Timber Construction
for Architects and Builders

McGraw-Hill Construction Series: *M. D. Morris, Series Editor*

Godfrey
PARTNERING IN DESIGN AND CONSTRUCTION

Matyas et al.
CONSTRUCTION DISPUTE REVIEW BOARD MANUAL

Minsk
SNOW AND ICE CONTROL MANUAL FOR TRANSPORTATION FACILITIES

Nichols and Day
MOVING THE EARTH—THE WORKBOOK FOR EXCAVATION, 4TH ED.

Palmer et al.
CONSTRUCTION INSURANCE, BONDING, AND RISK MANAGEMENT

Peurifoy and Oberlander
FORMWORK FOR CONCRETE STRUCTURES

Rollings and Rollings
GEOTECHNICAL MATERIALS IN CONSTRUCTION

Steinberg
GEOMEMBRANES AND CONTROL OF EXPANSIVE SOILS IN CONSTRUCTION

Timber Construction for Architects and Builders

Eliot W. Goldstein, AIA

McGraw-Hill

New York   San Francisco   Washington, D.C.   Auckland   Bogotá
Caracas   Lisbon   London   Madrid   Mexico City   Milan
Montreal   New Delhi   San Juan   Singapore
Sydney   Tokyo   Toronto

Library of Congress Cataloging-in-Publication Data

Goldstein, Eliot W.
    Timber construction for architects and builders / Eliot W.
Goldstein.
        p.   cm.
    Includes bibliographical references and index.
    ISBN 0-07-024580-0
    1. Building, Wooden.   I. Title.
TA666.G64   1999
694—dc21   98-38782
                                                                          CIP

McGraw-Hill

A Division of The **McGraw·Hill** Companies

    2 3 4 5 6 7 8 9 0   FGR/FGR   9 0 3 2 1 0 9

ISBN 0-07-024580-0

*The sponsoring editor for this book was Larry S. Hager and the production
supervisor was Sherri Souffrance. It was set in Century Schoolbook by North
Market Street Graphics.*

Printed and bound by Quebecor / Fairfield.

This book is printed on acid-free paper.

Whenever "man," "men," or their related pronouns appear in this work, either
as words or parts of words (other than with obvious reference to named male
individuals), they have been used for literary purposes and are meant in their
generic sense.

Photographs are by the author unless otherwise credited.

To Risa

Contents

Chapter 4. Code Issues   **4.1**

Part 2   Timber Design and Detailing

Chapter 5. Timber Framing   **5.3**

Part 3   Timber Construction

Chapter 11. Estimating and Bidding

Chapter 12. Shop Drawings

Chapter 13. Timber Erection

Part 4   Timber Case Studies

Chapter 14. A New Timber Building

Construction
Series Preface

Construction is America's largest manufacturing industry. Ahead of automotive and chemicals, construction represents 14 percent of this country's Gross National Product. Yet it is unique in that it is the only manufacturing industry where the factory goes out to the point of sale. Every end product has a life of its own and is different from all others, although component parts may be mass produced or modular.

Because of this uniqueness, the construction industry needs and deserves a literature of its own, beyond reworked civil engineering texts and trade publication articles.

Whether management methods, business briefings, or field technology, it will be covered professionally and progressively in these volumes. The working contractor aspires to deliver to the owner a superior product ahead of schedule, under budget, and out of court. This Series, written by constructors and designers for the whole building team, is dedicated to that goal.

M. D. Morris, PE
Series Editor

Preface

Never could I have imagined where it would lead when, in the spring of 1992, we were interviewed, and ultimately hired, to design a new public library for a small town in northern New Jersey. What began as a routine architectural assignment turned out to be a unique opportunity to explore a construction language, that of heavy timber, to which I had been introduced in architectural school in the late 1970s, but which had laid dormant in my head ever since.

Before starting graduate school, I had spent the better part of a year helping one of my professors, Ed Allen, AIA, build his wood- and timber-framed house, and had learned a tremendous amount about what it took to transform lines on paper into built form. By the time I completed his course on timber design and engineering the following year, I was hooked on the subject of timber framing, especially timber trusses. I was fascinated by graphical analysis, by the idea that you could determine the forces in each and every truss member without calculation.

My first job out of architecture school was with a small firm in northern Vermont. I was hired specifically to help lead the design of an enormous wood-framed barn for a working dairy farm. While I learned more about cows and milking parlors than I ever cared to know, this project served as my introduction to large-scale timber construction. I designed a number of buildings over the next 12 years and considered using timber whenever the structure was to be left exposed to view. But not until the Montville Library did the preferences of my client and the imperatives of the building program align perfectly with my interest in timber construction. Here was the opportunity I had been waiting for, to create an entire world of timber, to shape a structure to which all other building systems would be subordinate. As I describe later in this book, realizing that building was an education in itself.

For years, I have been an avid reader of *Engineering News-Record* magazine. (When I was a child, I would read my father's copy; for the past 20 years, I've been reading my own.) When the Montville Library was nearing the midpoint of construction, I called the buildings editor, Nadine Post, to see if she was interested in writing about it. She was tied up at the time, but encouraged me to call back several months later. When I did, she asked if she could visit the building. Having observed that *ENR* focuses on large concrete and steel structures, I figured that any story on the library would be buried in its back pages. Much to my surprise and delight, Nadine's article about it ended up on the cover (although it would have been bumped from there, I'm told, had the Japanese earthquake in early 1995 occurred a day or two earlier). It couldn't get any better than this, I thought, until I was informed, 10 months later, than the editors had chosen me as one of their newsmakers of the year for my work on that project.

Several days after the celebratory dinner, I received a letter from a book editor in Ithaca, NY, Dan Morris. Knowing of the library, he asked whether I would be interested in writing a book about timber construction, aimed at architects and builders. I had written a number of articles, some even about timber, but I could not fathom the effort it would take to write a book on the subject, particularly while practicing architecture more than full-time. Don't worry, he said, you'll have plenty of time to complete it, and you don't have to write it all yourself. Find some contributors to help.

Little did I know how little I knew about timber construction. (Start again.) Little did I know how little I knew about *explaining* timber construction. Most of what I had learned about the subject had been absorbed at the blackboard (in architecture school), at the drawing board (in the office), or on the job site; the seeds that were planted in design studio nearly a generation ago had only just sprouted. Start by writing an outline. Study other books on the subject. Differentiate yours from theirs.

Not until I was well along with the writing did the book start to take on a life of its own. It happened, I think, when I realized that what was missing from the timber literature was not just a book *for* architects and builders, but *by* an architect and builder, with their concerns foremost in mind. The authors of the standard timber engineering texts will undoubtedly consider my approach simplistic. That's fine: My goal has been to complement their work, not compete with it.

With the completion of the manuscript, it seems the right time to acknowledge all those who have contributed in one way or another to this book.

I have been fortunate to have four outstanding contributors, each of whom is a master of his subject: Fred Severud, PE, a long-time skiing partner and a trusted collaborator on so many projects, including the Montville Library; Ben Brungraber, Ph.D., PE, who provided a wealth of information and feedback during the evolution of this book, and who, with his colleague Tedd Benson, is breathing new life into the field of traditional timber joinery; Stephen Smulski, Ph.D., a talented wood technologist and editor of the *Wood Engineering Handbook;* and Phillip Pierce, PE, an expert on the restoration of covered bridges.

As an architecture student at MIT, I was surrounded by great teachers, but none had as much influence on me as Ed Allen. While his initial design studio, his introduction to building construction, and his seminar on timber engineering gave me the skills I needed to work with wood, assisting him in the construction of his own home was the first opportunity I had to apply those skills creatively on a real building (and gave me the confidence, years later, to build my own home). Ed, who has reached thousands of architecture students and practicing architects through his many books, was the logical person to call for advice as I was deciding whether to proceed with my own. While supportive of the project, he warned me of the tremendous effort it would take to realize it. For that, I am grateful. If readers of my book find it even half as useful as any of Ed's books, I will consider it a success.

For the better part of two decades, I have been part of the firm founded by my father, James Goldstein, FAIA. For much of the first decade, I was an apprentice; during the second, I have been his partner. While the lessons I have learned from him would not fit within the covers of a book, those lessons informed nearly everything I wrote within the covers of *this* book. Chapter 10 on timber specifications started as an attempt to write down my understanding of his principles for organizing the project manual and each of its sections. I am indebted to him, also, for emphasizing the importance of coordination among all of the participants in the building process. I hope my book inspires him to write his own, so that he can share more of his half-century of experience with the many architects and builders who can benefit from it.

Although I did not realize it at the time, I was also surrounded by great teachers throughout the construction of the Montville Library. I would never have been able to write this book without guidance from the timber contractors, Norm Strauss and Bill Marakovits of Dajon Associates; the detailers, Mike Noonan and his colleagues at Enterprise Engineering; and the fabricator, Unadilla Laminated Products. And, were it not for a group of very supportive clients, I would never have had the opportunity to experiment with timber in the first place. Special appreciation goes to Florence Knoll Bassett; Alden and Doll Siegel; Steve and Suzanne Kalafer; the administrators of the Anheuser-Busch Brewery in Newark, NJ; representatives of the Township of West Orange, NJ; representatives of New Jersey Transit; and the Library Board of the Township of Montville, NJ. The lessons I learned on their projects are the foundations of this book.

To Dan Morris, I am particularly grateful, not just because of his assistance as my editor, but also for having encouraged me to embark on this project in the first place. Solely on the basis of a single example of my work as an architect, Dan had confidence that I would make a good author. With the book essentially complete, I hope I have turned out to be as good an author as he anticipated.

I have been fortunate to have two thorough and thoughtful readers, one to whom the subject was very familiar, and one to whom the subject was quite foreign: Keith Faherty, Ph.D., PE, a Professor at Marquette University (and a coeditor of one of the definitive wood engineering texts) and Roger Goldstein, FAIA, a Partner in Goody, Clancy & Associates of Boston (and my older brother). In addition to identifying instances where my message was unclear or incorrect, each helped me see the book from a fresh perspective.

Explaining timber construction is nothing compared to explaining theoretical physics. By showing, through example, that even the most abstract concepts can be put in simple terms, my younger brother Raymond Goldstein, Ph.D., has been an inspiration to me, especially in the development of the graphs and figures.

Special thanks go to Christine Ducker and M.R. Carey, editing supervisors at North Market Street Graphics, Lancaster, PA, for their guidance and advice during book design and production, to David Behrman, of Behrman House, for leading me through the intricacies and idiosyncrasies of book contracts, and to Sandra Gurvis for preparing the index. Thanks also to Laura Antonacos Berwind for assistance in producing the illustrations in Chap. 8, and to Randi Guller for assistance in finding appropriate chapter quotations.

I am grateful to my wife Risa for her encouragement to write this book (and not only because it feels so good to have it behind me). During the endless weekends and evenings I was holed up in my office, writing, her confidence in my ability to complete it never wavered (even when mine did). While it only takes one to write a book, the burden is carried by the entire family; thank you for carrying the burden so long and so well. And, as my toughest critic, thank you for deciding *not* to read the completed manuscript. To our children, Adam and Hannah, who have effectively been fatherless for most of the past 18 months, you'll be seeing a lot more of me now (whether you like it or not).

Timber construction has been evolving for thousands of years. While it would be nice to think that this book presents the subject in a *new* way, it will be sufficient if all it does is present it in a *different* way, one which makes its language more accessible to those to whom it has been foreign.

The contributions of each and every individual and resource I consulted during the development of this book are hereby acknowledged. While they deserve a share of the credit for instances where I got things right, I take sole responsibility for those where I got them wrong.

Eliot W. Goldstein, AIA
South Orange, NJ
September 1998

CHAPTER

4
3
2
1
?

Introduction

> *... although they could not be seen, the surrounding walls and the oversize beams made themselves felt, almost like something alive there in the darkness.*

ERIC SLOANE, *An Age of Barns*

Timber Framing Is a Language

This book is about a language. It is an old language, but one that is still evolving. It is a language that is widespread, but spoken fluently by relatively few. It is a language based on a few simple rules, but capable of profound and complex expression. It is the language of timber construction.

From the timber shrines of the Far East, to the stave churches of Norway (see Figs. 1.1*a* and *b*), the language of timber construction has been spoken for over a thousand years, in a variety of local dialects. The dialects of timber construction, like those of written languages, have evolved in response to local needs (see Fig. 1.2).

> The expression of a building language becomes architecture only when it is executed with skillful understanding. . . . Climate, materials, and technology enable a society to build structures, but they remain only modifying factors behind the significant choice of forms—forms whose meanings are derived from a particular set of cultural values.[1]

Yet, regardless of where the language of timber construction is spoken (see Fig. 1.3), the underlying principles are the same. This book is meant to introduce the components of the common language, to explain how they relate to one another, and to describe the process by which an architectural concept is realized in timber. Each chapter is devoted to

(a)   (b)

1.1   ". . . the posts or staves rise like the trees of the forest toward the dark ceiling, and humans coming from the enclosing horizontality of the stue are transported into a superior world." (Christian Norberg-Shulz, *Norwegian Wood.*) Two views of the 800 year old Gol Church, at the Norwegian Folk Museum, Oslo: (*a*) Building in context. (*b*) Detail of roof.

1.2   Gazebos, Mohonk Mountain House, near New Paltz, NY.

an essential part of the story. Although timber construction is essentially a type of wood framing, its design and construction process more closely matches that of structural steel than that of house building. I will focus on nonresidential timber buildings that are fabricated primarily off-site.

Purpose of This Book

This book is intended as an introduction, for designers, builders, and design students, to the world of nonresidential timber design and construction. Its primary objective is to familiarize you, the reader, with the architectural and structural language of timber. I will have succeeded if, on your next timber project, you are able to collaborate knowledgeably with the other members of the design and construction team—if you know what timber can and cannot do and why, and are able to evaluate different options from the perspective of a designer or builder.

Our focus is on nonresidential projects where the ends justify the means—where the spans are sufficient to justify the use of heavy timber. While plenty of houses (including log cabins, of course) incorporate timber, its use is not primarily structural. This book is limited to buildings where the choice is between timber, steel, and concrete, not between timber and dimension lumber, where the members are large enough to burn slowly in a fire and, therefore, qualify as heavy timber under the building code, and where the entire frame, or a significant part thereof, will be visible in the completed building (see Fig. 1.4).

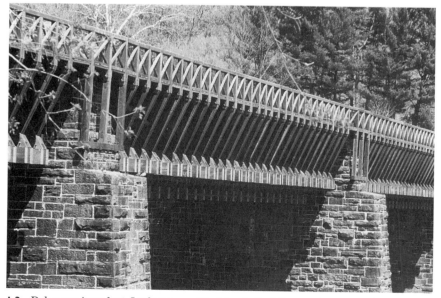

I.3 Delaware Aqueduct, Lackawaxen, PA. John A. Roebling, engineer, completed in 1849.

1.4 Classic framing and bracing in Vermont covered bridge.

Because it represents such a major portion of a building's cost and affects so many subsequent design decisions, the selection of a building's structural material is an extremely important decision. To assist in that selection, I will compare timber to concrete and steel.

This book is *not* intended to produce timber engineers; rather, it is meant to complement the definitive timber engineering texts already in print. Where those books contain pages and pages of detailed numerical tables, this book attempts to consolidate that information into a few simple graphs. Where those books present an enormous number of discrete choices, this book attempts to illustrate a few fundamental trends. Where those books cover each and every possible condition, no matter how unlikely, this book limits itself to those I believe are encountered most often. Where those books all rely on the same terms of art from the field of timber construction, I borrow, where appropriate, from other types of construction and introduce, where necessary to illustrate a point, some new terms of my own. Where the equations in those books contain every conceivable variable, the equations in this book have been stripped to the bone and simplified, to help distinguish "the forest from the trees" (sorry).

One of the most exciting aspects of working with timber is that the design possibilities are nearly limitless. From three-hinged glulam arches to long-span timber domes, from timber space frames to com-

1.5 Radial array of trusses with timber compression members. Becton Dickinson and Company corporate headquarters, Franklin Lakes, NJ. Kallmann McKinnell & Wood, architects.

posite trusses with timber and steel (see Fig. 1.5), the only constraint is the imagination of the designer. There is a world of design possibilities beyond the limited number of timber structures illustrated here.

To the extent that this book provides a conceptual framework (albeit a highly personalized one) for thinking about the entire process of timber design and construction, it may even be useful to the engineers already in possession of those other books. Architects and builders, on the other hand, may find this book sufficient without the others, unless they plan to actually engineer their timber structures themselves. However, no matter what your background or what your role in the process, make sure to acquire your own copy of the current edition of the rule book on which all timber books are based, the *National Design Specification for Wood Construction.*

It should not be surprising that the majority of the terms used in timber construction relate to specific components and their shapes. Rather than define those terms in a glossary arranged alphabetically, related terms are grouped and explained through composite illustrations (see Fig. 1.6).

Arrangement of This Book

This book consists of four parts, each containing several chapters covering different aspects of the design and construction of timber structures. The parts are 1, Timber Basics; 2, Timber Design and Detailing; 3, Timber Construction; and 4, Timber Case Studies. They are intended

	CHAPTER 1 Introduction
PART 1 **Timber Basics**	CHAPTER 2 Timber as a Construction Material
	CHAPTER 3 Preventing Wood Degradation
	CHAPTER 4 Code Issues
PART 2 **Timber Design** **and Detailing**	CHAPTER 5 Timber Framing
	CHAPTER 6 Fasteners and Connectors
	CHAPTER 7 Traditional Joinery
	CHAPTER 8 Timber Trusses
	CHAPTER 9 Lateral Bracing
	CHAPTER 10 Timber Specifications
PART 3 **Timber Construction**	CHAPTER 11 Estimating and Bidding
	CHAPTER 12 Shop Drawings
	CHAPTER 13 Timber Erection
PART 4 **Timber Case Studies**	CHAPTER 14 A New Timber Building
	CHAPTER 15 Covered Bridges

1.6 Organization of the book, showing icons for each chapter.

to parallel the process of education, design, and construction. (If the parts had been named according to their target audiences, they would have been entitled Students, Designers, Constructors, and the entire Building Team.) This book is intended to be read in at least two different ways: *in order* by those who are new to the field, and *according to subject* by those interested in specific topics.

Part 1, "Timber Basics," is intended to provide a basic understanding of wood to anyone who is unfamiliar with it as a construction material. It includes descriptions of timber's strengths and weaknesses, analyses of their implications with respect to the building codes, and discussions about the causes, effects, and prevention of timber degradation.

Part 2, "Timber Design and Detailing," is directed primarily at design professionals, but contains material that builders may also find useful. It covers the design and detailing of the various parts of a timber frame: the members themselves, some of the most common methods of joining them, various strategies for accommodating lateral forces, and the specifications necessary to assure that the design intentions are ultimately realized.

Part 3, "Timber Construction," is directed primarily at builders, but contains material that design professionals may also find useful. It is organized to follow the contractor's normal chronology: estimating, bidding, shop drawings (and their relationship to fabrication), and erection. It discusses not only the roles and responsibilities of timber contractors, but also their relationships to others involved in the construction process.

Part 4, "Timber Case Studies," is meant to put the lessons of the other three parts of the book into the context of actual projects. It, like each of the others, can be read in isolation but holds more rewards for those who have first read the other parts of the book.

Including this introduction, the book consists of the following 15 chapters:

Chapter 1, "Introduction," explains the purpose and organization of the book, and how it is intended to complement the existing literature on timber design and engineering.

Chapter 2, "Timber as a Construction Material," examines the structure of wood. Emphasis is placed on growth characteristics and moisture content, and their effect on structural and architectural properties. Timber grading is also discussed, because it provides a framework for classifying millions of unique members into a comprehensible set of structural categories, thereby enabling designers and builders to specify and procure with confidence. Other topics include brief analyses of the environmental consequences of timber construction and of the relationship between the direction of sawing and the requisite characteristics of saw blades.

Chapter 3, "Preventing Wood Degradation," looks at timber as a target of various "plagues": insects, water, fire, heat, chemicals, and corrosion. Topics include the circumstances under which each potential agent becomes destructive and the strategies recommended to prevent them. Under proper conditions, timber will remain serviceable for centuries; one of the purposes of this chapter is to help show how that objective can be achieved.

Chapter 4, "Code Issues," discusses heavy timber as a construction type, and the code implications of that classification. Topics include methods for limiting and slowing the spread of fire within and beyond the building, limitations on the height and area of timber buildings, and interfaces with other materials and systems.

Chapter 5, "Timber Framing," examines timber framing members, the loads on them, and their resulting behavior. To assist in the design of efficient timber structures, the concept of equivalent thickness is introduced. Comparisons are made between sawn timbers and glulams. Since timber framing is often left exposed to view in completed buildings, architectural issues are also addressed.

Chapter 6, "Fasteners and Connectors," is devoted exclusively to common approaches to the joining of timbers. In addition to a discussion of the advantages and disadvantages of various mechanical fasteners, this chapter illustrates how they must be arranged to avoid failure of the members where they are attached.

Chapter 7, "Traditional Joinery," explains the unique aspects of traditional timber frames, those in which members are joined without mechanical fasteners. Components of traditional timber frames are illustrated, as are the generic ways of joining their members. Problems related to shrinkage and to timber's directional properties, of concern in any timber framing system, are examined from the specific viewpoint of the traditional timber framer.

Chapter 8, "Timber Trusses," goes from the definition of a truss to construction and testing of a truss model. The general principles illustrated are then verified and quantified through the use of graphical analysis. Fasteners and connectors, introduced in Chapter 6, are discussed here in relation to the characteristics unique to timber trusses. Issues of continuity, deflection, and hierarchy complete the subject.

Chapter 9, "Lateral Bracing," introduces the various ways in which timber frames can be braced to resist seismic and wind loads. Resisting systems include diaphragms, diagonal braces, and shear walls, and combinations thereof. The behavior of timber structures under lateral load is compared to that of steel and concrete structures.

Chapter 10, "Timber Specifications," defines specifications as a particular type of instruction set, and identifies its general properties. Emphasis is given to the importance of thinking about and specifying the timber work in the context of constructing the entire building. The chapter identifies aspects where coordination between trades is critical and recommends appropriate language. Guidance is given on specifying for protection, appearance, and performance. The advantages and disadvantages of assigning timber construction to a separate contractor are compared to those of making it a part of the general contract.

Chapter 11, "Estimating and Bidding," provides a conceptual framework for translating the information in the contract documents—the working drawings, timber specifications, and general conditions—into a timber cost estimate and a timber bid. Rather than pricing the job simply as a collection of construction materials and the labor needed to install them, the reader is encouraged to look at the job holistically, from the implications of site access to the qualifications and responsibilities of the superintendent and project executive. Examples illustrate the necessity for reading between the lines of the bid package.

Chapter 12, "Shop Drawings," discusses what they should show and why, including their critical role in the planning and scheduling of the timber erection process. It examines the detailer's transformation of the architectural and engineering information into a set of detailed instructions for the fabricator and erector. As with timber specifications, timber shop drawings are examined in the context of the other systems and assemblies connected to or affected by them. Examples are given of situations where construction by others cannot happen without coordination with the timber contractor.

Chapter 13, "Timber Erection," surveys the entire erection process, from the loading of the trucks at the fabrication plant, through delivery, shakeout, protection, erection, decking, conditioning, and closeout. Other topics include temporary loads and bracing, preparations by others, and considerations related to preassembly.

Chapter 14, "A New Timber Building," is a detailed description of the design and construction issues associated with a recently completed heavy-timber library. It applies many of the principles covered earlier in the book, and explains how aspects that appear to be problems and constraints can be recast as design opportunities. Finally, it shows how a minor change in the physical relationship between the timber frame and other building systems can have a major impact on the ultimate architectural character of the building interior.

Chapter 15, "Covered Bridges," uses covered timber bridges to illustrate some general principles about maintaining, evaluating, and restoring existing timber structures. Because they are subject to frequent and substantial moving loads, are extremely narrow, are unheated, and are only partially protected from the weather, covered bridges are more vulnerable than virtually any other type of timber structure to structural distress.

Note

1. Holan, p. 62.

PART

I

Timber Basics

CHAPTER 2

Timber as a Construction Material

In one respect, timber is unique among structural materials. It is the only one which is the direct product of natural growth, and which is subject as it grows to structural actions comparable with those to which it is afterwards subjected in its use by man.

ROWLAND J. MAINSTONE, *Developments in Structural Form*

Introduction

Of all the common construction materials, timber has the most interesting range of potential life cycles: At one extreme are log cabins and tepees, made directly from trees with virtually no waste; at the other are contemporary glulaminated structures where the route from standing timber to building product is indirect and where the strengths of small members have been exploited to produce larger members with superior properties. When timber buildings are no longer serviceable, their timbers can be recycled for their structural properties—reused permanently as framing or temporarily as formwork—or for their heat content, as fuel.

The journey from tree to timber has gotten increasingly circuitous and the transformation increasingly profound. The imposition of a growing number of intermediate steps between the felling of a tree and the erection of selected parts of it has enabled this transformation and created opportunities for customization. Nevertheless, the performance of the finished product is still controlled to a great extent by the wood and wood fibers within it. For that reason, it is appropriate to take a close look at wood, the primary constituent of all timber construction.

It would be easy to focus on timbers as delivered—to ignore their past and future—but it would be a mistake. To build responsibly and well, both the designer and builder must understand in the broadest sense how timbers get to the job site and where they might go once they leave. (This, in spite of the fact that neither party has much control or influence over these events.) The designer can prepare better specifications and details and the builder implement them only if both understand:

- What wood is and what is unique about it
- What its strengths and weaknesses are
- How its properties are exploited in different timber products
- What its environmental properties are

Over time, man's control over the development and use of timber resources has widened. Long ago, trees destined for use as timbers were used essentially as found. We learned to saw in ways that maximized yield. Eventually, members were sorted and distributed among various end uses—framing, decking, and trim, for example—based on their visible characteristics. Visual grading systems were developed to codify the selection process. Since the grades conformed to market needs, they helped increase timber utilization. Since quality and price went hand-in-hand, and users could match products to their specific requirements, grading systems enabled consumers to control building costs. More recently, nondestructive testing has improved the quality of grading by measuring the performance of each piece, including the influence of its concealed characteristics. Enhanced understanding of tree growth has enabled the production of trees of more consistent quality. The advent of glulaminating made it possible to tailor the performance of each part of a member to its particular structural demands. The advent of composite timber products will make it possible to produce members with properties beyond those of wood alone. Where once we were pretty much at the mercy of nature, we are now capable of influencing the quality of the timber resource and of maximizing its utility.

The Structure of Wood

Although wood is a natural material with a wide range of natural variations, many aspects of its structure are similar regardless of tree size or species. All wood is composed of bundles of long hollow tubular cells or fibers. A transverse cut through a log reveals the large-scale aspects of that structure (see Fig. 2.1), but it takes a microscope to reveal its finer points.

Tree cross sections consist of several main parts. Around the perimeter is the bark. The outer bark acts as the tree's armor or skin, protect-

2.1 Cut end of an old timber. The ring spacing progresses from wide and weak when the tree was young to close and strong as it aged. Visible defects include checks and shakes. Delaware Aqueduct, Lackawaxen, PA. John Roebling, engineer.

ing it from infections, disease, and dehydration. The inner bark is the channel through which food is transported from the leaves.[1] Virtually everything to the inside of the bark is considered wood. At the interface between the inner bark and the outer edge of the wood is the cambium, which can be thought of as the tree's factory. Here, the food from the leaves above acts as the raw material for the production of new wood and bark.[2]

The production of wood fiber has three regimes. During the winter, the tree is essentially dormant. With the arrival of warmer weather and increased sunshine, wood production begins. Wood formed during this period consists of relatively large cells with thin walls, and is known as *springwood,* or less commonly but perhaps more precisely, as *earlywood.* Wood produced after growing conditions reach their peak is called *summerwood* or, again more precisely, *latewood.* The cells of summerwood are smaller and have thicker walls than those of springwood.[3] Summerwood is denser than springwood, and its cells are therefore stronger.[4] The higher the ratio of summerwood to springwood, the stronger the wood.[5] And, because they grow more slowly, trees "on north-facing slopes produce bigger, more erect, straighter grain and 'tougher' timber."[6]

Where growing conditions are fairly uniform throughout the year—as in the tropics—trees produce wood cells of relatively uniform size and wall thickness year round. Their relatively uniform density makes them ideal for carving. Only where growing conditions go through annual seasonal cycles does wood production alternate between springwood and summerwood.[7] The striations caused by the differences between these types of wood are what we know as *annual rings.* Dendrochronologists have determined that the widths of tree rings correspond to the growing conditions at the time of their production. For them, tree rings are indicators of past climatic conditions. For designers, tree rings are good predictors of wood strength and workability. The greater the density difference between the springwood and summerwood, the more difficult the wood will be to carve.

Functioning like drinking straws, wood fibers conduct sap from the roots of a tree up to its leaves. The material involved in this process, *sapwood,* encompasses virtually all of the wood in a young tree. When the tree reaches a certain size, the total transport capacity of the wood exceeds the sap requirements of the tree. The excess cells become inactive, and are classified thereafter as *heartwood.* Although somewhat of an oversimplification, the sapwood from then on can be thought of as a doughnut-shaped zone of constant area; the width of the sapwood will therefore be less near the ground, where the trunk diameter is greater, than near the crown.[8] Growth in the diameter of the tree is accompanied by enlargement of the inside diameter of the sapwood zone, bordering an ever larger area of heartwood.

Accompanying the transition from sapwood to heartwood is an increase in certain chemicals. These are referred to as *extractives* because, with solvents, they can be extracted or removed from the wood. Organic extractives affect some wood properties, including color, decay resistance, odor, and combustibility,[9] but not others, such as weight and strength. Heartwood is more resistant to decay than sapwood, which explains why, for example, all heart redwood makes superior siding. Sapwood, on the other hand, is more absorptive, enabling it to more easily accept decay-resistant preservatives.[10]

The cross section of a tree trunk is a like a plan detail in a set of working drawings; it is a snapshot of the two-dimensional relationships between the various components at one particular elevation, but does not tell the whole three-dimensional story. If trees were truly cylindrical, cross-sections would be identical regardless of the elevation at which they were taken. Trees, however, are really highly elongated cones; thus their rings—mistakenly thought of as circles since they are most often seen in cross-section—are really three-dimensional forms akin to *nested cones.* Some species, such as lodgepole pine, have less taper than others, but all species are tapered to some degree. All else being equal, the more pronounced the taper, the lower the total volume of lumber and the lower the percentage of long pieces of lumber a tree of a given height will yield. And, since trees are normally milled parallel to their longitudinal centerlines, those with more taper will yield more boards having *sloped grain*—grain that is not parallel to the faces of the member in which it is located.[11] Such members are more difficult to machine and more prone to warping.

Recall from high school math the conic sections, those geometric shapes produced when a plane in various orientations slices a cone. When the plane slices vertically through the cone, the intersection is a parabola. It should come as no surprise, then, that plain-sawn lumber—lumber sawn tangentially to the growth rings displays a series of nested parabolas along its length. The greater the taper of the tree, the closer and more pronounced the curves (see Fig. 2.2).

Moisture Content

Trees contain moisture in their cell cavities and cell walls. In a wood specimen, the ratio of this total moisture—free water plus bound water—to its oven-dry weight is called its *moisture content.* As felled timber dries, the free water in the cell cavities diminishes first. Loss of free water reduces the weight of a specimen, but not its size. The point where all the free water is gone but all the bound water remains is the *fiber saturation point,* and occurs at moisture contents between 25 and 32 percent.[12] Reductions in moisture content beyond that point result in shrinkage roughly proportional to the loss of bound water. Shrinkage, like strength, is proportional to wood density. The equilibrium moisture content of wood (EMC)—the moisture content it assumes in

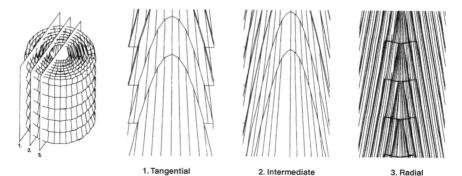

1. Tangential   2. Intermediate   3. Radial

2.2 A portion of a tree modeled on the computer as a set of nested cones. The figures revealed in sections cut through the model are similar to those of timbers sawn from corresponding portions of a tree. The slope of the cones in the model—corresponding to slope of grain in the tree (see Fig. 2.7)—is particularly evident in the slice cut radially.

service—is a function of the temperature and relative humidity of the surrounding environment (see Fig. 2.3).

In order of decreasing magnitude, shrinkage takes places in three different directions: circumferentially, radially, and longitudinally. Circumferential shrinkage, also referred to as *tangential shrinkage,* takes places along the growth rings, around the perimeter of the tree. Radial shrinkage takes place radially, perpendicular to the growth rings. Longitudinal shrinkage is that which occurs along the axis of the tree. On a scale of 1 to 100, relative shrinkages are as follows: tangential, 100; radial, 50; longitudinal, 1. This wide disparity explains why experienced designers are so careful in their specifying and detailing of timber construction.

The problem is not so much with shrinkage per se, but with differential shrinkage *within* a given member, *among* various members, and *between* timbers and other structural materials. For example, consider a structure that is partially balloon-framed—with continuous columns the full height of the building, and partially platform-framed—with interior columns interrupted at each floor and roof platform. In a three-story medium-span nonresidential timber-framed building, it would not be unusual for there to be a total of close to 50 in (1.25 m) of cross-grained wood in the floor and roof framing, and nearly 10 times that amount of

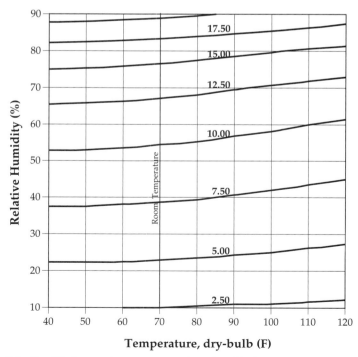

2.3   Equilibrium moisture content of wood, or EMC, in percent. (Based on Faherty.)

longitudinal fiber in the columns. As the building dries out in use, tangential shrinkage of approximately 2 percent can be expected, but longitudinal shrinkage may be as little as 0.02 percent. Therefore, even though longitudinal fiber predominates, both its total shrinkage and the longitudinal shrinkage differential between the balloon- and platform-framed portions of the building will be negligible.

The same cannot be said for tangential shrinkage. With its balloon framing, tangential shrinkage at the perimeter will total 0.33 in (8 mm) per floor, but will not accumulate from floor to floor. With its platform framing, shrinkage at the core will accumulate from floor to floor, totaling about 1 in (25 mm). The taller the building, the more pronounced will be the resulting difference. If the framing systems are in close proximity to one another, the shrinkage differential could result in noticeably uneven floors, or worse.

The moral of this story is to detail the structure so that differential shrinkage is kept to a minimum. There are several ways that this can be achieved:

- Do not mix balloon framing with platform framing.

- To the greatest extent possible, make sure all members are at or close to their equilibrium moisture contents.

- If possible, use members which are cut similarly (plain-sawn, quartersawn, etc.).

- Avoid mixing species with different shrinkage values.

- Use care when mixing bearing elements of concrete, masonry, or steel with those of timber.

Growth Stresses

Heavy timbers, by definition, have larger cross sections than boards of dimension lumber. Because dimension lumber is relatively thin, kiln-drying is quick and therefore economical. Because kiln-drying of heavy timber is slow and costly, it is rarely done. As a result, the moisture content of heavy timbers tends to be relatively high at time of delivery, with a significant drop over the course of construction and initial occupancy. This combination of large size and a large reduction in moisture content means that heavy timbers can develop serious shrinkage cracks. However, not all defects are the result of changes in moisture content.

The primary chemical components of dry wood—together comprising at least 75 percent of the wood by weight—are cellulose and lignin. As wood cells mature, lignin is deposited and polymerizes within their walls.[13] In general, cellulose makes up the walls of the wood fibers, while the lignin binds them together.[14] Just as the bundling of cellulose fibers enhanced their collective structural performance in ancient reed boats, lignification of cellulose fibers enhances their structural performance in trees. Once the lignin is removed, wood fibers act like extremely slender columns, strong in tension, but extremely flexible. The resulting material is ideal for making paper.

Just as the thickness of a rubber band decreases as the rubber band is stretched,[15] the length of each wood fiber decreases as the fiber expands radially during lignification. "The shrinkage in length of each cell is small, but the cumulative effect is sufficient to produce a considerable tensile stress longitudinally on the outside of the woody cylinder of the trunk."[16] Not surprisingly, the tensile force is highest at the outer ring, where the column of fibers is longest. The tension in the wood at the perimeter of the tree induces compression of the wood near the center. While the tree is standing, these forces are in balance.

Felling and sawing disturbs this balance and redistributes the forces, causing the outer wood to shorten and the inner wood to lengthen. With the release of these growth stresses, wood warps longitudinally (see Fig. 2.4). The type of warpage is a function of the location and orientation of the timber within the cross-section of the tree. In a plain-sawn (tangentially sawn) member, the outer *face* shortens more than the inner face, causing an outward *bow;* in a quartersawn (radially sawn) member, the outer *edge* shortens more than the inner edge, causing an outward *crook* or *hook*. Where the member is not parallel to grain or where there are variations in grain orientation along its length, the member's reaction to the release of growth stresses will vary along its length, causing *twist*.[17] At any given cross-sectional aspect ratio, pieces symmetrical in both directions about the axis of the pith will tend to react more symmetrically to the release of growth stresses, and will therefore warp less than members that are not. The release of longitudinal stresses results not only in distortion of logs and timbers, but also in end cracking, with which most carpenters are familiar. End cracking starts as soon as the log is cut, and lengthens during subsequent drying.[18]

Although large longitudinal growth stresses can cause problems in heavy timbers, they are of substantial structural benefit in standing timber. Winds put the windward side of a tree into tension and the leeward side into compression. In the absence of growth stresses, there would be a linear transition from one to the other. However, the perimeter tension arising from growth stresses means that the tree is essentially prestressed: Compression in the leeward face induced by the wind is partially offset by tension induced there by the growth stresses. "By this method the tree roughly *halves* the maximum compressive stress . . . and so *doubles* its effective bending strength. It is true that the maximum tensile stress has been raised, but the wood has plenty in hand in this respect."[19]

Other growth characteristics include shakes and checks. *Shakes* are fiber separations between or parallel to the rings,[20] which may result at

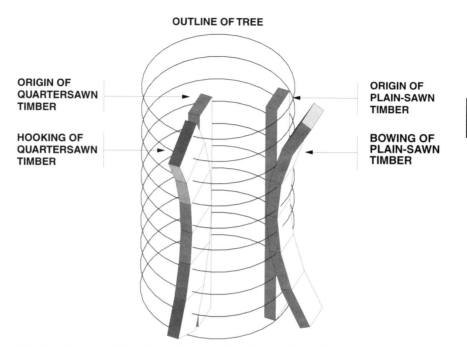

OUTLINE OF TREE

ORIGIN OF
QUARTERSAWN
TIMBER

ORIGIN OF
PLAIN-SAWN
TIMBER

HOOKING OF
QUARTERSAWN
TIMBER

BOWING OF
PLAIN-SAWN
TIMBER

2.4   Growth stress distortions. Compare these longitudinal distortions to the cross-sectional seasoning distortions at corresponding locations in Fig. 2.6.

areas of high shear stress when growing trees are bent under snow or wind loads.[21] *Checks* are fiber separations oriented perpendicular to the annual growth rings, and result from the stresses associated with tangential shrinkage during seasoning. We shall examine checking in more detail later.

Density Variations

When old-timers say that timbers today are not as good as the ones they used to get, they are absolutely correct. Not only is it nearly impossible today to find members in the large sizes that used to be common, but if found, they may not be as strong as their predecessors. Why is this so?

Overlaid on the density variation between springwood and summerwood is a long-term variation between juvenile wood and mature wood. Trees, like people, grow more rapidly when they are young than when they get older. Rapid growth means rings that are wider and, therefore, less dense. Lower density translates into lower strength.

Thus, a tree's ability to resist bending under dynamic (wind) loads is a function not only of its girth, but also of its age. As they age, trees get stronger both because they get bigger and because the mature wood at the perimeter is denser than the juvenile wood at the core. (For this reason, glulams can be thought of as assemblies of small second-growth pieces meant to mimic the performance of large old-growth sawn timbers.) Due to greater density, an older tree will be stronger than a younger one of identical species and girth. Within a given species, a large member cut from a small old tree will be stronger than the same member cut from the heart of a larger younger tree. By decreasing wood density, ideal growing conditions actually *reduce* structural properties. Timbers in old buildings are valuable not only because they are often larger than those available today, but also because they are stronger. Thus, it may be structurally advantageous in new construction to reuse existing timbers from old-growth trees rather than use new material from contemporary tree farms. Improvements in forest management also mean that wood density standards should be based decreasingly on naturally grown samples and increasingly on those from managed forests.[22]

The transition from juvenile wood to mature wood also has implications in terms of cutting strategy. From a given log, the strongest timber of a given width is likely to be the one whose edges are of equal age and as old as possible. All else being equal, the structural properties of members with edges of different ages will be controlled by the properties of the younger material, the weakest link in the chain. For heavy-timber construction, try to use members whose diagonal dimensions are as close as possible to the diameters of the trees from which they were sawn. "We do not grow big trees to make small products."[23]

Knots

That portion of a branch embedded within the trunk of a tree or within a member sawn therefrom is called a *knot*. All knots affect grain orientation and, therefore, wood strength.

The cross section of a branch is similar to but smaller than that of the main stem or trunk of the tree. As with a tree, a branch's cross section does not tell the whole story. From ordinary experience with knots in plain-sawn framing lumber, it would be natural to conclude that branches are basically cylindrical where embedded within the trunk, but such is not the case. In fact, branches start very small where they spring from the pith near the center of the main stem, and grow conically outward.[24] Only the portion of a living branch that projects from a tree approximates a cylinder.

The branch can only live as long as its tissue is connected to that of the trunk. During that time, the interconnected cambia add growth to both the trunk and branch simultaneously.[25] For this reason, the intersection between the trunk and a branch looks, in radial cross section, somewhat like a mitered joint. The older the branch, the greater the number of rings it shares with the trunk and, therefore, the larger the mitered area.

Once the branch dies, everything changes. Subsequent growth of the trunk goes around the perimeter of the branch. Whereas the living branch is structurally continuous with the tree, the embedded portion of a dead branch is more akin to a dowel. Because a branch weakens after it dies, it is likely to fall from natural causes or be removed during pruning. If the resulting stub is short enough or the subsequent growth of the trunk is great enough, the entire remaining portion of the branch will eventually lie within the trunk.

Knot classifications relate to the physiological relationships between the branch and the trunk, and their geometric consequences. Physiologically, where the tissues of the trunk and branch are interconnected, knots are *intergrown;* where the tissue of the trunk has grown around the long axis of the dead branch, knots are *encased.* Geometrically, the former are known as *tight knots* and the latter as *loose knots.*[26] To those involved in timber construction, the geometric terms are both more descriptive and more practical than the physiological ones.

Timber strength is related not only to the grain variations resulting *from* the knots, but also to the geometric relationship of the sawn member *to* the knots. Because the direction of sawing is roughly perpendicular to the long axes of the branches, knots in plain-sawn members are generally round or oval. Tight knots in such members are encircled by grain deviations the severity of which diminishes with distance from the knot. Because the direction of sawing is roughly parallel to the long axes of the branches, knots in quartersawn members look like spikes, and are consequently known as *spike knots.* Where the axis of a knot coincides with the width of the piece, the spike will spread across most of it, drastically reducing its strength. By its very definition, plain sawing encompasses the perimeter of a log in fewer members than quarter sawing. A given number of knots, therefore, will be distributed among a higher percentage of the plain-sawn members, but will have less impact on their strength than the limited number of spike knots will have on the few quartersawn members in which they are located. Given two identical trees, one quartersawn and one plain-sawn, the properties of the quartersawn members will be at the extremes—either knot free (i.e., clear) and extremely strong, or spiked and extremely weak—while those of the plain-sawn members will be at the mean—more likely to have knots, so moderately strong or weak depending on their locations.

There is an interesting reversal of structural properties associated with branches and knots. Like a moment connection in steel, structural continuity is what enables the trunk to support the branches. And, as in that case, continuity is achieved by enlarging the connection between the parts. Moment resistance in steel is achieved through welding or through the introduction of gussets. In trees, by contrast, successive growth increments at the branch/trunk interface cause the grain to take the form of a set of nested escutcheons; the size and strength of the joint is a function of the number of shared rings.

While the branch is alive, these grain variations are not problematic. On the contrary, without them, the branch could not be supported. For construction purposes, however, what matters is not just the size of the knot, but its area of influence. Grain distortions around tight knots tend to be more pronounced and extend farther than those around loose knots. Therefore, tight knots can reduce timber strength more than loose knots of similar size. Even though it seems counterintuitive, the hole left after a loose knot falls out is less detrimental to the strength of the piece than the area of distorted grain around a tight knot. Put simply, the local *increase* in structural capacity developed in a tree to support an appendage projecting from it—the branch—ultimately translates into a *decrease* in the structural capacity of the timber sawn from that tree.

A knot behaves like a secondary tree oriented at an angle to and embedded within the primary one. When it dries, it shrinks. The significant cross-grain shrinkage of a tight knot is resisted by the negligible longitudinal shrinkage of the surrounding wood; after enough shrinkage, the knot may become checked.[27] The shrinkage of a loose knot is, by contrast and by definition, unrestrained; after enough shrinkage, the knot may fall out.

Although the formation of limbs is a natural process, their ultimate diameters and embedded lengths are somewhat within the control of foresters. Through management of stand density and pruning, they can influence limb growth, and to the extent that structural properties are a function of knot type and size, wood strength (see Fig. 2.5).[28]

Creep

The structural behavior of wood under load is a function of load duration. Under short-term loading, wood is *elastic:* Upon removal of the load, the deformation disappears. Under long-term loading, however, a significant portion of the deformation is *plastic* (i.e., nonrecoverable). The elasticity comes from the cellulose; the plasticity, from the lignin. Wood's property of deforming plastically in relation to load duration is referred to as *creep.*[29] Creep deflection increases with load, moisture content, and temperature.[30]

To the designer, creep can be accommodated by assuming that the capacity of a timber will vary inversely in relation to the duration of the loads likely to be placed on it. On the one hand, exposed timber columns in a public lobby should easily accommodate the lateral loads exerted by the occasional individuals leaning against them, as should the timber deck in a wood shop under the impact of a dropped tool. On the other hand, floors subjected to heavy live loads typical of libraries and warehouses, and roofs under continuous loads from heavy snow or ponded water must be engineered with creep very much in mind.

The problem of ponded water is of particular concern. If the roof structure has no slope or so little slope that it will be canceled out by

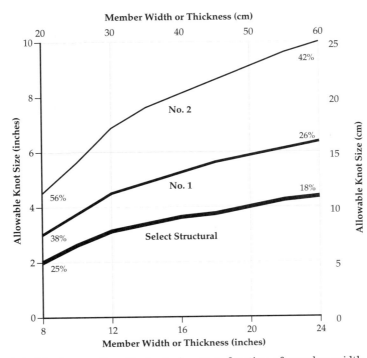

2.5 Maximum allowable knot size as a function of member width, labeled in percent (based on *WWPA Grading Rules,* Beams & Stringers, Wide Face).

deflection under dead and live loads, ponding will occur. The deflection it causes will increase the volume of ponded water, causing greater deflection and greater ponding.[31] Through this process of continuous positive feedback, ponding can lead to progressive collapse.

Other Defects

Wood is never "perfect"; grain variations and other growth characteristics are always present, though in varying amounts. While swirling figures may enhance the appearance and value of decorative members, they reduce the performance of structural ones. Therefore, even though these and other characteristics are a natural consequence of the growth and seasoning of wood, they are known as *defects*.

Defects occur during one or more of these stages: growth, seasoning, or production. Growth characteristics are those that happen during the growth of the tree and, as such, are subject to minimal control. Seasoning and production defects are those that arise after the tree has been felled; as such, they can be controlled, though to varying degrees. The size, magnitude, and distribution of defects and the intended use of the member in which they are located determine their impact on structural performance.

Distortion of member cross section is a common seasoning defect. The magnitude of the distortion is a function of a member's reduction in moisture content. Members sawn while wet—*surfaced green*—shrink more after milling than those sawn after drying—*surfaced dry*. The shape of the distortion is a function of the aspect ratio of the member and of its original location within the tree's cross-section. It is a consequence of the fact that wood shrinks more tangentially than radially.

Since distortion is a natural consequence of the seasoning process, it effects every sawn timber to some degree; therefore, it is helpful to have a basic understanding of how it works. Every member shrinks tangentially, radially, and longitudinally. The width of a plain-sawn member is roughly tangential, while the width of a quartersawn member is roughly radial. It should not be surprising, then, that the former will shrink more in width, and the latter, in thickness, at least for small members taken from large trees.

Things get more interesting as the members get larger, as we go from dimension lumber to heavy timber (see Fig. 2.6). Consider a quartersawn timber beam as wide as the radius of a tree: Tangential shrinkage will dominate at its outer edge, while radial shrinkage will dominate at its inner edge. Shrinking will produce *tapering;* its cross section will eventually look something like the elevation of a column with entasis. Now consider a plain-sawn beam with its centroid on a

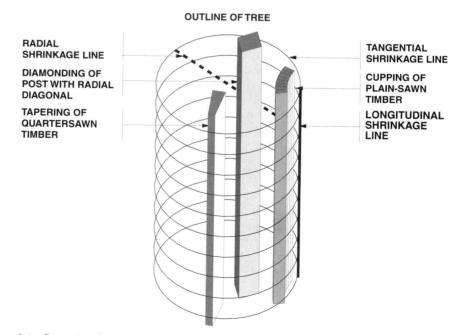

OUTLINE OF TREE

RADIAL
SHRINKAGE LINE

DIAMONDING OF
POST WITH RADIAL
DIAGONAL

TAPERING OF
QUARTERSAWN
TIMBER

TANGENTIAL
SHRINKAGE LINE

CUPPING OF
PLAIN-SAWN
TIMBER

LONGITUDINAL
SHRINKAGE
LINE

2.6 Seasoning distortions. Compare these cross-sectional distortions to the longitudinal growth stress distortions at corresponding locations in Fig. 2.4.

diameter of the log: The outer face will shrink more than the inner one, resulting in *cupping*. Finally, consider a post having one of its diagonals oriented along a radius: That diagonal will shrink less than the opposite one, resulting in *diamonding*.[32] Since shrinkage in the tangential direction—along the rings—is greater than that in any other, the longest ring within a member's cross-section will shrink the most. It follows, therefore, that the curvature of a cupped timber will be opposite to that of the rings within it, and that members centered on the pith will distort symmetrically, while those which are not, will not.

If you are ever in a heavy-timber building during its first heating season, you might occasionally hear a loud noise not unlike a gunshot. Soon thereafter, you will discover that one of the beams or columns has what looks like a long crack along one of its faces. What, you wonder, is going on here?

The answer is *checking*. Sawn timbers are so large that seasoning—in a kiln or in the air—takes a very long time. Consequently, the moisture content of most sawn timbers is relatively high at time of delivery and installation. Fibers close to the surface dry out faster than those in the interior, but the drier outer fibers are restrained from shrinking by the less-dry inner fibers. The greater the differential between the respective moisture contents, the higher the shrinkage stresses.[33] When the stresses get high enough, the fibers separate with a loud report.

Dimension lumber can be dried more uniformly and more completely than heavy timber because it is relatively thin. Glulaminated members are composed of dried dimension lumber; therefore, they are far less prone to checking than sawn timbers of equivalent size. When glulam checking does occur, it is often at the glue line of the bottom lamination, the layer with the greatest percent of its surface area in contact with the surrounding air.[34] If you think of the balloon-frame/platform-frame scenario discussed above as an example of the effects of shrinkage on the macro scale—at the scale of an entire building—it is easy to see that the glulam/sawn timber example discussed here is an example of the same phenomenon on the micro scale, at the scale of the individual timber element. In both cases, compounding or accumulation of shrinkage can be prevented through careful design.

Although the large dimensions typical of heavy timbers make seasoning defects virtually inevitable, the presence of such defects need not preclude use of the pieces containing them. Crook, for example, can be thought of as natural camber; it is desirable in a horizontal framing member as long as the crown is oriented up, the load is large enough to produce a deflection eliminating the crown, and the member is free of other warp. Similarly, a little twist is rarely of concern in a column, as long as there is some play in either its top or bottom anchorage.

Sawn timbers are rarely straight. Indeed, one of the attractions of glulams is that lamstock is sufficiently thin and flexible that some types of warpage can be squeezed out during the laminating process, producing straight and relatively stable members. Similarly, bowing of decking may disappear in the process of fastening it to its supports, nailing from one end to the other. Problems occur in use when the defects affect structural performance, detract from appearance, or interfere with construction. Examples include:

- Columns that are bowed or hooked
- Beams that are twisted or cupped
- Timbers whose loose knots have fallen out
- Deckboards so twisted that they cannot engage the tongues and grooves of their neighbors

Slope of Grain

The strength of wood is a direct consequence of its directional and cellular structure. The long and slender wood fibers stacked essentially end to end for the entire height of the tree produce great strength along the grain in both tension and compression. Across the grain, i.e., against the thin walls of the hollow fibers, wood is relatively weak. Its strength across the grain is higher tangentially than radially. *Radial* compressive loading is resisted by the thin alternating "sheets" of springwood and summerwood; since all of the growth rings are loaded simultaneously, radial compressive strength is controlled by the weaker springwood. By contrast, tangential compressive loading is resisted by the alternating "walls" comprising the growth rings. The weaker springwood walls are not loaded until and unless the stronger summerwood walls fail. Thus, tangential compressive strength is controlled by the stronger summerwood. It is clear, then, that the three orientations important to understanding wood shrinkage—axial, radial, and tangential—are also important to understanding wood strength.

The difference in orientation between the grain of the wood and the longitudinal axis of the member in which it is located is known as the *slope of grain* (see Fig. 2.7). The tangent of that angle—the deviation of the grain from the axis of the member over a certain distance—is expressed as a fraction, with 1 in the numerator. Slope of grain can also be thought of as the units of distance along the piece over which the grain deviates 1 unit out of line. Thus, an angle of 6° equals a slope of grain of 1/10, while an angle of 14° equals a slope of grain of 1/4. The larger a member's slope of grain (i.e., the smaller the denominator), the lower its axial and bending strengths.

Slope of grain can arise naturally during growth of the tree or can be the result of its orientation during sawing. Natural slope of grain can vary significantly along the length of a member, depending on the tree's straightness and taper, among other things. Slope of grain result-

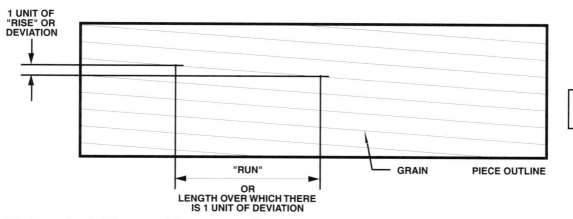

2.7   Slope of grain. Whereas roof slopes are typically given with respect to a run of 12 (i.e., 2 in 12, 6 in 12, etc.), slope of grain is typically given with respect to a rise of 1 (i.e., 1:8, 1:12, 1:15).

ing from sawing (which I refer to as *artificial slope of grain*) affects all natural slope of grain to the same degree. In trees with significant taper, artificial slope of grain can be minimized by sawing parallel to the bark, rather than to the pith. The same is true of *wane,* the wood missing at the corner of a member sawn too close to the curved surface of a log.[35] Because natural slope of grain is so variable within a single piece and among different pieces, its positive impact on members loaded across the grain is ignored. (See Table 2.1.)

One way to think about the grain orientation is to imagine a longitudinal axis connecting an entire column of wood fibers. It is then easy to see that the grain can be at an angle to one or more of the edges of the member. In the latter case, cross-grain should be evident on two adjacent edges, and the actual composite slope of grain will be greater than either of the individual slopes.[36]

TABLE 2.1   Comparison of Selected Aspects for Three Different Timber Grades

	Select structural	No. 1	No. 2
Characteristics, beams and stringers*			
Slope of grain: middle third	1/15	1/11	1/6
Slope of grain: balance of piece	1/12	1/8	1/6
Wane as fraction of any face	1/8	1/4	1/3
Grain	*medium*	*medium*	*medium*
Design values (psi), Douglas fir–larch†			
Extreme fiber stress in bending	1600	1350	875
Tension parallel to grain	950	675	425
Compression parallel to grain	1100	925	600
Compression perpendicular to grain	*625*	*625*	*625*

*WWPA *Grading Rules,* 1995, p. 149–153.
†WWPA *Grading Rules,* 1995, p. 247.
SOURCE: *WWPA Grading Rules.*

Grading

For structural purposes, clear straight-grained wood is the ideal. The greater the number and severity of defects, the lower the strength and cost of the member. This general relationship between timber quality and timber strength has been recognized for thousands of years, and was first formalized in grading rules nearly 250 years ago.[37]

Although no two timbers are identical, all contain a combination of growth characteristics, and all of these characteristics come from the same limited set. Grading rules provide a framework for classifying timbers with an infinite number of combinations of growth characteristics into a finite number of categories, based on the effects of those characteristics on their performance. Furthermore, they are part of a quality control system extending from the mill, via the supply chain, to the building site. Knowing the properties associated with each structural grade, the designer can *specify* with confidence. Correlating the grade stamp on each member with the timber specifications, the builder can *purchase* with confidence. In theory, anyone with a set of grading rules could fell a tree, saw it, and grade it. In practice, however, that is rarely done. There are many reasons for this, but the most important are the complexity of the rules for each species, the number of species, and the economies of scale associated with the grading process. Leaving grading to experienced graders working in sizable mills increases the quality of the grading, while lowering its unit cost.

Aspects of concern to the grader include wood density, slope of grain (including the effect of knots), shake, checks, wane, and decay.[38] In general, these aspects correspond to specific timber properties: overall strength, localized stiffness, shear resistance, integrity, volume, and soundness. In addition to grade, grade stamps typically indicate species, moisture content at time of surfacing, mill, and the name of the supervising agency.[39] To illustrate how selection criteria vary among different structural grades within a given group of species, consider a selected group of characteristics and design values given in the *Western Lumber Grading Rules* (see Table 2.1). Among the softwood species it covers are Douglas fir, lodgepole pine, and ponderosa pine, all of which are commonly used in sawn form in heavy-timber construction. Criteria constant across all of the indicated grades are shown italicized.

The relationship between timber characteristics and timber strength is obvious; the greater the slope of grain or wane, or the larger the knots, the lower the design values. Note that compression perpendicular to grain is constant across all three grades, while compression parallel to grain is roughly equal to it in the lowest grade. Why is this so? Slope of grain is typically a local phenomenon. As such, it represents a local improvement in perpendicular-to-grain performance and a local degradation in parallel-to-grain performance. Both compression design values are controlled by the worst conditions along a piece. Since the improvement is local rather than global, it does not affect the *overall* perpendicular-to-grain design value. Therefore, only the parallel-to-grain design value is affected by such a defect.

The strength differences between springwood and summerwood and between juvenile and mature growth are the bases for grain-related grading rules. Early in their lives and early in their growing seasons, trees in temperate climates grow more rapidly than they do late in their lives and in their growing seasons. This rapid growth is manifest in large and relatively weak thin-walled wood fibers. The higher the percentage of thick-walled summerwood or the greater the number of growth rings per unit of radial measure, the stronger the wood. This relationship is codified in the *Western Lumber Grading Rules* in the form of three categories of specific gravity: medium grain, close grain, and dense. The distinctions between them include those shown in Table 2.2 for Douglas fir.

Saw Blades and Wood Grain

Wood properties which differ across and along the grain necessitate saws tailored to their orientations of use. These are known as crosscut and rip saws, respectively. Since using the wrong tool endangers the user, the tool, and the work itself, a basic understanding of what happens at the wood/saw interface is useful for timber designers and builders alike. Nevertheless, all saws have certain common characteristics, independent of their use orientation:

The blades consist of thin steel sheets with teeth along one edge.[40] In hand saws, the cutting edge is usually straight, but may be slightly convex. The blade of a circular saw is a flat disk, the entire perimeter of which is toothed.

Provisions are made to minimize binding. Particularly when cutting through standing timber or where the workpiece flexes under the saw, the inside faces of the kerf or channel created by the teeth tend to squeeze against the outside faces of the blade. Tapering the blade away from its teeth reduces binding. Handsaw blades that taper

TABLE 2.2   Minimum Growth Ring Counts for Different Density Classifications

Category*	Rings/in (min.)	Rings/in (alternative min.)
Medium grain	4	3 or less, if summerwood averages 33% or more
Close grain	6	5, if summerwood averages 33% or more
Dense	6 and 33% summerwood	5 or less, if summerwood averages 50% or more

*WWPA *Grading Rules,* 1995, p. 202.
SOURCE: *WWPA Grading Rules.*

toward their back edge are *taper ground;* circular saw blades that taper toward their center are *hollow ground.* The width of the kerf being a function of the lateral geometry of the teeth, bending or setting the teeth slightly beyond the plane of the blade further reduces binding. Since the benefits of taper and set are similar, a taper ground blade requires less set than one that is flat ground (i.e., untapered).[41]

Provisions are made to remove cutting waste. Sawing is a subtractive process: The sum of the lengths or widths of the sawn pieces is slightly less than the dimensions of the original member. The reciprocation of the handsaw and the revolution of the circular saw not only cut the wood, but remove the waste that accumulates in the notches or gullets between the sawteeth.

The *differences* between crosscut and rip saws are found primarily in the details of their teeth. Crosscut teeth are filed at an angle to the work, creating knives that, because of their set, slice across the wood fibers in two parallel rows. Friction from the blade wears down the tiny uncut ridge left between the rows.[42] Ripping teeth, by contrast, are filed parallel to the work, creating chisels that chip off the ends of the wood fibers. Their set results in the removal of laterally adjacent wood chips by alternate teeth.[43] The condition of the blade is especially important during crosscutting:

> In early times it was discovered that a knife blade must be free from nicks and notches to cut well. Then it could be pushed against a piece of wood and a shaving whittled off. At about the same time it was noticed that if the nicked knife were drawn back and forth across the wood, it would tear the fibers apart, making saw dust.[44]

Environmental Factors

Construction materials do not exist in a vacuum. Regardless of how they are produced, transported, erected, utilized, and eventually disposed of, they have environmental consequences. One common way of comparing the environmental impacts of different construction materials is through life-cycle analysis, a process that examines both the direct and indirect impacts on the environment at each stage of their life cycles.[45] In the case of the structural components for a timber building, those stages could include growth, harvesting, transportation, manufacture, treatment, transportation, preparation, installation, disposal, recycling, and reuse or remanufacture. Similar analyses should be performed for the metals used in connectors and fasteners, the chemicals used for treatments and coatings, and the adhesives used for laminating.

It is important to recognize that timber was used for building long before the advent of the tools and manufacturing equipment typical of today's construction industry; that it has been and often continues to be worked with hand tools. Seen in this context, the environmental implications of timber construction can best be thought of as having two separate tracks, depending on the methods of harvesting, manufacture, fabrication, erection, and disposal. Unfortunately, unaware of any studies of the "low tech" track, I will confine my discussion to the mechanized processes more likely to be employed in contemporary large-scale heavy-timber construction.

Of the three dominant structural materials—wood, steel, and concrete—wood is the only one that is renewable. And, although steel will eventually break down in the presence of moisture, wood is the only one that is biodegradable. Furthermore, until the advent of structural adhesives and the large laminated panels and assemblies they enabled, wood was the only structural material the sizes of whose pieces were constrained by the sizes of the raw materials—steel members can be rolled to any length and welded up to any size, while the size of a concrete pour is essentially unlimited.

Environmental comparisons among structural materials are made increasingly difficult by the development of composites and by the spread of recycling. An example of the former is the fiber-reinforced glulam, which enjoys the benefit of lower wood volume for a given span, at the costs—monetary and environmental—associated with integrating high-performance plastic fibers into their construction. An example of the latter is the rapid growth in the use of scrap in the manufacturing of structural steel, drastically reducing the energy required during manufacturing. Also, because of code provisions permitting the use of wooden roof structures in many steel and concrete buildings, not all of their structural frames are composed purely of steel or concrete. Comparisons are further complicated by the fact that, with the limited exception of those using traditional joinery and employing wood foundations, most timber buildings employ steel and concrete in their connections and foundations, respectively.

Timber's advantages include its renewability, the energy content of its waste, its fire performance (compared to steel), its workability (minimizing the energy used in fabrication), and its lightweight cellular structure (minimizing the foundations required to support it, the energy needed to erect it, and the materials, if any, needed to insulate it). Offsetting factors include the environmental implications of any nontimber components of the structural system and of any nonstructural components—such as upright sprinkler heads above suspended ceilings—that the building code requires be used in conjunction with a timber frame.

Summary

Just as word processing has enabled the manipulation of word content, timber engineering has enabled the manipulation of wood content.

Nevertheless, the underlying properties of heavy timber materials are those of the wood and wood fiber of which they are composed. The unique structure of trees gives timber many of its unique properties.

Local strength is a function of the frequency of the springwood/summerwood bands or rings across the cross section of the tree.

- Since trees grow more slowly near the end of the growing season than near the beginning, the summerwood is denser and stronger than the springwood.

- Since trees grow more slowly as they age, the mature wood around their perimeters is denser than the juvenile wood at their cores.

- Since the competition for sunlight within yesterday's old-growth forests was greater than within today's managed forests, the timber resource is being replaced more quickly today, but with weaker material than in the past.

Performance of a given timber is a function not only of the density of the wood within it, but also of the orientation of the wood fibers with respect to its long axis; the more closely they are aligned, the higher the strength. Decay resistance is higher in the inactive heartwood than in the active sapwood. Shrinkage is highest circumferentially, somewhat lower radially, and negligible longitudinally; the distortions that result from shrinkage will therefore be a function of the proportions of a member's cross section, and of its location and orientation within the tree, prior to sawing. Warpage results when the growth-related prestressing forces in a member are not balanced as they were in the symmetrical cross sections of the original standing timbers. Because it is caused by shortening of outer fibers and lengthening of inner ones, its effect on a given piece is a function of how it is sawn. Member performance is also affected by the types, sizes, orientations, and spacing of knots, and by the grain variations associated with them. Creep is related to the magnitudes and durations of the applied loads.

Grading rules enable timber's infinite range of natural variations to be translated into a limited number of grades. Timbers are specified and purchased by grade, with the grade stamp as evidence of contract conformance. In general, higher grades are characterized by higher wood density, straighter grain, and smaller and fewer defects than lower grades. The rules for timbers relate primarily to their structural properties, while those for finish materials relate primarily to their appearance. Because timber construction represents a fusion of architecture and structure, it is worthwhile, whether you are a designer or a builder, to familiarize yourself with the grading rules pertaining to your project, so that you know what to expect of its timbers.

The directional properties of timber have influenced the evolution of saw blades. Binding of the blade can be minimized by tapering it or by setting its teeth. Waste is transported away from the cutting surfaces by notches between the teeth, and expelled from the blade as it clears the workpiece. Crosscut teeth, like miniature knives, slice across the wood fibers, while ripping teeth, like miniature chisels, chip away at their ends.

In getting from the forest to the site, timber is involved in three separate processes: growth, seasoning, and production. Defects can arise during any of them. Historically, our control over growth was limited to distributing the trees during planting and pruning them as they grew, but our control is increasing all the time. By contrast, our control over seasoning has long been substantial, enabling checking to be minimized. Finally, our control over production is virtually total, not only minimizing waste, but assuring that sawn members are graded to enable their highest and best use.

What happens to timber once it reaches the job site depends on whether the construction it will be part of will be temporary—formwork, shoring, or bracing—or permanent. Though temporary members need only satisfy structural criteria, permanent members must also satisfy architectural ones. That, plus the abuse to which temporary members are often subjected, typically makes them unsuitable for reuse in the permanent construction.

What happens to the scraps generated during sawing, fabrication, and erection, and to the timbers within structures to be demolished, depends on their sizes, lengths, compositions, and conditions. From any environmental perspective, one subsequent use is more desirable than another if it can be achieved by adding less energy or producing less pollution. Consequently, structural reuse is the most desirable approach, and resawing to enable reuse is next. Way down the list are burning or landfilling the waste. Notwithstanding the importance of environmental issues, they represent only one set of selection criteria among many that should be considered when deciding whether to use heavy timber for a building's superstructure.

Notes

1. Weiner, p. 79.
2. Weiner, p. 79.
3. Faherty, p. 1.3.
4. Breyer, p. 140.
5. Faherty, p. 1.3.
6. Courtenay in Mark, p. 184.
7. Breyer, p. 140.
8. Panshin, p. 28.
9. USDA FPL, *Wood Handbook,* p. 2.4.
10. CWC, *Wood Reference Handbook,* p. 85.
11. Telecon with David Kretschmann, USDA/FPL *Research Engineer.*
12. CWC, *Wood Reference Handbook,* p. 86.
13. Panshin, p. 293.
14. Faherty, p. 1.4.
15. Gordon, p. 159.
16. Panshin, p. 293.
17. Faherty, p. 107.
18. Panshin, p. 298.
19. Gordon, p. 282.
20. Breyer, p. 155.
21. Panshin, p. 296.
22. Jozsa, p. 12.
23. Jozsa, p. 12.
24. Panshin, p. 291.
25. Panshin, p. 289.
26. Panshin, p. 291.
27. Panshin, p. 334.
28. Jozsa, p. 25.
29. Panshin, p. 233.
30. Faherty, p. 1.15.
31. Breyer, p. 33.
32. Smulski, "Detailing for Wood Shrinkage," p. 54.
33. AITC, *Tech Note 11,* p. 1.
34. AITC, *Tech Note 11,* p. 1.
35. Harris, p. 534.
36. Panshin, p. 231.
37. *WWPA Grading Rules 95,* p. 3.
38. Faherty, p. 2.11.
39. Smulski, "Lumber Grade Stamps," p. 70.
40. Graham, p. 166. According to Courtenay in Mark (p. 183), steel-edged blades were used widely as far back as the Middle Ages.
41. Miller, p. 4.
42. Graham, p. 170.
43. Graham, p. 172.
44. Graham, p. 169.
45. CWC, *Case Study,* p. 2.

CHAPTER 3

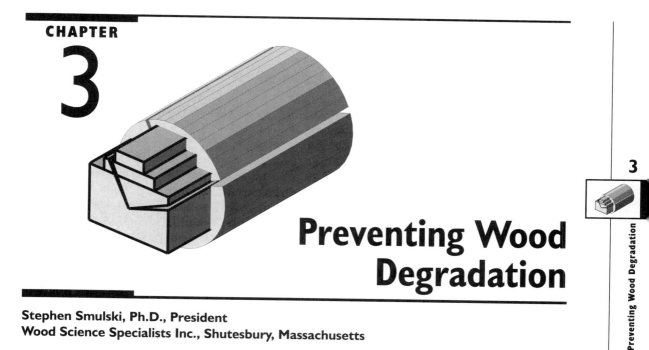

Preventing Wood Degradation

Stephen Smulski, Ph.D., President
Wood Science Specialists Inc., Shutesbury, Massachusetts

*The following trees are to be rejected: those that grow in cemeteries,
roads, temples, ant-hills . . . those that are stunted, those that are
joined with others, those that are infested with creepers, that have been
struck down by lightning and storm, those that have fallen by
themselves, that have been broken by elephants, that have withered,
are burnt by fire and that contain bee-hives . . . One should go to the
desired tree and offer worship to it with food and flowers.*

6TH CENTURY INDIAN TEXT[1]

Introduction

Up to 10 percent of the timber harvested each year in the United
States is used to replace wood that has degraded in service. This need
not be the case. Damage to wood products in buildings and other structures by decay fungi, insects, sunlight, and other destructive agents is
entirely preventable. The presence of degraded wood points to inappropriate use of products, building design flaws, poor workmanship, and
neglected maintenance.

The serviceability and structural integrity of wood products do not
degrade simply with the passage of time. Loss of function or strength
requires a causal agent that is present, most typically, because of elevated wood moisture content. As such, the most important consideration in protecting wood products from deterioration is to keep them dry.
As evidenced by 3500-year-old wooden furniture and statuary recov-

ered from the desert tomb of Egyptian King Tutankhamen, wood that is kept dry and protected from direct exposure to the elements will last indefinitely.

The principles and practices for preventing degradation of wood products in service are well established and, if followed, will ensure durable, long-lived wooden buildings and structures. Design features, construction practices, and finishes that promote water shedding protect wood products from the deleterious effects of exposure to the elements. In situations where wood cannot be kept dry, preservatives defend wood products against attack by decay fungi, insects, and other organisms. Fire retardants reduce wood's combustibility and slow the spread of flames across its surfaces. Incorporating the appropriate protective measures into the design and construction of a wood-frame structure is key to maintaining the serviceability and structural integrity of wood products for the life of the structure.

How Wood Products in Buildings Get Wet

Experience has shown that more than 95 percent of the performance problems with wood products in buildings and other structures occur because the wood gets wet. Wood products in structures can be wetted by ground water; piped water; condensation; and dew, rain, and snow melt. In some cases, water naturally contained in sawn timbers at the time of erection is to blame.

Ground Water

Soil surrounding building foundations and floor slabs virtually always contains water in both liquid and vapor form. Moisture can enter into basements and crawl spaces by gross flow and capillary suction of liquid water, and by diffusion of water vapor. Liquid water can leak into basements and crawl spaces through shrinkage and settlement cracks, joints, utility cutouts, and other penetrations in foundation walls and floor slabs. It can also be drawn by capillarity from soil into the micropores inside concrete and masonry that lacks dampproofing or waterproofing. Water vapor can move through walls and footings that lack dampproofing or waterproofing, and through floor slabs that lack a vapor retarder. Upon reaching an exposed surface, the water evaporates, sometimes leaving behind a telltale crystalline efflorescence or surface bloom. Evaporating water increases the relative humidity of air inside the basement or crawl space, which, in turn, can raise the moisture content of sills, girders, joists, and subflooring to mildew- and mold-susceptible levels. Often, liquid and capillary leakage are intermittent and associated with seasonal fluctuations in ground water level and heavy rain.

Gross leaks in basements and crawl spaces tend to be noticed and repaired before significant damage is done. In most cases, entry of liquid water into basements and crawl spaces can be eliminated by installing perimeter drains, by sealing cracks and other points of entry, by applying dampproofing or waterproofing to the exterior of foundation walls, by backfilling with free-draining soil, by grading soils so that they slope away from the foundation, and by installing gutters and downspouts along eaves. Dampproofing is often confused with waterproofing. Dampproofing materials such as bituminous liquids, cementitious coatings, and surface bonding mortar retard vapor transmission and capillary movement of water through concrete and masonry, but are considerably less effective in stopping liquid water. Only waterproofing compounds such as bituminous and elastomeric membranes, and bentonite coatings prevent the passage of liquid water through concrete and masonry.

Basement and crawl space moisture problems sometimes arise not because of gross leaks, but because of capillary water and water vapor. Transport of water into a basement or crawl space via capillary suction or diffusion of vapor through walls is controlled by applying dampproofing to a foundation's exterior. Movement of moisture into a basement or crawl space via capillarity and diffusion through the floor is prevented by installing a vapor retarder—most commonly 8 or 10 mil polyethylene—under the slab or over exposed soil in a crawl space. In addition to making a basement warmer, rigid foam insulation placed against foundation walls and under floor slabs also blocks capillary and diffusion ingress of water.

Sills, girders, and other framing members in direct contact with concrete and masonry are especially prone to mildew, mold, and decay because capillary water can travel unimpeded and undetected from the soil through the foundation and directly into the wood. Wetting of wood by capillary water is prevented by inserting a "capillary break" of metal or plastic between wood and concrete and masonry. The circle of dampness that sometimes shows up around the base of a foundation owes to capillary movement of water from the soil through the footing and up into the wall. Known as "rising damp," this can be stopped by placing polyethylene sheeting over the footing before walls are cast or built.

Piped Water

The labyrinth of pipes inside a wood-frame building that provides water and, sometimes, heat, occasionally leaks. Forceful plumbing leaks are typically discovered and corrected immediately. Though wood may be temporarily saturated with water as a result, it dries out before fungi can get established. Slow, persistent leaks hidden inside walls and floors, however, can go undetected for long periods, and can elevate the moisture content of wood to fungi- and insect-attractive levels. Condensation dripping from cold water supply lines onto wood often leads to decay, especially in humid basements and crawl spaces. Prevention entails insulating cold water supply lines. Providing access to critical connections for purposes of inspection and repair should be considered during the plumbing layout design phase.

Condensation

Condensation of water vapor from warm, moist air that leaks inside walls, floors, ceilings, attics, basements, and crawl spaces can create conditions favorable to mildew, mold, decay fungi, and insects. Water vapor that diffuses into these spaces can do the same. Condensation-caused problems are often an unintended consequence of the use of energy-efficient building materials and construction techniques. While improving occupant comfort and reducing heating costs, increased insulation, gasketed doors and windows, continuous vapor retarders, air sealing measures, and air infiltration barriers also substantially reduce the number of air changes per hour in energy-efficient buildings. Unless active mechanical ventilation is provided, water vapor generated by occupants' activities and released from other interior sources is carried into walls, floors, ceilings, attics, basements, and crawl spaces on convection currents of air flowing through joints between materials; penetrations in walls, floors, and ceilings; and other hidden air leakage paths. When warm, moist air entering into these spaces is cooled below the dew point, the excess moisture (the amount above 100 percent relative humidity) is deposited as condensation on framing, sheathing, and other cold surfaces.

Significant sources of indoor moisture usually include soil moisture migrating through foundations and floor slabs; showering and bathing; humidifiers; clothes dryers that are not vented to the outside; indoor storage of firewood; and backdrafting of gas-fired appliances (Table 3.1). Seemingly minor sources such as occupant respiration and perspiration, cooking, dishwashing, floor mopping, plants, aquariums, and seasonal desorption from building materials, when considered in toto, can make a sizable contribution.

In northern climates, condensation within walls, floors, ceilings, attics, basements, and crawl spaces can occur during both the heating and cooling seasons. Condensation on the inside of windows and within walls, floors, ceilings, and attics happens mostly during the heating season, and usually because of excessively high indoor relative humidity. Warm, moist air from living areas entering into these spaces is cooled below the dew point; the excess moisture is deposited as frost or ice that later wets framing and sheathing to fungi-favorable levels when it melts. In summer, water vapor held in hot, humid outdoor air entering into basements and crawl spaces can condense on cooler framing and subflooring, creating conditions irresistible to fungi.

In southern climates, condensation is a problem primarily during the cooling season. In this case, the culprit is hot, humid outdoor air that seeps into walls, floors, ceilings, attics, basements, and crawl spaces cooled to below the dew point, most often by air conditioning. The prob-

TABLE 3.1 **Indoor Moisture Sources**

Source	Estimated release (pints)
Occupant generated	Variable evaporative loss
Aquariums	
Bathing	
Tub (excluding towels and spillage)	0.12
Shower (excluding towels and spillage)	0.52 per 5 min
Clothes washing (lid closed, standpipe discharge)	Virtually 0 per load
Clothes drying	
Dryer vented outside	Virtually 0 per load
Dryer vented inside or indoor line drying	4.7–6.2 per load (more if gas dryer)
Combustion	
Unvented kerosene space heater	7.6 per gal of kerosene burned
Cooking	
Breakfast (4 persons)	0.35 (+ 0.58 if gas range)
Lunch (4 persons)	0.53 (+ 0.68 if gas range)
Dinner (4 persons)	1.22 (+ 1.58 if gas range)
Simmer in 6-in pan at 203°F for 10 min	<0.01 if covered, 0.13 if uncovered
Boil in 6-in pan for 10 min	0.48 if covered, 0.57 if uncovered
Dishwashing	
Breakfast (4 persons)	0.21
Lunch (4 persons)	0.16
Dinner (4 persons)	0.68
Firewood stored indoors (1 cord green wood)	400–800 per 6 mon
Floor mopping	0.03 per sq ft
Gas range pilot light (each)	≤0.37 per day
House plants (5–7 average plants)	0.86–0.96 per day
Humidifier	To 120+ per day (average 2.1 per h)
Pets	Fraction of human adult weight
Respiration and perspiration (4 persons)	0.44 per h
Refrigerator defrost	1.0 per day average
Sauna, steambath, whirlpool	To 2.7 per h
Naturally occurring	
Combusion	
Gas-fired appliance exhaust backdrafting	To 6720+ per year
Desorption of building materials	
New construction	10+ per day average
Seasonal	6.3–16.9 per day average
Dew, rain, or snow melt penetration	Variable
Piped water leaks	Variable
Roof leaks	Variable
Seasonal high outdoor absolute humidity	64–249 per day
Soil moisture migration	To 105 per day

SOURCE: Adapted from Angell and Olson, 1988.

lem is especially prevalent in basements and crawl spaces under air-conditioned rooms. Condensation in cooling climates forms on the outside of windows in air-conditioned rooms.

Condensation-caused problems can be largely avoided by reducing indoor relative humidity by eliminating potential sources of indoor moisture through judicious design and construction; by placing vapor retarders against the warm side of walls, ceilings, and floors; by venting clothes dryers, heating appliances, and kitchen range, bath, and

3

Preventing Wood Degradation

other exhaust fans directly to the outside; by providing the needed ceiling insulation and roof ventilation; and, to a limited degree, by dehumidification.

Wetting by the Elements

Siding, trim, windows, and doors are routinely wetted by dew, rain, and snow melt. Protecting these and other exterior wood products from fungi and insects requires that building exteriors be designed and constructed to shed water. Wide eaves, and gutters and downspouts, prevent water running off of a roof from cascading directly down walls, or splashing back onto walls from a lower roof, deck, or the ground. Siding and trim should be at least 8 in off the ground to minimize these splashback effects. Appropriate flashing must be installed in roof valleys and at roof/wall intersections to prevent water that seeps past shingles and siding from entering inside walls. A cricket should be built on the upslope side of any chimney that does not penetrate the roof at the peak. Flashing must be placed around windows and doors to prevent water from entering into walls at these and other openings. Likewise, the exposed tops and ends of heavy-timber beams and arches that terminate at or project beyond the eaves must be sloped and flashed to deter absorption of water. Because they draw in water more than 10 times as fast as side grain, wood's end grain surfaces are especially vulnerable to decay. Repeated cycles of swelling and shrinking occasioned by entry of water into end grain is why exterior coatings usually crack and peel first at joints. Liberal application of a water-repelling coating or caulking to both end grain surfaces before a joint is closed is effective in retarding absorption of water. The back and ends of siding and trim should be backprimed, that is, finished with a water-repelling coating before being installed to prevent water from being drawn by capillarity into joints and the overlap between courses. Siding, trim, windows, and doors should be finished with opaque film-forming finishes that combine high water repellency with high permeability to water vapor. Opaque finishes are necessary to prevent degradation of wood surfaces by ultraviolet light from the sun. Foundation plantings should be trimmed at least 12 in away from walls to provide the airflow needed for drying, and sprinklers positioned so that water is not sprayed onto siding and trim.

Under most circumstances, exterior wood products dry sufficiently between wettings from the combined effects of water-shedding designs, good construction practice, use of film-forming finishes, and sun and wind. In climates where wettings are frequent and intense, however, exterior wood products may not dry sufficiently between wettings to deter fungi and insects. In this case, either naturally decay-resistant species or preservative-treated wood should be used for siding, trim, and other exterior products.

Water Built-In During Construction

The sawn timbers used in heavy-timber construction are often in the green or water-saturated condition when installed. Normally, the water is slowly released to the air inside the building as timbers dry. Under certain circumstances, however, water in the timbers can create condensation problems or permit the timbers to decay. An 8-ft-long hemlock nominal 8×8, for example, contains about 9 gal of water when green (about 97 percent moisture content). When it eventually dries to the 12 percent moisture content typical of framing in heated buildings, it will contain about 1 gal of water, the other 8 having been released to the air or surrounding materials. The total water given off by a green timber frame can count in the hundreds of gallons, with the greatest release occurring during the first heating season. Severe window condensation that damages sashes, sills, and walls is often the result. The problem can be avoided by dehumidifying a timber-frame structure during the first heating season.

Decay can develop in green timbers that are partially or fully boxed in with framing, sheathing, or finish materials shortly after erection. This severely retards the drying of timbers and can keep the wood at a decay-susceptible moisture content for years. Decay of boxed-in timbers is a problem mostly with woods that lack natural resistance to decay, such as hemlock, southern pine, and red oak. Solutions include using partially air-dried timbers and green timbers treated with borate preservatives. The problem does not exist when glulam timbers or parallel strand lumber are used in heavy timber construction because these and other engineered wood products are manufactured at about 12 percent moisture content.

Degradation of Wood Products in Buildings

Wood products in service can be degraded by biological organisms, excessive heat, sunlight, corroding metal, chemicals, or a combination thereof. Degradation of wood products can result in either structural failure or service failure. Structural failure involves a substantial loss of wood strength that threatens a wood product's ability to safely support loads. Service failure refers to a reduction in the functional utility or esthetics of a wood product. Extensive decay in a timber girder, for example, represents a structural failure, whereas excessive vibration of the floor supported by a glulam timber is a service failure.

Natural Decay Resistance of Wood

As seen on the end of a log or stump, a tree consists of a central core of (usually) darker heartwood surrounded by a shell of light-colored sapwood (Fig. 3.1). Only the heartwood of a tree may possess appreciable

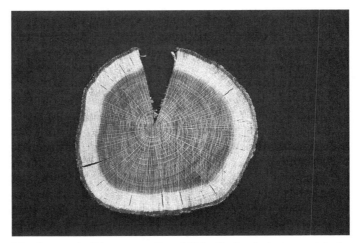

3.1 As seen in this cross-sectional disk of red oak, trees consist of a light-colored shell of sapwood surrounding a core of (usually) darker heartwood. Only the heartwood of a tree may have appreciable natural resistance to decay; regardless of the species, the sapwood has none. (Wood Science Specialists Inc.)

natural resistance to decay because of chemicals called *extractives* that form in the heartwood of some species. Regardless of the kind of wood, sapwood lacks extractives and has no natural resistance to decay. The natural decay resistance of heartwood varies widely, not only among different kinds of wood, but among different pieces of the same wood, depending on the toxicity and concentration of the extractives (Table 3.2). The heartwood of some kinds of wood—western red cedar, redwood, bald cypress, and white oak, for example, is highly resistant to decay. As a result, the heartwood of these species has been used tradi-

TABLE 3.2   Relative Heartwood Decay Resistance of Some Construction Woods

Resistant or very resistant	Moderately resistant	Slightly or nonresistant
Bald cypress (old growth)	Bald cypress (young growth)	Hemlocks
Cedars	Douglas fir	Oak
Chestnut	Larch, western	red and black species
Oaks	Pine, eastern white	Pines (other than
Bur	Southern pine	eastern white
Chestnut	longleaf	longleaf, and slash)
Gambel	slash	Spruces
Oregon white	Tamarack	True firs
Post		
White		
Redwood		

SOURCE: Adapted from Forest Products Laboratory, 1987.

tionally for siding, shingles, shakes, sills, and other exterior and ground contact uses. The heartwood of most construction woods, however, has only slight-to-moderate natural resistance to decay and must be protected from direct wetting or treated with preservative to prevent decay.

Because so many variables are involved, it is impossible to say just how long the heartwood of any kind of wood will last when used outdoors or in ground contact. The most important determinants of wood's service life outdoors are the decay hazard posed by the local climate and the way in which the wood is used (Fig. 3.2). All other things being equal, exterior wood will last longest in the arid Southwest, shortest in the warm, humid Southeast, and somewhere in between in the cooler, drier Midwest and Northeast. While vertically oriented siding, trim, and columns readily shed water, exposed beams, window sills, and other horizontally oriented members let water pond, encouraging decay. Wood in ground contact will succumb to decay and insects faster than wood above grade. The service life of wood products used outdoors but above grade can be extended by decades by judicious use of

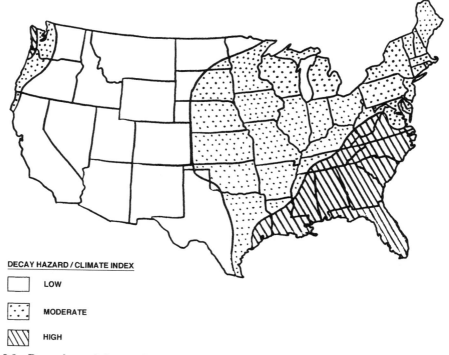

DECAY HAZARD / CLIMATE INDEX

	LOW
	MODERATE
	HIGH

3.2 Decay hazard for wood exposed above ground, by climate index. (Adapted from Scheffer, 1971.)

design features, construction practices, and finishes that promote water shedding. Likewise, preservative treatment can lengthen the service life of wood in contact with concrete, masonry, or the ground by 50 years or more.

Dramatic changes in the forest resource in the last few decades have altered the natural decay resistance of wood products, and greater reliance must be placed on keeping wood dry or treating it with preservative to avoid decay. Historically, heavy timbers were sawed entirely from the heartwood of large-diameter, old-growth trees, and gave long service life even in high decay-hazard environments. Today, however, the trees from which timbers are being manufactured are much smaller in diameter. As a consequence, sawn timbers often have decay-susceptible sapwood at their corners. (See Fig. 3.3.) Also, the decay resistance of the heartwood these second- and third-growth trees is less than that of the heartwood of old-growth trees. Increasingly, engi-

3.3 Sapwood on the corner of this southern pine timber rotted because water leaking through the roof was trapped between the timber and decking; the more decay-resistant heartwood is still intact. (Wood Science Specialists Inc.)

neered wood products like glulam timbers and parallel strand lumber are being used in heavy-timber construction. As these products may be almost all sapwood, they lack decay resistance. Only preservative-treated engineered wood products should be used in exterior and other high decay–hazard applications.

Weathering of Unfinished Wood

Direct exposure to the elements triggers in the surface of unfinished wood irreversible changes in color and physical appearance due to the combined effects of sunlight, water, and biological organisms in a process known as *weathering*. During the first several weeks of exposure, virtually all woods begin to develop a yellow or brown color as naturally occurring water-soluble extractives in the wood rise to the surface and are altered by sunlight and water. Light-colored woods tend to darken, while darker woods lighten. Ultraviolet light breaks down lignin, the natural adhesive that binds wood fibers, which are made largely of cellulose. Within a few months, the wood turns a pleasing silvery gray as more and more of the lignin and extractives are washed away by rain, leaving behind a surface of light-colored cellulose. In most regions of the United States, however, the silvery gray color is temporary, and the wood deepens to dark gray, black, or brown as mildew colonizes its surface. Wood retains the much sought after silvery gray color only in regions with low rainfall and in coastal climates where salt air inhibits the growth of mildew.

With each cycle of wetting and drying, surface wood fibers swell and shrink. Over the next 1 or 2 years, microscopic checks open between fibers; fibers begin to lift from the surface; and the grain raises. Eventually, loosened fibers are eroded from the surface by rain, wind, and freeze/thaw actions. The process of weathering slows markedly after about 2 to 3 years, proceeding almost imperceptibly from then on.

The rate at which unfinished wood erodes varies widely depending on climate, severity of exposure, and characteristics of the wood, including density, growth rings per inch, percent earlywood and latewood, and growth ring orientation. In general, unfinished softwoods erode at a rate of about 0.25 in per century, while the rate for hardwoods is about 0.125 in per century. Wood in regions with higher annual rainfall and temperature weathers faster than wood in areas with lower rainfall and temperature. Wood weathers fastest on a building's southern exposure, somewhat slower on the western and eastern exposures, and slowest of all on the northern exposure. Quartersawn hardwood lumber and vertical- or edge-grained softwood lumber weathers more evenly than plain-sawn hardwood lumber and flat- or slash-grained softwood lumber, largely because the bands of softer, more easily eroded earlywood are substantially wider in the latter.

Fungal Degradation of Wood Products

Wood products in service can be affected by various types of primitive plants known as *fungi*. Unable to produce their own food, fungi feed instead on carbohydrates stored inside wood cells or on the cell wall substance itself. Mushrooms that sprout from tree trunks, stumps, and other infected wood are fungal "fruits." They release millions of dust-size "seeds" called *spores* that are scattered helter-skelter by air currents. When the conditions of the wood surface they eventually settle on are right, spores germinate, sending out threadlike filaments called *hyphae*. Enzymes secreted by hyphae break down organic matter so fungi can use it for food. Two broad groups of fungi attack wood: nondecay fungi that merely discolor wood, and decay fungi that cause it to rot.

Nondecay Fungi

Nondecay fungi include mildews, molds, and sap stains. These fungi feed on sugar and starch stored inside wood cells, on airborne organic detritus, and on the organic ingredients found in certain finishes. In the process, wood surfaces and finishes are discolored.

Mildews

Occurring outside and inside buildings, most mildews are black, but reds, greens, blues, and browns are possible. Masses of dark spores and hyphae give mildews their characteristic splotchy look. Merely discoloring the surface they grow on, mildews have no appreciable effect on wood itself.

Mildews appear most often on unheated, projecting parts of buildings that cool quickly after sunset, like eaves, decks, and porch ceilings. North-facing walls and those shaded by foundation plantings, trees, and other obstructions that restrict sunlight and airflow are also candidates. Mildew's location often mirrors a building's dew pattern. Absent where siding crosses "hot spots" over framing and other thermal bridges, mildews may thrive where dew persists over cooler, insulated bays in between. Mildews on the outside of buildings can be largely avoided by finishing wood products with coatings containing a mildewcide.

Mildews occur indoors most frequently in crawl spaces, basements, baths, and other areas prone to high relative humidity. They also show up in places with poor air circulation such as closets and closed-off rooms. Mildews can appear whenever the relative humidity of air near a surface exceeds 70 percent. Spores and musty odors emitted by mildews growing indoors can trigger allergic reactions in susceptible persons.

Thermal bridges that lead to "hot spots" outside create "cold spots" inside. Exterior corners are notoriously mildew prone because of poor air circulation inside and heat-robbing windwashing outside. In summer, water vapor from warm, humid outdoor air entering crawl spaces and basements below air-conditioned rooms may condense on cooler framing and subflooring, creating conditions irresistible to mildews, as well as molds and decay fungi. Moisture condensed as frost or ice from heated air leaking into walls and attics in winter likewise wets framing and sheathing when it melts, encouraging mildews.

Preventing mildew inside buildings lies wholly in controlling indoor air moisture levels and condensation potential through proper site drainage and dampproofing, and use of soil covers, vapor retarders, insulation, and ventilation as ambient conditions call for.

Molds

Molds need a wood surface moisture content of about 20 percent to get started. Although most are green, black and orange molds are not uncommon. Color comes from spores strewn across surfaces. Though hyphae reach deeper into wood, discoloration in softwoods tends to be limited to the surface of the sapwood. It can usually be planed, sanded, or even brushed off. Brown, gray, or black patches penetrate more deeply into hardwoods and cannot be machined away. Discoloration aside, mold's effect on wood is generally inconsequential. Flourishing in damp crawl spaces and basements, and in poorly vented attics, molds form a living veneer on framing and sheathing. Molds are controlled with the same methods used for mildews.

Some molds are tolerant of wood preservatives. This explains the fuzzy growths occasionally found between timbers and boards in banded shipments of solid-piled preservative-treated wood. Molds die once the wood dries, but can be washed off beforehand with a dilute solution of bleach or a commercial fungicide.

Sap Stains

Discoloration of wood by sap stain fungi happens almost exclusively in logs and freshly sawn timbers and lumber. As a precaution, rough wood products are often dipped in a fungicide immediately after sawing. Staining fungi invade only the sapwood portion of logs, timbers, and lumber, and are most troublesome in softwoods. The steel gray to blue-black discoloration they cause in softwoods—called *blue stain*—and the brown hues in hardwoods, are due to pigmented hyphae that permeate cells in the sapwood in search of stored carbohydrates. Inactive sap stains are routinely found in timbers and lumber. The deep-reaching stains are indelible. In finding food, sap stains destroy certain wood cells. As a consequence, wood loses a little bit of its strength and toughness, and becomes substantially more permeable and more absorbent to liquid water. Thus, conditions conducive to the growth of decay fungi exist for longer periods in blue-stained wood that gets wet in service, increasing the risk of rot.

Decay Fungi

While discoloration by mildews, molds, and sap stains is only an appearance problem, decay fungi threaten the structural integrity of wood products. Wood decays or rots because decay fungi eat the actual wood cell-wall substance. Before decay fungi can colonize wood, four requirements must be met: an oxygen supply, adequate temperature, a supply of sufficient moisture, and a food source (wood). Infection can be prevented by eliminating any one of the requirements.

Because fungi take oxygen from the air, it is not possible to prevent decay of wood products in service by limiting oxygen. Wood that is saturated with water, however, does not contain enough oxygen to support decay fungi. This is one reason why stored logs are continuously sprinkled with water or floated in a pond prior to sawing. It also explains why untreated wood foundation piles and piles supporting freshwater and marine structures do not rot below the low waterline. Even the cottage industry that salvages water-logged old-growth trees and logs from the muck of swamps, rivers, and lakes for conversion into specialty timbers and lumber takes advantage of this fact.

Limiting decay by controlling temperature is generally impractical because, like most living things, fungi thrive in the 40 to 100°F range. Even at subfreezing temperatures, many fungi do not die; they simply go dormant. The high temperatures used in kiln-drying lumber kill fungi and insects present in green wood, and, in effect, sterilize lumber. Because it is uneconomical to kiln-dry members larger than about 4 × 4 in, timbers are sold at a green or air-dry moisture content and may already be infected with fungi or insects when incorporated into a structure.

Moisture content is the critical factor determining wood's susceptibility to decay. Hence, the most effective "method" of preventing fungal deterioration is to keep wood dry. Wood moisture content must exceed 28 percent, and liquid water must be present in cell cavities before fungi can infect wood. Once established, some fungi can carry on their destruction at a moisture content as low as 20 percent. When moisture content falls below this level, all fungal activity ceases. This is one reason why construction lumber is dried to 19 percent moisture content or less. With the moisture content of wood indoors over most of the United States cycling annually somewhere between 6 and 16 percent, it is too dry for most fungi to get started.

Wood can be made highly decay resistant by eliminating it as a food source for fungi and other organisms by impregnating it with a preservative. Use of preservative-treated wood is often the only option for preventing decay and other biological degradation in wood products embedded in soil, submerged in seawater, or used in other hazardous environments.

Decay fungi fall into three major groups: brown rots, white rots, and soft rots.

Brown Rots

Brown rots are so-named because infected wood turns dark brown (see Fig. 3.4). Most commonly colonizing softwoods, brown rots consume cellulose, hardly touching the darker lignin. Mycelia appear as white sheet-like or fluffy growths on wood surfaces. Brown-rotted wood shrinks excessively and splits across the grain as it dries. Friable and crumbly, surfaces then show brown rots' hallmark cubical checking.

Water-conducting fungi are a special type of brown rot that show up infrequently in the Southeast, Northeast, and Pacific Northwest. Sometimes called dry rot fungi, the name unfortunately suggests that dry wood can decay. What contractors, building inspectors, and others routinely mislabel "dry rot" is almost always, in reality, wood that got wet, rotted, and dried out before discovery. Unique in their ability to

3.4 Brown rots most often attack softwoods and are easily recognized by the characteristic cubical checking that results. (Wood Science Specialists Inc.)

pipe moisture from the soil over long distances through root-like rhizomorphs, water-conducting fungi wet otherwise dry wood in advance of their attack. Infecting softwoods and hardwoods, their light-colored mycelia look like large, papery, fan-shaped sheets. Dirt-filled porches, damp crawl spaces, and wood in ground contact are avenues for entry.

White Rots

White rots impart a white, gray-white, yellow-white, or otherwise bleached appearance to wood (see Fig. 3.5). Most often infecting hardwoods, they feed on both cellulose and lignin. In advanced stages of decay, white-rotted wood is spongy, has a stringy texture, and lacks the cubical checking of brown-rotted wood. A thin black line often marks the advancing edge of incipient white rot in hardwoods. Ironically, this partially decayed wood, called *spalted wood,* is coveted by woodworkers for its unique figure.

Soft Rots

As the name suggests, soft rots impart unusual softness to wood. Their attack is generally limited to a shallow layer of wood on a product's surface that, when scraped away, reveals sound wood. These fungi prefer hardwoods, but attack softwoods as well. When dry, soft-rotted wood is similar in appearance to brown-rotted wood, but with a finer pattern of cubical checking. Wood products that are constantly wet or repeatedly wetted and dried—bridge timbers, freshwater piles, utility poles,

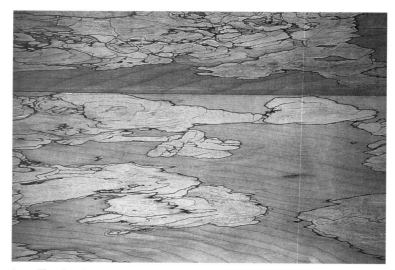

3.5 The thin black lines that mark the leading edge of white rot damage in hardwoods give this so-called *spalted wood* a unique figure prized by woodworkers. (Wood Science Specialists Inc.)

and railroad crossties, for example—are favored by soft rots. Rarely found inside buildings, soft rots occasionally degrade wooden shakes and shingles on heavily shaded roofs in wet climates.

Bacteria

Bacteria are unique in that they can degrade wood that is water saturated. Living trees, logs stored in ponds before milling, sinker logs recovered from rivers and lakes, and foundation piles are occasionally affected. Fortunately, bacterial degradation of wood proceeds slowly, and strength loss is usually slight. However, timber foundation piles embedded in wet soil for decades can show appreciable strength loss. Living hardwood trees—especially oaks—growing on flood plains or other wetlands are infrequently infected with bacteria. Timbers sawed from infected trees may emit a rancid odor and are often prone to excessive checking, splitting, and ring separation as they dry in service.

Detecting Decay

In its incipient and early stages, decay is difficult to detect. Confirmation of infection requires that wood be examined with a microscope for the presence of fungal hyphae, boreholes, and other cellular damage. Strength loss, however, can be appreciable even in the early stages of infection. As decay advances, wood's luster fades. Surfaces become dull and discolored. A musty odor is often evident. The rate at which decay progresses depends on wood moisture content, temperature, and the specific fungus. In the advanced stages of decay wood is visibly discolored, spongy, and musty. Surfaces may be stringy, shrunken, or split across the grain. Cottony masses of hyphae, called *mycelia,* as well as fruiting bodies, may be present. Decay extends deep into wood; strength loss is significant.

Measuring wood moisture content with a moisture meter is useful in determining whether decay is active in wood that is suspiciously wet or discolored, but otherwise looks okay. If the moisture content is 20 percent or below, there is no active decay present. If it is between 20 and 28 percent, existing decay can continue its attack. At moisture contents above 28 percent, conditions are ripe for a new infection to get started, or for an established one to continue.

The pick test is also useful in detecting decay. Here, the soundness of wood is judged from the way a large splinter breaks when pried from it with an awl or ice pick. Sound wood emits a sharp crack as the splinter is pried up. The splinter is typically long, with one end still attached to the wood. Sometimes it breaks in the middle over the tool, but the fracture will still be splintery and fibrous. A splinter pried from wood with incipient decay lifts quietly from the surface and almost always fails brashly directly over the tool, with both ends anchored to the wood. The

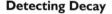

pick test is highly subjective; natural characteristics of sound wood can produce misleading results. Accurate interpretation comes only with experience and consideration of other clues.

Decay hidden inside timbers can be revealed by examining turnings ejected from a bored hole. Discolored, wet, and musty shavings signal decay. Plug the hole with a preservative treated dowel.

Dealing with Decayed Wood

The first and most important thing to do once decay is discovered is to locate the source of the water. Check for the obvious—roof and plumbing leaks, and missing or punctured flashing. Look for stains and drip tracks symptomatic of ice dams. Are eaves wide enough to prevent water from cascading down walls? Are gutters poorly maintained or missing? Do finish grades slope toward or away from the foundation? Are foundation cracks admitting water? Is untreated wood in direct contact with concrete, masonry, or soil? Check to see if crawl spaces have soil covers, and if ventilation and/or insulation is present, adequate, and properly placed. The same goes for attics. Peeling and blistering paint sometimes signals excessively high indoor relative humidity, inadequate interior ventilation, or a missing vapor retarder. Water stains on framing and sheathing inside walls suggest condensation. To make the remedy permanent, you must cure the disease—water infiltration, not just treat its symptoms—mildew, mold, and decay.

Once the source of water has been eliminated, remove as much decayed wood as is practical and economical. This is especially important with girders, columns, and other structural members whose load-carrying ability may have been compromised. There is no known way of accurately determining the remaining strength of decayed wood left in place, nor is there a way to restore strength short of reinforcement or replacement. Cut back rotted members to sound wood, keeping in mind that difficult-to-detect incipient decay can extend well beyond visibly rotted areas. When a partially decayed structural member cannot be replaced, reinforce it with a "sister" anchored to sound wood. Decayed wood absorbs and holds water more readily than sound wood, so let rotted areas of members not removed dry out before making repairs and closing in.

In damp crawl spaces or other places where water is likely to appear, replace decayed members with preservative-treated wood. The major model building code agencies—BOCA, ICBO, and SBCCI—require that treated wood be used for sills and sleepers on concrete or masonry in ground contact, for joists within 18 in of the ground, for girders within 12 in of the ground, and for columns embedded in the ground supporting permanent structures.

Treatments for Decayed Wood Left in Place

Dormant fungi can be reactivated when dry, infected wood is rewetted. Infected, but otherwise serviceable wood left in place should be treated with a spray application of waterborne borax-based preservative that will not only kill active fungi, but guard against future infection as well. Because of the decay hazard posed whenever wood bears on concrete or masonry, solid borate rods are often inserted into holes bored near contact areas. Should wood ever get wet, the rods dissolve and ward off infection.

In instances where replacement is not an option, decayed wood can be stabilized with epoxy. Epoxies consist of resin and hardener that are mixed just before use. Liquids for injection and spatula-applied pastes are available. After curing, epoxy-stabilized wood can be shaped with regular woodworking tools and painted. Epoxies are useful for consolidating rotted wood, restoring lost portions of members, and for strengthening weakened structural members. In the last case, they are used to bond concealed metal reinforcement inside holes or channels cut into hidden faces. Epoxies are not preservatives and will not stop existing decay or prevent future infection.

Wood-Destroying Insects

Nearly two billion dollars is spent annually in the United States to treat, repair, and replace wood in buildings and other structures that has been damaged by wood-destroying insects (Table 3.3). Infestations can be avoided by employing proper design and construction practices, coupled with regular postconstruction inspections.

Termites

Termites are by far the most economically important pest plaguing wooden structures. About 95 percent of all damage is done by subterranean termites which nest underground. Other termites of local significance include drywood termites found along America's southern border from California to Florida, dampwood termites of the coastal Pacific Northwest, and a recent exotic arrival in several Gulf states, the Formosan termite.

Lacking the natural antifreeze of other insects, subterranean termites cannot hibernate during freezing weather and must remain active year round. This is why they are concentrated in the Southeast, having expanded northward only since the early 1900s with the widespread adoption of central heating. Rarely seen because they shun light, subterranean termites favor locales with wetter soils and a readily available source of wood from which they derive their diet of cellulose.

TABLE 3.3 Habits of Common Wood-Destroying Insects

	Attacks	Preferred % wood moisture content	Exit/bore holes (in)	Galleries and frass	Can reinfest	Typical source of infestation	Remarks
Termite	Softwood Hardwood Sapwood Heartwood Old wood New wood*	>8	Seldom seen; uses existing gaps and cracks	Messy, packed with excrema, soil, wood fragments	Yes	Enters heated buildings from soil	Seldom seen except when swarming
Carpenter ant	Softwood Hardwood Sapwood Heartwood Old wood New wood*	>15	Seldom seen; uses existing gaps and cracks	Clean with "sandpapered" walls; insect parts and shredded wood nearby	Yes	Enters heated and unheated buildings from soil	Often seen foraging inside buildings though nest is in nearby tree or stump
True powderpost beetle	Hardwood Sapwood New wood*	6–30	Round $\frac{1}{32}$ to $\frac{1}{8}$	Loosely packed with very fine powder	Yes	Brought into buildings in infected timbers, flooring, firewood, etc.	Hardwoods coated with film-forming finishes are safe from attack
False powderpost beetle	Hardwood Sapwood New wood*	6–30	Round $\frac{3}{32}$ to $\frac{9}{32}$	Tightly packed with fine-to-coarse powder that tends to clump	Yes, but rarely does	Brought into buildings in infected timbers, lumber, etc.; can enter from outside	Common in tropical hardwood products
Anobiid beetle	Softwood Hardwood Sapwood Old wood New wood*	13–30	Round $\frac{1}{16}$ to $\frac{1}{8}$	Loosely packed with fine powder and tiny pellets	Yes	Enters damp crawl spaces, basements, and attics from outside	30 holes per sq ft indicates well-established infestation
Old house borer	Softwood Sapwood New wood*	10–30	Oval $\frac{1}{4}$ to $\frac{3}{8}$	Tightly packed with fine powder and tiny pellets; walls with ripplemarks	Yes	Brought into buildings in infected timbers and lumber; can enter from outside	Larvae inside wood make ticking, clicking, or rasping sound
Carpenter bee	Softwood Sapwood Heartwood Old wood New wood*	>15	Round $\frac{1}{2}$	Clean, but with brooding chambers and stored food	Yes	Enters from outside	Often seen, attracted to only unpainted or lightly stained wood

*"New wood" is less than 10 years old. "Old wood" is more than 10 years old.
SOURCE: Adapted from Smulski, 1992.

In nature, termites feast on dead trees, stumps, and other woody debris. In buildings, they will attack virtually any wood- and cellulose-based building material including timbers and lumber, sheathing, flooring and siding, cardboard-forming tubes, wastepaper-based cellulose insulation, and even the paper faces of gypsum wallboard. Though attracted by odors given off by moist or decaying wood, termites will attack wood at a moisture content as low as 8 percent, which is typical of the year-round average moisture content of wood indoors across most of the United States. No species of untreated wood is immune to attack, but wood pressure impregnated with preservative is an effective deterrent.

Termites usually enter buildings through existing gaps at or below grade. In the absence of shrinkage and settlement cracks in foundations and slabs, and gaps at electrical, plumbing, and septic penetrations, they build shelter tubes of soil and digested wood up the sides of exposed foundations to reach the wood above. Metal shields inserted between foundation and framing will not stop termites from entering, as they will simply build shelter tubes up and over such inconveniences. Shields do force shelter tubes to be built where they are more easily visible, prevent termites from entering sills directly through cracks in foundations, and provide a capillary break between concrete and wood.

Poor building practices that attract termites to buildings include burying stumps, cut-offs, and other wooden debris during backfilling; failing to remove grade stakes and wooden or cardboard concrete forms; placing untreated wood in contact with concrete or soil; and leaving soil exposed in crawl spaces. Termites' high soil-moisture needs can be met by omitting gutters and downspouts, by backfilling with poor-draining soils, or by failing to provide adequate foundation and site drainage.

Termites are serious pests. Unchecked, their thoroughness in excavating tunnels in wood can lead to structural collapse. Because termites hollow out the interior of wooden members without breaking through the surface, there are few visible signs of their presence. Symptoms include shelter tubes, blistered or puckered wood surfaces, crushed and collapsed wood at framing bearing points, and fine soil lining the edges of cracks in foundations and slabs. Termite-damaged wood resounds with a dull thud when tapped with a hammer. When broken open, termite galleries are characteristically messy and often filled with an oatmeal-like mixture of fecal matter, soil, and chewed wood. (See Fig. 3.6.)

Prevention measures in new construction include treating soil with a termiticide before footings, foundation, and floor slab are placed; installing a seamless metal shield between foundation and sill; using preservative-treated wood wherever framing is in contact with concrete and masonry or soil; and capping backfill with a layer of specially screened sand through which termites can tunnel only with great difficulty. Infestations in existing buildings can be treated by pressure-injecting pesticide into soil surrounding the foundation and beneath the floor slab, or by fumigation.

Carpenter Ants

Found all across America, carpenter ants are a problem primarily in the Northeast and Northwest. Like termites, they live mainly under-

3.6 While tunneling through this window header termites left their signature oatmeal-like mixture of fecal matter, soil, and chewed wood. (Wood Science Specialists Inc.)

ground. Unlike termites, carpenter ants do not actually eat wood, but tunnel in it only as a place to live. The distinction may seem trivial, but it is not. Wood that is CCA-treated, for example, although immune to termite attack, is still susceptible to destruction by carpenter ants because they do not ingest the tainted wood.

Carpenter ants may nest underground, in live and dead trees, in stumps, in stored timbers and lumber and firewood, and inside buildings. Ants nesting outdoors hibernate during freezing weather, while those inside heated buildings may be active year-round. Foods include plant juices, "honeydew" secreted by aphids, insects, and household food scraps.

Ants enter buildings via the same routes as termites, but do not build shelter tubes. No untreated wood is safe from ant attack. These hexapods prefer wood whose moisture content is 15 percent or higher, and are especially attracted to decaying wood. As owners of structural insulating panel buildings have learned, carpenter ants will tunnel in these panels' foam cores. Panel makers now treat the foam with a pesticide and recommend that panel edges near grade be capped with a seamless metal shield that ants cannot gnaw through.

Carpenter ants excavate an irregular maze of tunnels in wood that are free of debris, with signature "sandpapered" walls. Extensive tunneling can weaken structural members, but ant infestations are usually detected long before damage becomes serious. A sure sign of activity is the coarse shreds of wood, called *frass,* and occasionally insect parts, that ants dispose of through joints and cracks outside of nests as they tidy their tunnels.

Inviting trouble spots in buildings include eaves and walls wetted by roof leaks, ice dams, condensation, and overflowing or leaky gutters. Plumbing leaks and within-wall condensation can raise wood moisture content to ant-attractive levels. Exposed soil, wood in ground contact, inadequate ventilation, and poor foundation and site drainage create crawl space and basement moisture conditions conducive to ants.

Carpenter ant infestations are difficult to avoid but relatively easy to detect via routine inspection. Once a nest has been located, usually by observing the path by which foraging ants return to it, it can be treated with insecticide.

Powderpost Beetles

Though collectively called powderpost beetles because each reduces wood to a fine powdery frass, true powderpost beetles, false powderpost beetles, and anobiid beetles are each distinct in their preferences and destructive habits.

True Powderpost Beetles

Belonging to the genus *Lyctus,* true powderpost beetles occur throughout the United States and are second only to termites in the dollar damage done. From nests in dead trees, they infest hardwood logs, timbers, and lumber at sawmills and storage yards. Occasionally seen adults lay eggs only in the large earlywood pores of ring-porous hardwoods like oak and ash that are less than 10 years old.

After hatching, *Lyctus* larvae limit their attack to sapwood with a moisture content of 6 to 30 percent, where they tunnel extensively in search of stored starch. Hidden under a veneer of unaffected wood, galleries are loosely packed with talcum powder-like frass that sifts from drying checks and small, round bore holes. Adults emerge from wood after 1 to 2 years. Infestations tend to die out naturally as the carbohydrate content of wood drops over the first few years, but reinfestation can occur if favorable food and moisture conditions persist. If exit holes or adult beetles are not seen within five years of a building's construction, chances are they will never show.

True powderpost beetles most commonly enter buildings as eggs or larvae in new hardwood timbers, flooring, and millwork. Tropical hardwood products are frequently a source of infestation because of inadequate wood storage and drying practices in the country of origin. *Lyctus* may lurk in salvaged timbers, firewood, and in antiques recovered from unheated buildings. Once coated with a film-forming finish, hardwoods are safe from attack, as the coating clogs pores where adults want to lay eggs. In many cases, damage is limited to a single piece of flooring or trim, so removal of the affected item solves the problem. Exit holes in timbers are usually seen long before structural integrity is threatened. Remedies include in-place treatment with insecticide or simply allowing the infestation to die out naturally.

It is difficult to determine whether *Lyctus* activity is ongoing. One way is to vacuum up all frass and mark existing bore holes. The reappearance of frass and new holes confirms activity. Fresh frass is bright and cream colored like new wood; old frass sifting from an inactive infestation is yellow or brown.

False Powderpost Beetles

False powderpost beetles, or bostrichids, nest in dead trees across America. Attacking primarily the sapwood of hardwoods less than 10 years old, they occasionally infest softwoods as well. Both adults and larvae inflict damage, preferring wood whose moisture content is near the upper end of the 6 to 30 percent range over which they operate.

Bostrichids chew small, round holes in wood in search of starch. Larval tunnels are tightly packed with fine-to-coarse frass which tends to clump and does not easily sift from holes. Tunnels made by adults are frass-free. Unlike most beetles, adults bore into wood to lay eggs. Larvae feed for about 9 months before emerging as adults. Since bostrichids prefer wetter wood, most infestations die out with the departure of the first crop of adults. Reinfestation is possible when conditions are right.

As with most true powderpost beetle infestations, the source of the majority of bostrichid problems comes hidden in hardwood products installed during construction or later brought into a building. Active bostrichid infestations can be detected and treated with the same techniques as those used for *Lyctus* beetles.

Anobiid Beetles

Anobiid beetles make their natural home in dead trees. While concentrated in the Southeast, they can be found in buildings in the northeastern, north central and Pacific coastal states as well. Anobiids attack the sapwood of softwoods and hardwoods, regardless of age. Attracted to wood at 13 to 30 percent moisture content, as well as to decayed wood, anobiids most often infest framing in damp crawl spaces and basements.

Rarely seen adults deposit eggs in drying checks, on rough-sawn wood surfaces, and in joints. Tunnels excavated by larvae are loosely packed with fine powder and lemon-shaped fecal pellets that feel gritty when rubbed between the fingers. Adults emerge after 3 or more years from small, round exit holes from which frass freely sifts. Anobiid infestations develop so slowly that the few exit holes present may go unnoticed for 10 or more years. Thirty or more holes per square foot indicates a well-established infestation. Reinfestation is routine.

Unlike *Lyctus* and bostrichid beetles, anobiids rarely enter buildings via infected wood. Adults fly in directly from the outside, attracted by the moist conditions found in crawl spaces and basements lacking soil covers and proper ventilation, and foundation and site drainage. The same techniques for detecting and treating *Lyctus* infestations are used for anobiids.

Old House Borer

A type of long-horned beetle found in the mid-Atlantic states, the misnamed old house borer, primarily infests wood that has been in service for 10 years or less. Adult beetles do not destroy wood. It is in the wormlike larval stage that this insect damages softwood structural framing members in attics, crawl spaces, and basements.

Larvae emerge from eggs adults deposit in drying checks or in joints between framing members and immediately bore into wood. The large larvae often make a faint ticking sound when tunneling. Though their attack is limited to the sapwood portion of a member's cross section, it is so thorough that the sapwood beneath a thin, intact surface layer may be completely pulverized. The large, oval larval tunnels are tightly packed with fine powder and rod-shaped fecal pellets. Walls are characteristically ripple marked, looking like sand that has been lapped by waves. Larvae may feed in wood for several years before emerging as adults through large, oval exit holes chewed through the surface of infested wood or covering materials such as sheathing, siding, flooring, and gypsum wallboard. Exit holes are often the first and only visible symptom. The majority of infestations die out once adults emerge.

Most old house borer infestations are built into buildings during construction through use of softwood timbers and lumber infected during drying or storage, or through use of salvaged timbers or lumber. Beetles can infest or reinfest wood older than 10 years, providing its nutritional content is still high and its moisture content exceeds 10 percent. It is difficult to gauge whether an old house borer infestation is active. Two signs of ongoing activity are the sounds made by tunneling larvae, and the reappearance of fresh frass on cleaned surfaces.

Carpenter Bees

Found throughout the continental United States, carpenter bees look like and are the same size as the familiar bumblebee. But while a bumblebee's abdomen is covered with yellow hair, a carpenter bee's is hairless, shiny, and black. Because these distinctive insects and the large round entrance holes they chew in wood are easily seen, they are more of a nuisance than a serious wood-destroying pest. Like carpenter ants, bees do not eat wood.

Carpenter bees usually attack exterior eaves trim, porch ceilings, siding, exposed timbers, and, occasionally, door and window trim. Their interest is limited to unfinished or lightly stained wood only; painted wood is shunned. Tunnels are usually excavated in easy-to-chew softwoods like redwood, bald cypress, and western red cedar, though harder Douglas fir and southern yellow pine are susceptible. The frass-free tunnels may extend parallel to the grain for 1 to 3 ft or more. Adults emerge about 2 months after eggs are laid. Bees overwinter in the tunnel, and may reuse it year after year.

Live bees, fecal stains around the entrance hole, and fresh frass below confirm an active infestation. Carpenter bees can be controlled by spraying pesticide into the tunnel then, after waiting several days so that all bees using the tunnel are exposed, plugging the hole.

Thermal Degradation of Wood Products

Wood will experience a gradual loss of strength when subjected to intense heat for long periods or to fire. Under ordinary circumstances of use, heavy-timber products in buildings are seldom exposed to the prolonged high temperatures necessary to cause thermal degradation and the resultant loss of strength. Sometimes, though, heavy-timber products are exposed to potentially damaging temperatures. Sawn timber, glulam timber, or parallel strand lumber framing, for example, can be subject to hot air escaping from ovens or kilns, or to steam or vagrant heat from industrial processes. Even intense sunlight falling on a black roof membrane can raise the temperature of heavy-timber decking to susceptible levels for hours at a time.

Exposure to High Temperature

The stiffness and strength of wood decrease as its temperature is raised, and increase as its temperature is lowered. For this reason, extended exposure to subzero temperatures—as a heavy-timber bridge in a northern climate experiences—typically poses no problems. But prolonged exposure to high temperature is another matter. Whether wood suffers thermal degradation depends mainly on how hot the wood gets and how long it stays hot. Below about 200°F, the immediate effects of short-term exposure are fully reversible. But long-term exposure of wood to temperatures above 150°F can produce irreversible strength loss because of thermally driven chemical changes that take place in the wood. The amount of strength loss is affected by temperature, time of exposure, heating medium (air versus water, for example), wood moisture content, wood species, and member cross-sectional dimensions. Because of a synergistic effect between moisture content and temperature, strength loss is greater and occurs faster in wood of high moisture content exposed to high temperature. In general, wood's bending strength is reduced more than its bending stiffness, and its impact resistance lowered more than its bending strength.

The possibility of permanent strength loss in structural wood products due to sustained exposure to elevated temperatures should be addressed as part of the design process. Allowable design values for timbers and lumber given in the *National Design Specification for Wood Construction* are "applicable to members used under ordinary ranges of temperature and occasionally heated in use to temperatures of up to 150°F." The NDS advises that design values may have to be adjusted for each 1°F difference in temperature from a reference value of 68°F "when wood structural members are cooled to very low temperatures at high moisture content, or heated to temperatures up to 150°F for extended periods of time."

Exposure to Fire

Though combustible, sawn timbers and glulam timbers have an outstanding record of performance during fire exposure. The excellent fire safety performance of heavy-timber construction is owed to the fact that the strength of a burning wood member is reduced very slowly and in proportion to the reduction in cross-sectional area lost to fire. This gradual loss of strength during a fire is characteristic of heavy-timber products, and arises because wood's thermal properties render such members self-insulating. Consequently, there often is sufficient time for occupants to evacuate a burning timber-frame building and for firefighters to extinguish the flames before any structural damage is done. As long as subsequent engineering analysis confirms that the unaffected cross sections are large enough to carry the design loads, surface-burned members are simply cleaned and refinished.

The fire endurance of heavy-timber products arises because wood burns in a two-step process that involves first, pyrolysis, then flaming combustion (Fig. 3.7). On being exposed to an outside source of extreme heat (400°F or higher) or to open flame, wood's chemical constituents—cellulose, hemicellulose, and lignin—decompose in a process called *pyrolysis.* During pyrolysis, combustible gases are released from the wood, leaving behind the black char residue known familiarly as charcoal. On mixing with oxygen in the air, the combustible gases ignite into open flames. Eventually, enough heat is generated by the burning of gases evolved from the wood that the pyrolysis/flaming combustion reaction becomes self-sustaining.

Under some circumstances, however, heavy-timber products are self-extinguishing. As the layer of char that forms on the surface of the member thickens, it insulates unaffected wood below from the heat needed for pyrolysis. The insulating properties of the char layer, combined with wood's low thermal conductivity and high specific heat, slow the rate at which the zone of active pyrolysis progresses deeper into the

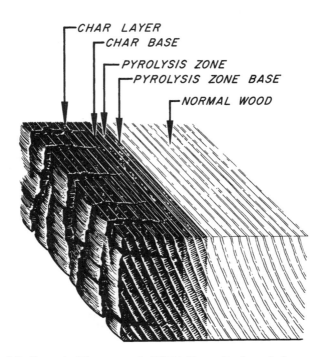

3.7 Impact of fire on wood. (USDA Forest Products Laboratory, Madison, WI.)

wood. Eventually, the heat conducted inward may be insufficient to cause pyrolysis, and unless an external source of heat or flame is reapplied, the fire stops. Thus, the rate at which the wood chars is the most important factor contributing to the endurance of heavy-timber products during a fire. The rate of charring is typically about 1 to 1.5 in per hour. Charring rates are lower for high-density woods, and decrease as wood moisture content increases.

Chemical Degradation of Wood Products

The cellulose, hemicellulose, and lignin composing wood have high natural resistance to being dissolved or otherwise degraded by most chemicals. As such, chemical degradation of wood in service is an uncommon occurrence. Wooden tanks of white oak, Douglas fir, redwood, and bald cypress, for instance, have a tradition of use in the brine processing of onions, olives, pickles, and other acidic or alkaline foodstuffs that will corrode metal vats. When wood does fall victim to chemical degradation, it may be due either to direct exposure to chemicals or, more likely, to contact with corroding metal.

Heavy-timber products can be at risk of chemical degradation by both mechanisms. Sawn timber, glulam timber, and parallel strand lumber framing in factories, storage depots, and other industrial buildings may be exposed to caustic solids, liquids, or aerosols used in or released from manufacturing processes. Likewise, wood in contact with corroding metal fasteners or hardware can suffer localized damage from by-products of the corrosion process. In both cases, the degradation usually takes place gradually, and is occasioned most often by chronic exposure to low levels of corrosive material and exacerbated by high temperature, high relative humidity, and high wood moisture content.

Attack by Alkalis and Acids

Despite its high natural resistance to attack by most chemicals, wood can be degraded by strong acids (pH < 3) and especially by strong alkalis (pH > 11). (The pH scale ranges from 0 to 14. Substances with a pH less than 7 are acidic; those with a pH over 7 are alkaline. Substances of pH 7 are neither acidic nor alkaline, but neutral.) Wood's moderate resistance to alkalis such as sodium hydroxide, for instance, is exploited commercially in pulping wood for making paper and by furniture makers for bleaching the color from wood.

Strong alkalis or alkaline salt solutions in contact with wood first degrade hemicellulose, then lignin. Because cellulose is only mildly affected, alkali-degraded wood surfaces have a white, bleached appearance. Having been separated from one another by disintegration of the layer of lignin that formerly cemented them together, surface fibers are raised, loose, friable, and easily broken free by gentle rubbing. Depending on the duration and conditions of exposure to alkali, the layer of degraded fibers can extend well below the surface. Painters using chemical strippers containing strongly alkaline sodium, calcium, and magnesium hydroxide to remove multiple layers of oil-base paint have learned the hard way just how easily wood can be damaged by these aggressive products.

Sawn timbers, poles, and planks in a bulk chemical storage structure may suffer chemical degradation because of direct contact with lime, fertilizer, or other caustic solids. The surfaces of wooden walls against which road deicing salts are heaped, for example, frequently become fuzzy and friable. The culprit is salt dissolved by rain or high humidity that diffuses into water in the wood. When the wood dries out, the salt recrystallizes inside the wood, causing its fibers to rupture. Repeated cycles of wetting and drying advance the damage deeper into the wood, and can substantially reduce member strength. Wooden road salt storage structures should be lined with polyethylene or built atop a concrete kneewall to limit contact between salt and wood.

Acids break down cellulose and hemicellulose, and in doing so, reduce both wood's tensile and bending strengths. Wood attacked by acid turns dark brown and is friable and brittle. Glulam arches spanning an indoor swimming pool, for example, may be exposed to vapors containing dilute hydrochloric acid originating from chlorinated water. Preventing damage in this case entails finishing members with chemical-resistant surface coatings and venting the enclosure.

Contact with Corroding Metal

Heavy-timber products may experience localized chemical degradation at points of contact with corroding metal fasteners and hardware. Chemicals released or formed as part of the corrosion process substantially weaken wood surrounding the corroding metal. As the wood degrades, its grip on the shanks of nails, drift pins, and other friction fasteners lessens. As a result, fasteners may withdraw and connections between members may loosen to reduce the structure's rigidity or threaten its structural integrity. Because affected wood turns dark brown or black, it is sometimes mistakenly thought to be decayed. When examined microscopically, however, no fungi will be found. Known to shipwrights and boat builders as "nail sickness," and to railroaders as "spike kill," the phenomenon occurs only in wet wood. At risk are heavy-timber products subject to frequent or prolonged wetting such as those used in bridges, marine vessels and structures, and other exterior applications. Green timbers inside buildings whose drying is severely retarded by surface coatings or persistent high relative humidity, for example, are sometimes affected.

So-called crevice corrosion occurs when iron leached from rusting fasteners or hardware reacts with water to form ferrous hydroxide. This compound in turn breaks down cellulose, weakening and embrittling wood. Iron also reacts with tannins and other extraneous chemicals naturally present in wood to stain it a characteristic black or blue-black. The situation can be worsened by airborne pollutants such as chlorides and sulfur oxides that can create acidic conditions (hydrochloric, and sulfurous and sulfuric acids, respectively) in wet wood abutting rusting metal. Nongalvanized and electroplated iron fasteners and hardware are especially prone to crevice corrosion and should never be used in heavy-timber products in exterior applications, or in indoor uses where relative humidity is constantly high, such as in swimming pool enclosures and skating rinks. Only stainless steel, hot-dipped galvanized, or similar corrosion-resistant fasteners and hardware should be used under high moisture conditions.

A second type, galvanic corrosion, happens when dissimilar metals are used in wet wood. Here, water in the wood acts as a conductor to allow electrons from one kind of metal to flow to another kind, forming, in effect, a simple battery called a *galvanic cell*. The metal giving up electrons is called the *anode* (positive terminal), and is the one that corrodes. The metal accepting electrons—the cathode (negative terminal)—is protected from corroding by the steady supply of new electrons. By-products of the corrosion process, in conjunction with metal ions dissolved in rain, freshwater, and saltwater, may form acidic or alkaline compounds that degrade wood around fasteners and hardware. The zinc coating on galvanized nails driven through copper flashing on the end of a glulam timber, for instance, will slowly corrode as electrons flow from the zinc (anode) to the copper (cathode). The iron shank thus exposed will eventually be eaten away as well. Options for preventing galvanic corrosion in heavy-timber products include using the same metal for all bolts, nuts, washers, connector plates, etc.; isolating dissimilar metals with nonconducting gaskets; and using fasteners and hardware of less reactive metals like stainless steel.

Durability of Adhesives in Engineered Wood Products

Only waterproof, structural adhesives that meet the rigorous strength and durability requirements of ASTM D2559 Standard Specification for Adhesives for Structural Laminated Wood Products for Use Under Exterior (Wet Use) Exposure Conditions are used to manufacture the glulam timber and parallel strand lumber used in heavy-timber construction. Under ASTM D2559, adhesives must meet minimum criteria for shear strength, resistance to delamination during accelerated exposure to wetting and drying, and resistance to deformation (creep) under static load. The phenol-formaldehyde, resorcinol-formaldehyde, and phenol-resorcinol-formaldehyde adhesives used in making these products have an excellent record of long-term performance and durability in the field—over 50 years in the case of glulam timber. Because these adhesives are waterproof, parallel strand lumber can be pressure-treated with waterborne preservatives after manufacture. Glasslike when cured, these adhesives are not attacked by biological organisms, do not emit formaldehyde or other vapors, and do not soften or melt even when exposed to direct flame. Since the strength of adhesion between these adhesives and wood is greater than the cohesive strength of the wood itself, test blocks bonded with these adhesives routinely show over 75 percent wood failure when loaded to destruction. Glueline quality is continuously monitored during the manufacture of glulam timber and parallel strand lumber, with strength and durability tests conducted daily on samples taken directly from production.

Preservative Treatment of Wood Products

Supplies of naturally durable woods are too small to meet today's demand at an ecologically and economically acceptable price. Thus, woods lacking decay resistance are impregnated with a preservative toxic to fungi, insects, and other organisms that can extend service life by 30 to 50 years or longer.

Treating Wood with Preservatives

Both nonpressure and pressure processes are used to introduce preservatives into wood. Nonpressure methods such as brushing and spraying are usually limited to field treatment of wood during construction, or remedial treatment of wood in place. Millwork makers dip window and door parts in a combined water-repellent preservative. Lasting from days to weeks, soaking of poles, posts, and piles is really nothing more than extended dipping. Penetration of preservative into wood with nonpressure methods is minimal; only a thin layer of wood near the surface is treated with preservative. Thus, the amount of protection gained with nonpressure methods is unpredictable.

The most effective processes are those in which preservative is driven into wood under pressure, hence "pressure-treated wood." Though variations abound, two basic pressure processes are the full-cell or Bethell process, and the empty-cell or Rueping/Lowry process. With both, wood cell walls are saturated with preservative. After treatment, cell cavities are filled with preservative in the Bethell process, but nearly empty in the Rueping/Lowry process.

Green wood is generally dried to around 20 percent moisture content before treatment. Otherwise, water saturating its cell walls inhibits absorption of preservative. The surfaces of timbers of difficult-to-treat woods like Douglas fir are often incised—punctured with knives or needles as they pass between spiked rollers—to promote absorption and penetration of preservative.

In the full-cell process wood is placed inside a sealed vessel and a vacuum drawn to remove air from its hollow cells. After being flooded with preservative, the vessel is pressurized to drive the solution deep into the wood. Pressure is later released, excess liquid pumped to a holding tank, and treated wood removed for air- or kiln-drying.

No initial vacuum is pulled in the empty-cell process. As the flooded vessel is pressurized, air inside wood cells is compressed. After pressure release, the expanding air, aided by a small applied vacuum, kicks preservative out of the cell cavities leaving them practically empty.

Generally, lumber up to 1 in thick is completely penetrated by preservative. Depending on the type of wood, 2-in-thick lumber may or may not be fully penetrated. Only the shell of timbers 4 in and thicker will be penetrated; the core will not. Also, while the sapwood of virtually all woods is easily penetrated, the heartwood of most resists penetration. For all intents and purposes, only the sapwood of treated wood has protection against decay greater than what nature already handed the heartwood. What is important is that a zone of untreated wood may be exposed if timbers are cut to length, notched, bored, or otherwise machined after treatment. Fungi can also gain access to untreated wood as timbers check and split in service. Repeated rapid swelling and shrinking of timbers wetted by rain, for example, can carry surface checks into the untreated core of large members. Water trapped in these fissures allows fungi to destroy timbers from the inside out.

Depth and uniformity of penetration, and amount of preservative retained by wood, determine success of treatment. Measured in pounds of preservative per cubic foot of wood (pcf), retention varies with the preservative, product, wood species, and intended use (Table 3.4). Retention standards set by the American Wood-Preservers' Association are enforced through chemical analysis of treated wood by third-party agencies that periodically inspect treating companies. The type and retention of preservative, as well as the inspection agency, are identified in the treater's quality mark stamped on preservative-treated wood.

Of countless compounds suggested and tested as preservatives, only a handful have the safety, effectiveness, permanence, and economy that make them commercially important. In addition to creosote, preservatives fall into two classes, oil-borne and waterborne, depending on whether they are carried into wood in an organic liquid or water.

3

Preventing Wood Degradation

Creosote

In use since the 1850s, creosote is highly effective against fungi, insects, and marine borers. Impregnated into crossties, marine piles, and bridge and highway timbers in a full-cell process, creosote may exude from products into surroundings. Organic in origin, it eventually biodegrades. Utility and building poles, freshwater piles, fence posts, and industrial wood block flooring are treated in an empty-cell process that yields a clean, nonbleeding surface.

Dark brown to black, creosote-treated products are rarely appropriate for indoor use. Freshly treated wood emits harmful vapors that eventually disappear. Gloves must be worn when handling creosote-

TABLE 3.4 **Recommended Preservative Retention for Sawn Timbers and Lumber**

	Retention (pcf)		
	Above ground	Soil and fresh water	Marine
Creosote	8.00	10.00	25.00
Pentachlorophenol	0.40	0.50	NR*
Waterborne			
CCA	0.25	0.40	2.50
ACA	0.25	0.50	2.50
ACC	0.25	0.40	NR*
ACZA	0.25	0.40	2.50

*Not recommended
SOURCE: Adapted from American Wood Preservers' Association, 1997.

treated timbers. Creosote products cannot be painted; coal tar pitch, urethane, epoxy, and shellac are acceptable sealants. Creosote crossties last about 30 years; utility poles may survive 60.

Oil-Borne Preservatives

Carried in organic solvents such as liquefied isobutane, oil-borne preservatives useful in heavy-timber construction include pentachlorophenol, and copper and zinc naphthenate.

Potent against terrestrial pests, pentachlorophenol, or penta, extends wood's service life by 20 to 40 years. In use since the 1930s, most is borne in heavy oil to treat utility and building poles, fence posts, and highway timbers. Tinted light to dark brown, penta products glue and finish reasonably well after the noxious oil carrier evaporates.

Because penta can migrate to form surface deposits, gloves must be worn when handling these products. Urethane, latex enamel, shellac, and varnish are effective sealants. Penta can move into surrounding soil, but because it binds tightly to soil, groundwater contamination is unlikely. Penta slowly breaks down into biodegradable compounds.

Only licensed pesticide applicators may use EPA-regulated creosote and penta to treat wood. For contractors, builders, and others who need to treat untreated wood exposed by machining on-site, over-the-counter oil-borne preservatives are available. Copper naphthenate is a common ingredient in the green-tinted preservatives, while zinc naphthenate is found in many clear products. Because penetration and retention with self-treating are minimal, preservative should be liberally applied to untreated surfaces. Always wear eye protection, avoid skin contact, and follow label directions when using over-the-counter preservatives.

Waterborne Preservatives

Most of the treated building poles, timbers, lumber, and engineered wood products such as glulam timber and parallel strand lumber used in heavy-timber construction are protected with one of several preservatives carried in water, such as chromated copper arsenate (CCA), ammoniacal copper arsenate (ACA), acid copper chromate (ACC), and ammoniacal copper zinc arsenate (ACZA). Though Douglas fir and other western woods are commonly treated with ACA and ACZA, and southern yellow pine with CCA, wood treated with any one of these compounds has pretty much the same characteristics.

CCA

The workhorse of the waterborne stable is CCA. Since use began in the 1930s, three basic formulations, types A, B, and C, which vary by amount of chromium, copper, and arsenic, have evolved. The type C, or oxide form, is preferred. Before and during a modified full-cell treating process, CCA is water soluble. During the first day or two of air-drying after treatment, CCA is rendered insoluble in water in a process called *fixation*. During fixation the preservative reacts chemically with the wood, permanently bonding itself to the cell walls. For this reason, CCA does not leach from wood in service. Because it is applied as a water solution, no vapors are ever emitted.

In addition to resisting attack by fungi and termites, CCA thwarts marine borers. CCA-treated wood is not completely immune to insects that, like carpenter ants, do not ingest the wood. Also, CCA products are susceptible to surface mold if solid piled when wet. Treaters guarantee CCA-treated wood for 40 years and consider 100 possible.

CCA-treated wood has a blue-green tinge. Products are used mainly in exterior situations where decay hazard is high, and whenever wood is used in ground contact or against concrete and masonry. Products have a clean surface and can be used both indoors and where skin contact is frequent without being sealed. Because wood is saturated with water during treating and seldom kiln-dried afterward, it is almost always still wet during construction. Shrinkage of large timbers must be accounted for during design and construction as a result. Through-bolts properly tightened at the time of installation may loosen as timbers shrink and will need to be retightened a few weeks later.

CCA products finish reasonably well. Water repellents, semitransparent stains, and other finishes should not be applied until the surface of treated products is thoroughly dry. Oil-base semitransparent stains perform best on CCA-treated wood. A water repellent containing fungicide should always be applied to minimize the warping and checking that affect CCA-treated wood exposed to the elements. With water repellents so important, the latest development in pressure treating is to impregnate wood with preservative and water repellent simultaneously.

Copper in CCA, ACA, and ACC is corrosive to uncoated metal. In above-grade construction, use stainless steel or hot-dipped galvanized fasteners. Joist hangers, framing anchors, and other hardware should be corrosion resistant. Type 304 and 316 stainless steel, Type H silicone bronze, ETP copper, and monel fasteners are required for below-grade applications like wood foundations.

Borates

A waterborne preservative that is being used increasingly to treat sawn timbers is borax in the form of disodium octaborate tetrahydrate. Borates protect wood from most fungi and wood-eating insects. Borate-treated wood is unchanged in color, noncorrosive to fasteners, and can be readily glued and finished. Nontoxic to people and animals, borates also increase wood's fire resistance.

Borates are applied by dipping diffusion. Green timbers and lumber are immersed in a hot, aqueous borate bath, then removed and solid piled. Over a few weeks' time, the preservative naturally diffuses into the water in the wood. Sapwood is completely penetrated, as is the heartwood of some woods. Dry wood is treated in a full-cell pressure process. Spray application is used for treating wood in place. Borates are most commonly used for treating timbers for post-and-beam construction and logs for log structures.

Borates remain water soluble, however, and can leach out of treated wood that gets wet. Until a way to render them insoluble after treatment is developed, borate products should not be used where they are exposed to direct wetting. Leaching of borates from the exterior of log homes is all but eliminated by applying a water repellent every couple of years.

Precautions Regarding Treated Wood

To help contractors and builders use treated wood wisely, the treating industry publishes an EPA-approved Consumer Information Sheet for creosote, pentachlorophenol, and inorganic arsenical pressure-treated products. The CIS is available from treated-wood retailers and AWPA. Though recommended uses vary with preservative type, some caveats found on the CIS apply to all treated wood products:

- Use treated wood only where such protection is needed.
- Wear a dust mask and goggles when machining treated wood.
- Wash hands before eating after handling treated wood.
- Wash work clothes separately, and before reuse.
- Do not burn treated wood; dispose of by burial or ordinary trash collection.
- Do not use treated wood for cutting boards, countertops, silage or fodder bins, or where it could become a component of animal feed.

Fire Retardant Treatment of Wood Products

The fire resistance of heavy-timber products can be enhanced by pressure impregnation or surface application of fire retardants. Of importance is that these chemicals do not render wood fireproof. They do, however, substantially reduce the rate at which flames spread across its surfaces. As such, fire retardants assist in containing the spread of fire and improve the chances that it will be brought under control before damage becomes extensive.

The fire retardant formulations in use today are proprietary, differ widely in chemistry and ingredients, and are waterborne. Common active ingredients include zinc chloride, mono- and diammonium phosphate, ammonium sulfate, aluminum trihydrate, and boric acid, or a combination thereof. Regardless of whether they are pressure impregnated into wood or merely brushed on, these and other chemicals inhibit the spread of flames across the surface of wood during fire exposure by altering the mechanism of thermal degradation. Most do so by lowering the temperature at which pyrolysis begins (about 400°F for untreated wood). With this approach, flame spread is slowed because fewer combustible gasses are released from the wood and more self-insulating char is produced. Some fire retardants limit the advance of flames by decomposing into nonflammable gasses that dilute the combustible gasses released by pyrolysis. Others increase the thermal conductivity of wood so that heat is conducted away from the surface to the interior, thereby cooling the surface and slowing the advance of pyrolysis. One type, called *intumescent coatings,* expand when heated to form a layer of glasslike foam that prevents combustible gasses from escaping from the wood and stops oxygen from reaching them. Fire retardants are often formulated with multiple chemicals to create a synergy among these effects.

Surface Treatments

Fire retardants can be applied topically to sawn timber, glulam timber, and parallel strand lumber by brushing or spraying. A superficial surface coating at best is achieved with either method. The use of surface-applied fire retardants on heavy timber products has pretty much been limited to members in existing buildings. The amount of fire retardant brushed on or sprayed on must be carefully monitored to ensure that the required level of flame spread resistance is achieved.

Pressure Impregnation

The pressure impregnation processes employed for treating heavy-timber products with fire retardant are essentially the same as those used for introducing preservatives into wood. As with preservative-treated wood, the treating and posttreatment kiln-drying conditions, penetration, and retention standards for fire retardant–treated wood are set by the American Wood-Preserver's Association. Because some fire retardant chemicals may be leached from wood by water, three performance rating categories of fire retardant–treated products are recognized by AWPA. Products intended for indoor use are identified as Interior Type A and Interior Type B; those for outdoor use are designated Exterior. Heavy-timber products pressure impregnated with fire retardant in conformance with AWPA standards must have a flame spread index of 25 or less when tested under ASTM E84 Standard Test Method for Surface Burning Characteristics of Building Materials. By comparison, the flame spread index for untreated wood ranges from 75 to 135.

Certain properties of wood can be affected as a result of pressure impregnation with fire retardant. Potential side effects include reduced stiffness and strength, increased hygroscopicity (wood's ability to adsorb moisture from the air), and increased potential for corrosion of fasteners and hardware. Experience has shown that an interaction between fire retardant chemicals and heat during the treating process and posttreatment kiln-drying can induce in wood a 5 to 10 percent loss of stiffness and a 10 to 20 percent loss of strength. As a consequence, allowable design values for fire retardant–treated timbers and lumber are lower than those for untreated products. Due to the proprietary nature of fire retardants, the magnitude of stiffness and strength loss varies widely among formulations. As such, the formulator or treater must be contacted directly to obtain the proper adjustments to be applied to allowable design values.

Like wood itself, some fire retardant chemicals are hygroscopic; that is, they are able to adsorb moisture from the air. Thus, some fire retardant–treated products will achieve a higher moisture content than untreated wood will when exposed to the same ambient temperature and relative humidity. Since wood's stiffness and strength decrease as its moisture content increases, allowable design values for heavy-timber products treated with hygroscopic fire retardants have to be adjusted downward according to the formulator's or treater's recommendation. The ability of a fire retardant to adsorb moisture from the air, coupled with its potential for reacting with metals, can, under some circumstances, lead to corrosion of nails, lag screws, through-bolts, connector plates, and other fasteners and hardware used with heavy-timber products. Although most current fire retardant formulations are both low in hygroscopicity and noncorrosive, the possibility for metal corrosion should be discussed with the formulator or treater, especially if treated heavy-timber products will be exposed to high relative humidity, high temperature, saltwater, or used outdoors. In these cases, stainless steel or other types of corrosion-resistant fasteners and hardware are required. As fire retardants may discolor or otherwise affect surface finishes applied to treated products, the formulator or treater should be consulted for finishing recommendations to ensure compatibility.

Protective Finishes for Wood Products

Finishes are applied to the surfaces of heavy-timber products in most cases, regardless of whether they are used indoors or outdoors, because of the protection these coatings provide. Paints, solid-color stains, clear coatings, water repellents, semitransparent stains, and other finishes enhance esthetics, retard uptake of liquid water and water vapor, reduce seasonal dimensional changes, limit surface checking, and prevent weathering. For certain uses, however, it may be neither necessary (from a performance standpoint) nor desirable (from an esthetic perspective) to finish certain heavy-timber products. For instance, there is no compelling reason to finish penta-treated bridge timbers because the oil carrier imparts water repellency to the wood. Likewise, a designer desiring a rustic look in a post-and-beam residence may opt to leave the roughsawn surfaces of exposed timbers in their natural state.

Finish Types

Finishes for wood products are classified as either film-forming or penetrating. Both types may be either waterborne (also known as *latex*) or oil-base, and formulated for interior or exterior use. On drying, film-forming finishes—primers, paints, solid-color stains, and clear coatings—coalesce into a thin continuous sheet that sits on top of the surface to which they are applied. Of all coatings types, film-forming finishes provide the greatest protection to wood by virtue of the physical barrier they pose to the elements. These finishes are a two-edged sword, however, being equally effective in deterring the entry of water into dry wood and retarding its escape from wet wood. Thus, untreated exterior wood products finished with primers, paints, solid-color stains, or clear finishes that get wet are susceptible to coatings failure and decay. Penetrating finishes do not form surface films, but rather are absorbed into the wood. Water repellents and semitransparent stains fall into this category. Penetrating finishes allow water vapor to escape from wet wood and are effective in repelling liquid water and controlling surface checking, but do not stop wood from weathering.

The type of finish, the properties of the wood to which it is applied, local climate, and directional exposure influence the performance and longevity of coatings applied to exterior wood products. All other things being equal, paints last longest, followed closely by solid-color stains, with semitransparent stains, water repellents, and clear coatings all about an equally distant third. All other things being equal, finishes last longest on vertical grained surfaces of low-density softwoods whose moisture content stays below about 16 percent. Coatings failures occur fastest in geographical regions with higher annual temperatures and rainfall, and on a building's southern exposure. Finishes used on heavy-timber products inside buildings are not subject to the elements, of course, and are renewed typically when faded, abraded, or a color change is desired.

Application of Finishes

Finishes may be applied to heavy-timber products at the factory or in the field. Most fabricators of glulam timbers, for example, will factory-apply primers and top coats on request. Afterward, members are wrapped with kraft paper or other protective coverings to prevent surface marring during shipment and erection. Surfaces of smooth-sided heavy-timber products to be finished on-site should be lightly sanded just prior to applying primers and other finishes to remove contaminants and reactivate surfaces for good adhesion. Such surface preparation is, of course, not possible with rough-sawn timbers. Best adhesion is achieved on both smooth and rough wood when finishes are worked into surface irregularities by brushing. Spray- and roller-applied finishes should be back-brushed for this reason.

Film-Forming Finishes

Primers

Primers are a mixture of a polymeric binder and pigment particles carried in water or an organic solvent. Applied to bare wood in advance of paints and solid-color stains, primers create a smooth base for top coats to adhere to, and serve as a chemical barrier between the wood and top coats. Best performance is obtained with primers containing a fungicide to stop mildew and a stain-blocker that prevents water-soluble extractives in wood from discoloring top coats. Referred to colloquially as "cedar bleed," the phenomenon is more accurately termed *extractive staining* because it happens in redwood, Douglas fir, southern pine, and other woods as well.

Paints

Paints are opaque coatings that obscure both wood's color and surface texture. In addition to their decorative function, paints protect wood surfaces by blocking the sun's ultraviolet rays, repelling liquid water, retarding the adsorption of water vapor, and preventing erosion by wind-driven particles and rain. Containing typically from 50 to 70 percent solids by weight borne in water or an organic solvent, paints consist of a polymeric binder that forms the protective film, pigments that lend color, and myriad additives that impart other desirable properties. For each coat applied, the solvent evaporates, leaving behind a dry film 2 or 3 mil thick (1 mil equals 0.001 in). Oil-base paints generally have lower permeability to water vapor and are less flexible over time than waterborne paints. As a consequence, oil-base paints are more susceptible to cracking and peeling occasioned by swelling and shrinking of the underlying wood. Performance and longevity of paints can be enhanced by treating bare wood with a "paintable" water repellent prior to priming. Research and field experience show that paints' best performance is obtained when two top coats of 100 percent acrylic latex paint with fungicide are applied over a primer containing fungicide and stain-blocker. Depending on local climate and severity of exposure, such a system should give 7 to 10 years of good performance before refinishing is needed.

Solid-Color Stains

Containing the same ingredients as paints, solid-color stains are essentially thin paints. With a solids content by weight of about 30 to 40 percent, the one-coat dry film thickness is usually less than 1 mil.

Solid-color stains are frequently the finish of choice for rough-sawn wood because they mask wood's color, but accentuate its surface texture. The thinness of the film permits both liquid water and water vapor to pass more easily into and out of the wood. Though some solid-color stains can be applied directly to bare wood, most benefit from being applied over a compatible primer containing fungicide and stain blocker. Application of a "paintable" water repellent prior to priming further improves performance and longevity. As with paints, best performance is achieved with a three-coat system of primer containing fungicide and stain blocker and two top coats of 100 percent acrylic latex solid-color stain with fungicide. Depending on the environment, this system should give 3 to 7 years of satisfactory service before maintenance is required.

Clear Coatings

Essentially paints without pigment, varnishes, urethanes, and other clear coatings are intended primarily for indoor use. When used outdoors, the sun's ultraviolet rays pass through these transparent finishes and degrade lignin on the surface of the wood. In usually 2 years or less, even the toughest marine finishes are likely to embrittle, crack, and peel with exposure to sun and rain. Additives that block or absorb ultraviolet light slightly lengthen the service life of clear coatings. Best performance outdoors is obtained by applying a "paintable" water repellent to the bare wood, followed by three or more coats of finish. Varnishes, urethanes, lacquers, and shellac are commonly used indoors on smooth-surface wood products to accent the wood's grain and natural beauty. Despite their poor water resistance, shellac-based knot sealers work well on exterior wood as long as they are covered with paint or solid-color stain. Some clear coatings may develop a yellow tinge over time in a process known as *ambering* as a result of being oxidized by oxygen in the air and exposed to sunlight. Coatings formulated with ultraviolet light blockers or absorbers are less likely to discolor.

Penetrating Finishes

Water Repellents

Used on exterior wood that is otherwise unfinished, water repellents retard absorption of liquid water by wood, causing dew, rain, and snow melt to bead on the surface where it can harmlessly evaporate. Effective in limiting warping, surface checking, and end splitting in exterior wood, these penetrating finishes do not prevent wood from turning gray with exposure to ultraviolet light, nor stop adsorption of water vapor. Without water repellent, repeated rapid swelling and shrinking can carry surface checks into the untreated core of preservative-treated timbers, inviting internal rot. Most water repellents consist of paraffin wax dissolved in turpentine, mineral spirits, or other organic solvents, along with a drying oil and fungicide to discourage mildew. Others, marketed as water-repellent preservatives, also contain an oil-borne preservative such as tributyltinoxide (TBTO) or iodo propynyl butyl carbamate (IPBC) to deter decay. Applied liberally by brushing or spraying, water repellents are readily absorbed into smooth and rough wood. Because water repellents applied to otherwise unfinished wood remain effective for only 1 or 2 years, they must be reapplied regularly. So-called paintable water repellents that contain less wax are applied directly to bare wood before the primer to enhance the performance and longevity of paints and solid-color stains by reducing extractive staining, and blistering and peeling.

Semitransparent Stains

Containing 10 to 20 percent solids by weight, semitransparent stains generally include binder, pigment, water repellent, ultraviolet light inhibitors, and fungicide dissolved in an organic solvent and, sometimes, in water. An excellent choice for rough-sawn or weathered wood, semitransparent stains impart color to wood without hiding its natural texture. If used on smooth wood, only a single coat of semitransparent stain should be applied initially. Otherwise, a film may form; surfaces may appear glossy and later begin to flake because of smooth wood's inability to absorb the second coat. Additional coats can be applied once smooth surfaces have become more absorptive after weathering for 1 or 2 years. Semitransparent stains provide to wood slightly more protection from the elements than water repellents. Reapplication is usually necessary every 3 to 4 years.

Summary

Under certain circumstances, wood products can be destroyed by fungi, damaged by insects, consumed by fire, eroded by weathering, weakened by high temperature, and disintegrated by chemicals and corroding metal. Yet, in spite of these shortcomings, sawn timbers, glulam timbers, and other heavy-timber products are used worldwide in residential, commercial, and industrial buildings and structures because of their strength, economy, and beauty. The principles and practices for overcoming wood's liabilities are well established and, if followed, will ensure good performance and longevity of heavy-timber buildings and structures. Though the necessary preventive measures differ depending on the agent of degradation, the most important consideration in creating durable, long-lived heavy-timber buildings and structures is to utilize design features, construction practices, and maintenance regimens that keep wood dry.

Wood products in buildings can be wetted by ground water; piped water; condensation; and dew, rain, and snow melt; green-sawn timbers naturally contain substantial amounts of water. Techniques proven effective in preventing moisture-caused damage to wood in buildings include:

- *Foundation*—install perimeter drains; seal cracks and other points of entry; apply dampproofing or waterproofing to the exterior of foundation walls; backfill with free-draining soil; grade soils to slope away from the foundation; install gutters and downspouts along eaves; place vapor retarder under floor slab and over exposed soil; install metal shield as capillary break between wood and concrete or masonry; place polyethylene over footing before walls are erected to stop rising damp.

- *Piped water*—insulate cold water supply lines to stop condensation; make critical connections accessible for inspection and repair.

- *Condensation*—control indoor relative humidity with passive or active ventilation or dehumidification; seal potential points of air leakage in ceilings, walls, and floors; place vapor retarder on warm side of wall; insulate floors of air-conditioned rooms over basements and crawl spaces.

- *Wetting by the elements*—use wide eaves; install gutters and downspouts; hold wood 8 in or more off the ground; use flashing and caulking to keep water from entering into walls; apply protective finishes; back- and end-prime siding and trim; prune plants to ensure airflow around building; use naturally decay-resistant or preservative-treated wood; use corrosion-resistant fasteners.

- *Water built-in during construction*—use partially air-dried sawn timbers; use sawn timbers treated with borate preservative; use naturally decay-resistant wood.

Preventive measures for specific agents of degradation include:

- *Weathering*—apply opaque film-forming finishes.

- *Mildews and molds*—use exterior finish with fungicide; control indoor relative humidity with passive or active ventilation or dehumidification; insulate floors of air-conditioned rooms over basements and crawl spaces.

- *Decay fungi*—keep wood moisture content below 20 percent; use naturally decay-resistant or preservative-treated wood; sterilize salvaged timbers with heat or treat with borate preservative; replace or reinforce even moderately damaged members; conduct regular inspections.

- *Insects*—install metal "termite shield"; treat soil around foundation with insecticide; use naturally decay-resistant or preservative-treated wood; sterilize salvaged timbers with heat or treat with borate preservative; keep wood dry; conduct regular inspections.

- *Heat and fire*—reduce attic temperature with passive or active ventilation; account for anticipated thermal exposure during design; use fire retardant–treated wood; employ fire-resistive design features such as fire- and draftstopping, sprinkler systems, and fire-rated wall, ceiling, and floor assemblies.

- *Chemicals*—account for anticipated exposure during design; apply chemical-resistant finishes; use corrosion-resistant fasteners; use same metal for all fasteners, hardware, flashing, etc.; keep wood dry; conduct regular inspections.

Considerations in choosing treatments and finishes to enhance the performance and service life of heavy-timber products include:

- *Preservative-treated wood*—use wherever wood is likely to get wet; choose preservative type and retention level appropriate for decay hazard; require product to be pressure-treated; liberally treat all site-cut surfaces; account for shrinkage of timbers treated with waterborne preservatives during design; anticipate need for future retightening of through-bolts and lag screws; when appropriate, apply water repellent with fungicide on a regular schedule.

- *Fire retardant–treated wood*—choose low hygroscopicity and noncorrosive formulations; reduce allowable design values as per formulator's or treater's recommendations; select finishes according to formulator's or treater's recommendations; use corrosion-resistant fasteners and hardware.

- *Protective finishes*—select finish based on desired esthetics, smooth or rough wood, expected service life, and level of protection needed as dictated by local climate; use primers with fungicide and stain blocker and top coats with fungicide; apply water repellent with fungicide to otherwise unfinished exterior wood on a regular schedule.

As evidenced by Norwegian stave churches, Asian pagodas, and the colonial meeting houses and covered bridges of New England, heavy-timber structures designed, constructed, and maintained in accordance with these principles can last for centuries.

Note

1. As quoted in Galligan.

CHAPTER

4

Code Issues

The fire chief is not in charge of the fire, but of the fire department's effort to contain the fire. The fire sets its own agenda.

FRANCIS L. BRANNIGAN, *Building Construction for the Fire Service*

Introduction

If you've ever tried to build a fire, you know from that experience that the process starts with kindling: small dry sticks that readily ignite. The heat they give off is used to ignite larger branches, which in turn are used to ignite firewood. The smaller its ratio of surface area to volume, the longer a piece of wood will take to ignite. The greater its mass, the longer it will take to be consumed.

Wood has many desirable properties as a building material, but its combustibility is not one of them. As illustrated in the above example, the use of large members reduces but does not eliminate the importance of combustibility in the design of timber-framed buildings.

This chapter starts by looking at heavy timber as a construction type, defining the circumstances under which it can work for a given building. It has been estimated that fully three quarters of all nonresidential construction in North America "falls within the height and area limitations set by building codes for wood buildings."[1] Compartmentalization is discussed as a way of permitting larger timber buildings. Next is an examination of the fire performance of heavy timber, followed by a brief survey of the minimum dimensions and mandatory details that help reduce both timber's vulnerability to fire damage and the impact of such damage on other building elements.

Heavy Timber as a Construction Type

Building codes generally classify construction types by their degree of fire resistance. The greater the fire resistance of the structural frame, the larger the allowable building size. Light wood framing, being combustible, is at one end—the low end—of the fire-resistance spectrum, while reinforced concrete, being noncombustible, is at the other. Though combustible, heavy timber is considered *superior* to light wood framing, for its larger member sizes make it burn much more slowly. Though noncombustible, unprotected structural steel is considered *inferior* to concrete, for it loses strength at the elevated temperatures typical in fires. For the same reason, heavy timber is considered superior to unprotected steel. In fact, "model building codes have traditionally permitted heavy timber buildings to have at least the same allowable floor areas and heights as one-hour protected noncombustible construction."[2] Where the allowable area for light wood framing is x, the allowable area for steel is $2x$, and for timber, $3x$[3] (see Fig. 4.1). In general, the greater the allowable area, the higher the cost of the construction type with which it is associated. Except where the code permits roof construction to have a lower fire resistance than the rest of the structural frame, a building of several construction types is typically classified according to the least fire resistive of them.

Just as the code distinguishes between structural materials, it also distinguishes between use groups (see Fig. 4.2). The greater the fire hazard or the longer it will take occupants to leave, the smaller the allowable building area. The fire hazard relates to the type and quantity of fuel present. Nightclubs and the like are at the low end of the allowable areas in the use group spectrum, while churches, offices, and schools—whose occupants are assumed to be familiar with their surroundings—are at the high end. Where the allowable area for the former is x, the allowable area for the latter can be as high as $6x$.[4]

Although the fire resistance of a given material is inherent to it, various means are available to improve its performance in a fire. Its members can be protected with fire-resistive coatings or enclosures, or an automatic sprinkler system can be provided. Protection involves placing a barrier between the fire and the structural frame, while suppression attempts to put out the fire and/or cool the surfaces exposed to it. These approaches can be used separately or in combination. Protection of steel framing will increase a building's allowable area by 35 percent or more, while provision of sprinklers, regardless of construction type, may increase it by as much as 200 percent.[5]

Sprinkler systems result in allowable area increases only when they are automatic and installed throughout the building. The reasoning is simple: When elevated temperatures or products of combustion are sensed anywhere in the building, suppression will begin immediately, even when there are no occupants. Design measures that maximize

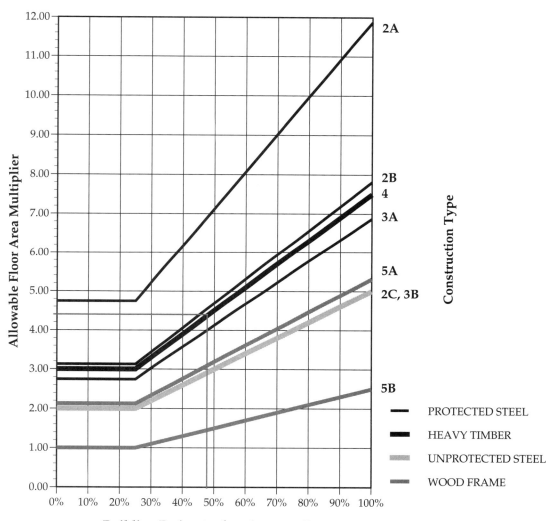

4.1 Area increases for street frontage. Determine the percentage of building perimeter fronting on a street or fire lane. Project up from the appropriate location on the x-axis to the sloping line associated with the construction type. Project across from there to find the allowable floor area multiplier. The product of this multiplier and the allowable base area derived from Fig. 4.2 is the allowable floor area of the overall building or of a fire zone within it. (Derived from *1993 BOCA National Building Code.*)

emergency access to the building are encouraged by the code, but the delay between the activation of the alarm and the commencement of firefighting operations prevents this approach from being quite as effective as sprinklering. Nonetheless, the potential increase in allowable area associated with a perimeter fire lane can be significant. Under one of the model building codes, a 2 percent increase in area is allowed for each 1 percent of the building perimeter, above 25 percent,

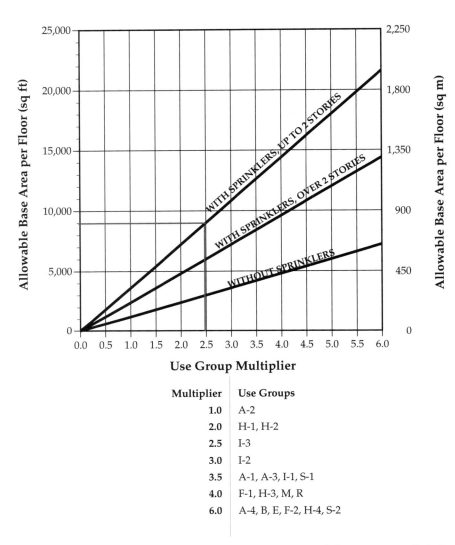

4.2 Allowable base areas for use groups. For the most restrictive use group, find the applicable multiplier in the table located under the graph. Project up from the appropriate location on the graph's *x*-axis to the sloping line associated with the building height and the presence or absence of sprinklers. Project across from there to the allowable base area. The product of this base area and the multiplier derived from Fig. 4.1 is the allowable floor area of the overall building or of a fire zone within it. (Derived from *1993 BOCA National Building Code.*)

which fronts "on a street or other unoccupied space . . . on the same lot or dedicated for public use."[6] In other words, a building with 100 percent of its perimeter with such frontage would be allowed a 150 percent area increase, or three quarters of the maximum increase resulting from sprinklering.

Even with sprinkler and street frontage increases, the floor area *required* for a given design and occupancy may be greater than the area *allowed* for the preferred construction type. In that case, the designer has four options:

1. Increase the allowable area by using a more fire-resistive construction type.
2. Decrease the area per floor by distributing the overall building area over more floors.
3. Divide the building into separate fire areas each within the maximum allowable size.
4. Employ some combination of the above approaches.

Let's take a closer look at each of these options. Since this is a book about timber construction, I'll assume that the construction type—option 1—is not negotiable. Option 2, redistributing the floor area, is primarily a function of allowable building height and secondarily a function of function—will the required spaces and adjacencies work in a taller building with smaller floors? Option 3 alone can make any construction type work, but a fuller explanation of it is in order.

Building codes contain a number of incentives for the construction of barriers to the spread of fire. If the distance between adjacent buildings—the *fire separation distance*—is increased, the codes permit the fire resistance of the exterior walls of both buildings to be reduced. Where land costs are high or properties are small, spacing buildings farther apart is impractical. In many urban areas, fire separation distances are, by necessity, zero.

Fire Walls

All else being equal, from the perspective of fire spread, it doesn't matter whether two adjacent buildings are owned by the same person. What matters is whether or not the fire barrier between them is adequate. If the barrier is adequate, the areas on each side of it are regulated essentially as separate buildings. Within a single building, the barriers are known as *fire walls* and the areas on each side of them are called *fire areas*. To prevent the spread of fire from one fire area to another, the fire wall: (1) must be highly fire resistant, (2) must to somewhat taller than the fire areas it abuts, and (3) must prevent the spread of fire from each side to the other, even in the event of total collapse of the side that is involved in the fire.

To satisfy this last objective, some codes mandate the use of *single* fire walls cantilevered from the ground and structurally independent of the adjacent fire areas,[7] others permit *double* fire walls, where the structural frame of each fire area is connected to and braces one of a pair of independent back-to-back fire walls.[8] In the former, collapse of either fire area leaves the fire wall standing; in the latter, collapse of either fire area takes one wall with it, leaving the other standing (see Fig. 4.3).

Assuming reinforced concrete or concrete block construction to achieve the applicable fire rating, a double fire wall will contain twice the mass of a single fire wall, all else being equal. Unfortunately, all else is not equal: As building height increases, the vertical reinforcing requirements of each type of fire wall differ markedly. A single fire wall is unbraced for its entire height, while a double fire wall is customarily braced at every floor. At a certain height, the stability of a single fire wall can only be assured if its thickness is increased, either for its entire length or at regular intervals through the use of pilasters. For single-story buildings, therefore, single fire walls tend to be more economical because of their efficiency of mass; for multistory buildings, double fire walls tend to be more economical because of their efficiency of reinforcing. And, separate and apart from structural integrity and fire resistance, in all buildings whose programs call for a high degree of interior openness, fire walls may be inappropriate.

It is not that openings in fire walls are prohibited, only that they are limited in size and number, and must have appropriate protectives to prevent the passage of smoke and fire. Under one of the model codes, "each opening through a fire wall shall not exceed 120 ft² (11 m²). The aggregate width of all openings at any floor level shall not exceed 25

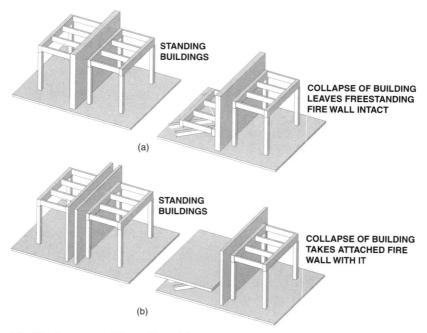

4.3   The two types of fire walls: (*a*) Single fire wall. (*b*) Double fire wall.

percent of the length of the wall."[9] Sprinklers permit the loosening of these constraints, but only if they are provided throughout both of the abutting fire areas. Protectives include two types, one—including rated fire doors and fire shutters—at openings used for the movement of occupants, and the other—including fire dampers in ducts—at openings used for the movement of air. Each opening protective must be self-closing and must have the appropriate rating in relation to the rating of the assembly in which it is located.[10] In general, fire doors are permitted to have slightly lower ratings than the walls in which they are located. The rationale for this is that "whereas combustibles may be placed against a fire resistance rated wall . . . a fire protection rated door assembly will not have similar combustibles placed against it because the opening must be maintained clear to allow passage through the door."[11] Notwithstanding this exception, I find it easier during preliminary design to assume that the rating of the doors matches that of the walls.

Fireblocking, Draftstopping, and Firestopping

Whether a building is constructed of steel or heavy timber, its walls and floors are typically framed with a series of parallel members, regularly spaced. If sandwiched between wall finishes or between floor or roof decking and ceilings, the voids within these assemblies resemble long and narrow channels. During a fire, these concealed spaces allow smoke and flame to spread out of sight of firefighters, and out of reach of their equipment. Junctures of wall, floor, and roof assemblies are particularly problematic, as they can enable the spread of fire far beyond the initial assembly.

Although weakened by high temperatures, steel framing is noncombustible. Timber framing, by contrast, will burn, even if slowly. For that reason, concealed spaces are prohibited in heavy-timber construction,[12] unless they are sprinklered.[13]

To prevent the spread of smoke and flame through concealed passages, two types of barrier are employed: *draftstopping* within large passages such as attics, and *fireblocking* within small passages such as floors, walls, and roofs.[14] Though some fireblocking—such as the top plate of walls in platform framing—is inherent to the building process,[15] most must be added. The precise locations, orientations, and conditions of such barriers vary from building to building, depending, among other things, on construction details at the interfaces between critical assemblies:

- Where construction is continuous horizontally between bays or vertically between floors, provide fireblocking at each column or floor line.

- Where the floor plane is interrupted by stairs, fireblock the walls along the stringers.
- Give special attention to the intersections at eaves, between floor and roof framing.
- Minimize the detrimental effects of shrinkage by postponing the installation of fireblocking until it and the framing have reached equilibrium moisture content. Otherwise, joints tight during installation will open up later, compromising the effectiveness of the barrier.
- Coordinate fireblocking with solid structural blocking; if their locations coincide, consider letting a single piece of blocking serve both purposes.
- In buildings with complex roofscapes, or with dormers or other roof structures, use particular care to identify potential interconnections among wall, floor, and roof assemblies; fireblock all of them.
- To verify that all fireblocking is in place and tightly fitted, thoroughly inspect the structure before any of it is hidden from view; add fireblocking wherever required.
- Use materials conforming to applicable codes, typically 2 in (5 cm) nominal lumber.

Whereas fireblocking is intended to slow the spread of fire through the structure, and draftstopping to slow its spread through building cavities, firestopping is provided to seal the joints between different building materials and systems. Our concern here is primarily with penetrations of the timber decking. Penetrations are generally permissible where they do not adversely affect the structural capacity of the decking. The requirements of the firestopping system are then a function of the size, insulation, and composition of the penetrating item, the width of the annular space around it, and the required rating, if any, of the penetrated building assembly.[16]

Allowable Building Height

The taller the building, the longer it is likely to take its occupants to vacate it in a fire. For this reason, building height is deemed an appropriate characteristic for regulation by the building and life-safety codes. The greater the fire resistance of the building construction, the longer the occupants are presumed to have available for safe escape. Nevertheless, except for protected noncombustible buildings where the members supporting more than one floor have ratings of 3 hours or more, building heights above 10 stories are generally prohibited.[17]

Life-safety design, like structural design, involves balancing loads and resistances. In structural design, the loads include the self-weight of the structure and the weights associated with occupancy, while the resistances are inherent to the structural materials and member sizes. In life-safety design, the fire loads include that of the structure (if any) and those associated with occupancy, while the resistances are inherent to the construction materials and member sizes. In structural design, the stress in a member is a function of the load P divided by the cross-sectional area of the member A:

$$\text{stress} = P/A$$

In life-safety design, I like to think of the urgency of evacuating a building as a function of the fire load L divided by the relative fire resistance of the structure R:

$$\text{urgency} = L/R$$

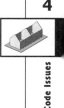

Just as the stress in a structural member can be reduced by increasing its size or strength or decreasing the load on it, the urgency of evacuating a building can be reduced by increasing the fire resistance of its members or decreasing the fire load.

Automatic sprinkler systems represent one proven means of suppressing fires. Their net effect is the same as reducing the fire load: The time available for egress is increased. For this reason, the codes allow height increases in sprinklered buildings.

A necessary condition for the issuance of a building permit is that the *proposed* height and area of the building's largest fire area do not exceed the *allowable* height and area given in the applicable code for the proposed construction type and use group. That is why the intelligent designer starts the structural selection process with an analysis of the applicable height and area constraints. No matter how compelling are the functional or esthetic arguments in favor of timber over steel or concrete, they are of no significance if the height or area is excessive, or if the fire walls needed to prevent such excesses cannot be arranged in an acceptable way.

Fire Separation Walls

Many buildings contain more than one use group. To accommodate mixed uses, building codes typically give the designer several different ways to conceptualize the building. In the first, entitled *nonseparation,* the mixed uses are not separated from one another; instead, the entire building is designed with the most restrictive construction type required for any one of them.[18] If, for example, the building in question is a hotel with a meeting room, the allowable height and area of the meeting room, a place of assembly, would control over that of the residential portions of the building. While this approach makes planning simple and flexible—it permits all of the use groups to be distributed freely throughout the building—it can make construction more costly, especially when the controlling use group occupies a relatively small portion of a building. *Under nonseparation, interior fire separations are avoided at the cost of a more fire-resistant building frame.*

The second approach, *separation,* involves the construction of fire separation assemblies between different uses. Assuming a given construction type, the ratio of the proposed and allowed areas is calculated for each use group. If the sum of the ratios is less than 1, the assumed construction type is permitted, but each use must be separated from every other use by fire separation assemblies having the appropriate rating.[19] *Under separation, the fire resistance of the building frame is minimized at the cost of interior fire separations.*

Whenever a building contains more than two use groups, you should study the implications of separation, nonseparation, and each of the scenarios between. A combination of nonseparation and separation is worth considering when the requirements associated with one of the uses are substantially more restrictive than those of the others. Examples include offices buildings (business uses) containing other uses, including meeting rooms (assembly uses), and police stations (business uses) containing other uses, including jail cells (institutional uses). Often in such cases, the costs of nonseparation are prohibitive, because the entire building would have to accommodate the requirements of a relatively small but highly restrictive use group. On the other hand, the functional limitations and inflexibility associated with complete separation can be equally unacceptable. The most practical approach may involve designing for the dominant or main use or for the second most restrictive use. And, if the most restrictive use is accessory to the main use and occupies no more than 10 percent of the building's floor area, it may not need to be separated from the main use.[20,21]

The allowable height and area for a given construction type can be thought of as the maximum size or capacity of each fire area of the building. If future modifications, such as additions, changes of use, or reductions in street frontage are contemplated, consider underutilizing that capacity in the initial design. Without such slack, changes like these may only be permitted if fire walls or fire separation walls are added within the existing building, a costly and disruptive operation under the best of circumstances.

Fire separation walls differ from fire walls in several important ways. Fire walls are continuous from the ground to, or slightly beyond, the roof; they may be erected before or after the structural frame. Fire separation walls, by contrast, run from the top of the floor to the underside of the floor or roof above.[22] If the ceiling is part of a rated floor/ceiling assembly, the wall must pass through it. If the ceiling itself is a rated assembly, the wall may stop at it.[23] In either case, the superstructure must be in place before the walls can be built. In a nutshell, in a building with multiple fire areas, the structure is delimited by fire walls, while in any building, fire separation walls are delimited by the structure.

The implications of the latter are critical in a building with timber framing. Each member passing through a fire separation wall represents a path—a fuse, if you will—for fire to take from one side of the wall to the other. The fire will wreak additional havoc along the way if the concealed spaces in the wall are framed with lumber that has not been fire treated. With large members, the fuse will burn slowly enough to pose minimal risk. However, though the goal is to construct the wall tightly around each penetrating member, open joints are inevitable there due either to the limitations of field construction or to the shrinkage of the timbers. Standards for the protection of these annular spaces are given in each of the model codes. Consider using resilient sealing materials to accommodate differential movement of the wall and the structural frame.

Although limited in size by the surrounding construction, fire separation walls do not exist in isolation; the members and assemblies that support them must have ratings to match.[24] Where the rating exceeds 1 hour, it will be necessary to protect the timber columns and beams comprising the supporting structure. If, however, the rating is just 1 hour, the members may meet the requirement without additional protection.[25]

Fire separation assemblies fall into three main categories: (1) exit enclosures, (2) shafts and elevator hoistways, and (3) mixed use separations. All require ratings of 2 hours or more,[26] unless the building has fewer than four floors—in which case the ratings of the first two categories drop *to* 1 hour[27]—or unless the building is sprinklered—in which case the ratings in the third category drop *by* 1 hour.[28] In many cases, therefore, a timber structure is likely to provide adequate fire resistance for the support of fire separation assemblies. Bear in mind, however, that timber *floor* deck is not rated, so compliance with the requirements for rating of the supports will be easiest to achieve if the fire separation walls are located over beams or joists.

Fire Performance of Heavy Timber

Unlike steel or concrete, timber framing adds fuel to a fire, between 7500 and 9050 Btu/lb (between 17,450 and 21,050 kJ/kg).[29] As each member is consumed, its structural capacity declines: "Wood, when exposed to heat and/or flame, forms a self-insulating surface layer of char. Although the surface chars, the undamaged inner wood below the char retains its strength and will support loads equivalent to the capacity of the uncharred section."[30] Therefore, to the extent the members are oversized (i.e., understressed), they will be able to accommodate fire-related reductions in cross-section without structural failure. Such extra thickness in a timber member serves a purpose similar to that of the concrete cover that protects reinforcing steel in reinforced concrete members: In a fire, it buys time. The fire resistance of members loaded to half their capacity is between 30 and 50 percent higher than that of fully loaded members, depending on structural function and length.[31]

Why, you might ask, isn't the fire resistance inversely proportional to the ratio of actual to allowable stress? Why, for example, doesn't a half-loaded member have *twice* the fire resistance of a fully loaded one? The primary reason appears to be that the members in question are exposed to fire on more than one face. Therefore, the area of the uncharred cross-section will diminish not in proportion to the first power of the fire duration, but rather to its second power. (See Fig. 4.4.)

Just as a timber frame is only as strong as its weakest link, it is only as fire resistant as its weakest link. Therefore, it makes little sense to understress the columns if the members in the roof trusses supported by them are fully stressed. Similarly, why worry about the section at midspan if the net section at the connections controls the design. The designer must view the building as a whole, recognizing that its behav-

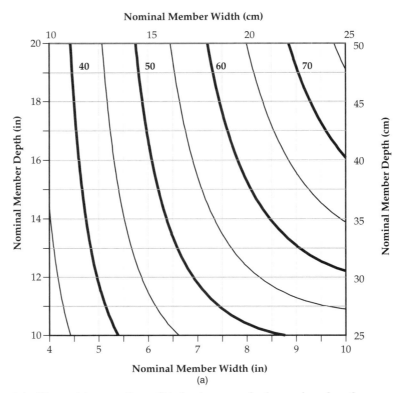

4.4 Fire-resistance ratings of timber beams and columns, based on the number of exposed faces: (*a*) Beams with four exposed faces. As the ratio of a member's actual load to its allowable load decreases, its fire-resistance rating increases, roughly in a straight line, leveling off when the ratio reaches 0.5. At that point, the increase in the rating, above that shown in these graphs, is 30% for slender columns and all beams, and 50% for other columns. (Derived from *Timber Construction Manual.*)

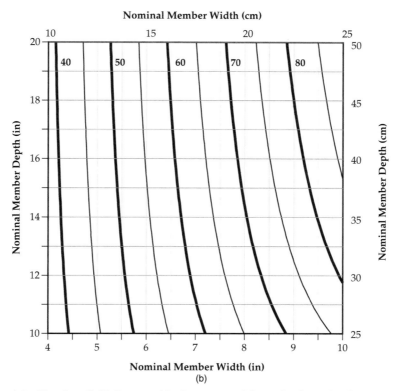

4.4 (*Continued*) (*b*) Beams with three exposed faces. As the ratio of a member's actual load to its allowable load decreases, its fire-resistance rating increases, roughly in a straight line, leveling off when the ratio reaches 0.5. At that point, the increase in the rating, above that shown in these graphs, is 30% for slender columns and all beams, and 50% for other columns. (Derived from *Timber Construction Manual.*)

ior under fire conditions will be much more complex than the behavior of each of its members. In addition to altering the distribution of loads among the members, a fire in an enclosed structure can produce enormous pressure, generating forces entirely different from those the structure was designed to accommodate.

Although heavy timber burns slowly, flames spread rapidly across its surfaces. Sprinklers help minimize the degree to which the timber structure contributes fuel and produces smoke during a fire. That is why sprinklers are appropriate, in most heavy–timber buildings, even where not required.

Mixed Construction Types

Timber construction is used not only by itself, but also in combination with other construction types. As in the case of mixed uses, buildings of mixed construction types are controlled by the requirements associ-

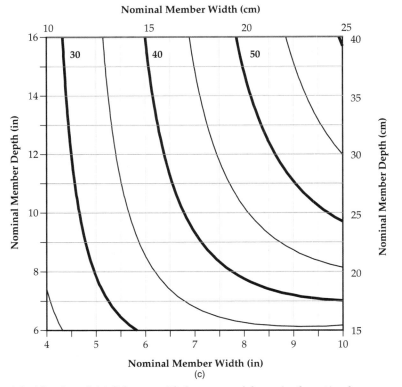

Nominal Member Width (cm)

Nominal Member Width (in)

(c)

4.4 (*Continued*) (*c*) Columns with four exposed faces. As the ratio of a member's actual load to its allowable load decreases, its fire-resistance rating increases, roughly in a straight line, leveling off when the ratio reaches 0.5. At that point, the increase in the rating, above that shown in these graphs, is 30% for slender columns and all beams, and 50% for other columns. (Derived from *Timber Construction Manual*.)

ated with the lesser type,[32] except when the fire resistance of designated members in specific locations is allowed to be lower than that of the rest of the superstructure.[33] This exception applies most commonly to roof construction, the required fire resistance of which decreases as its distance above the floor increases.

If you intend to use timber roof framing in an unprotected steel building, recognize that the steel is more vulnerable in a fire. Therefore, its fire performance will essentially control that of the entire building. The timber may enhance the architecture, but it will not enhance the effective fire resistance. While timber will *start* to lose its strength at or below 750°F (400°C),[34] steel will *finish* losing its strength after prolonged exposure to ordinary fire temperatures of 1000°F (540°C) or more. Temperatures in building fires tend to rise quickly, so the time interval between these critical points can be very short.

Steel's coefficient of thermal expansion is quite high; its length may increase more than 1 percent in a building fire. Steel members

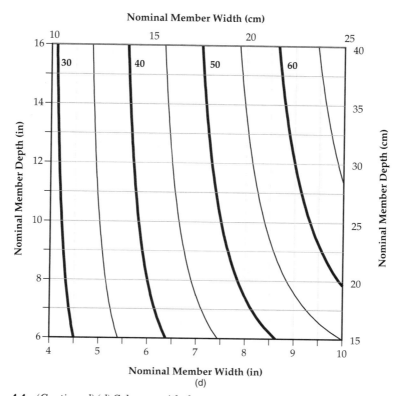

Nominal Member Width (cm)

4.4 (*Continued*) (*d*) Columns with three exposed faces, where unexposed face is the smaller side. As the ratio of a member's actual load to its allowable load decreases, its fire-resistance rating increases, roughly in a straight line, leveling off when the ratio reaches 0.5. At that point, the increase in the rating, above that shown in these graphs, is 30% for slender columns and all beams, and 50% for other columns. (Derived from *Timber Construction Manual*.)

restrained from elongating can buckle or overturn.[35] If allowed to lengthen, they may load surrounding materials in ways they cannot accommodate. Timber, on the other hand, distorts very little in a fire, minimizing the likelihood of such consequential damage.[36] Take this difference into account in the design and detailing of the interfaces between fire walls and structure, and between one structural material and another.

Steel Connectors

In a fire, the likelihood of structural failure may be obvious in a *steel* frame, even if it has a slow-burning timber roof as in the foregoing example. Less obvious, but equally likely, is structural failure of a *timber* frame having exposed steel connectors. Although they are generally small, steel connectors are critical to the integrity of a timber frame. Under the stress of a fire, even those connectors that carry no load under normal circumstances—such as side plates of bearing connections—could end up carrying substantial loads, some of which may not be predictable in direction or magnitude. Critical connectors—such as truss plates transmitting large tensile forces—are of particular concern.

The concern is twofold: (1) weakening of the steel itself and (2) conduction of heat by the steel connector and its steel fasteners to the timber surfaces with which they are in contact. The first issue involves materials and assemblies that are visible, while the second affects those that are concealed. The most effective ways to address these concerns are to protect the steel or sprinkler the building. These strategies apply to timber frames or trusses with steel connectors, and to composite timber and steel trusses.

Composite trusses often utilize timber for top chords and other compression members, and steel for bottom chords and other tension members. In a roof truss, a steel rod or cable might provide the only resistance to the outward thrust imparted by the timber framing of a sloping roof. A small tie will heat up and lose strength in no time.

Minimum Member Sizes

Not all framing members are of equal structural importance. Failure of a portion of the roof deck would normally result in limited localized damage, while failure of a main column could lead to progressive collapse of an entire structure.[37] In general, the structural importance of a framing member is roughly proportional to the load it carries.

The total load on a framing member relates to its total tributary area, which follows from its role and location. Because of gravity, columns accumulate vertical loads from the roof and any intermediate floors. Consequently, column loads are lowest near the roof and highest near the bottom floor. Furthermore, because roof loads tend to be lower than floor loads, column loads accumulate at an accelerating rate. The loads on rafters, girders, and joists are proportional to their spacing and span. They too can accumulate loads from perpendicular members, but usually pass these along to columns. In a frame with bays of constant area a at each level l and with two simply supported girders framing into each column at each level, the tributary area of each girder will be a and the load at the end of each girder will be proportional to $a/2$. By contrast, the tributary area of each column accumulated at the lowest level will be al. In a multistory building, therefore, the vertical loads that must be carried by the columns can dwarf the loads that must be carried by the girders.

Let's put these findings in perspective; in this and other chapters, we will examine three important relationships:

1. The relationship between member type and load

2. The relationship between load and timber size

3. The relationship between timber size and fire resistance

In general, members carrying heavy loads must be larger than those carrying light loads. The greater the supported load, the more serious the consequences of failure of the supporting member. Clearly, there is a correlation between member size and the consequences of member failure.

Because of the relationship between timber size and fire resistance, heavy-timber construction is distinguished from conventional wood framing primarily by the minimum nominal dimensions of its members. These are summarized in Table 4.1.

Another way to understand the significance of these sizes is to compare the resulting minimum cross-sectional areas as shown in Table 4.2.

If one of the objectives of setting minimum member sizes is to assure that a member's fire-resistance is proportional to the consequences of its failure, the current approach to sizing would appear to be misdirected. While it is based on the very rough correlation that exists between the minimum dimensions mandated for heavy-timber members and the relative tributary areas of and loads on those members, the actual dimensions required to satisfy structural considerations do not necessarily control the design. Under the current approach, final member size is the *larger* of:

1. The size that satisfies structural requirements
2. The minimum size that satisfies the requirements of the heavy-timber construction type (i.e., the size that possesses the required fire resistance).

Member size, in other words, is driven by one or the other. When structural considerations control the design, no additional allowance need be made for fire resistance. And when construction type controls the design, the member's fire resistance becomes, by default, a function of its degree of understress.

Even more disturbing is the fact that the hierarchy of minimum member sizes does not allow for differences between buildings. Consider, for example, a one-story building one bay wide having a repetitive structure consisting of a series of parallel bents. In this

TABLE 4.1 **Member Types and Minimum Sizes**

Minimum nominal dimension	Depth	Width
2 in (5.0 cm)	Roof deck	
3 in (7.5 cm)		Roof truss with sprinklers
4 in (10.0 cm)	Floor deck	Roof truss without sprinklers
6 in (15.0 cm)	Roof truss	Floor joist
	Roof-supporting column	
8 in (20.0 cm)	Floor truss	Floor truss
	Floor-supporting column	All columns
10 in (25.0 cm)	Joist, beam, or girder	

SOURCE: BOCA, 1996, p. 254.

TABLE 4.2   Minimum Nominal Sizes and Areas by Member Type

Member types	Minimum nominal depth		Minimal nominal width		Minimal nominal area
Roof truss with sprinklers	6 in (15.0 cm)	×	3 in (7.5 cm)	=	18 in² (112.5 cm²)
Roof truss without sprinklers	6 in (15.0 cm)	×	4 in (10.0 cm)	=	24 in² (150.0 cm²)
Column supporting roof	6 in (15.0 cm)	×	8 in (20.0 cm)	=	48 in² (300.0 cm²)
Joist, beam, and girder	6 in (15.0 cm)	×	10 in (25.0 cm)	=	60 in² (375.0 cm²)
Column supporting floor, floor truss	8 in (20.0 cm)	×	8 in (20.0 cm)	=	64 in² (400.0 cm²)

SOURCE: BOCA, 1996, p. 254.

construction, each column will carry only *half* the load of each truss, yet to be considered heavy timber, each column must be *twice* the size of each truss member. Since the failure of any column or truss will likely have similar consequences—the loss of an entire bay—it is difficult to justify the application of a hierarchy that assumes differing consequences. There has got to be a better way.

That better way would recognize that the portion of a timber's cross section destroyed by fire can no longer contribute to its structural performance. It would also recognize that a fully stressed member will fail more quickly in a fire than one that is understressed. As it is in the sizing of combined members—where cross section can be used to resist axial *or* bending stresses, but not both simultaneously—so should it be in the sizing of fire-resistant members. I propose a new approach, wherein final member size is the *sum* of:

1. The size that satisfies structural requirements

2. The minimum additional thickness, if any, required to achieve a designated level of fire resistance.

This approach is somewhat akin to that used for reinforced concrete, where the concrete cover below the rebars on the tension face is solely there to protect the steel from the fire. The difference in the case of timber is that the additional material would be sacrificial; the longer the fire duration, the less the remaining protection.

There are several additional advantages to this approach. First, it would enable the designer to generate a project-specific hierarchy of member importance, rather than being bound by a general and potentially inappropriate one. Second, it would require that the degree of understress be a primary consideration, rather than a secondary consequence of the ratio between the size mandated for structure and the size mandated by the code-defined construction type. Third, it would permit the benefits of sprinklering to be applied more consistently in the sizing of members. (There appears to be little justification for the current situation, where the presence of sprinklers affects the sizing of framing at the roof, but nowhere else.) Finally, it would be consistent with the recent trend away from prescriptive code provisions and toward those based on performance. (Obviously, such a change would have significant economic and esthetic implications.)

Proximity to Other Buildings

In matters of life safety, each building must be thought of in relation to adjacent buildings. When buildings are located in close proximity, fire spread between them is controlled largely by the fire resistance of their exterior walls. The smaller the *fire separation distance,* the greater the required fire resistance of the exterior walls[38] and of any structural members located within them.[39] In heavy-timber construction, load-bearing exterior walls are required to have a fire resistance rating of 2 hours or that associated with the fire separation distance, whichever is greater.[40] Because the highest rating recognized for heavy timbers is 1 hour, this means that structural timbers located within exterior load-bearing walls cannot meet the 2-hour rating associated with this construction type unless further protected. Such protection might consist of a coat of intumescent paint or a layer of fire resistant gypsum sheathing on the exterior faces of these members.

Conditions Necessary for a Fire

For a fire to start or be maintained, three elements must be present: fuel, oxygen, and heat (see Fig. 4.5). "Remove or reduce any single element below an established level, and the fire will cease."[41] In a building with a steel or concrete frame, the fuel must come from somewhere other than the noncombustible superstructure. In a building with a timber frame, by contrast, the superstructure itself represents a huge source of fuel.

By definition, heavy timbers have substantial minimum cross-sectional dimensions and are therefore inherently massive. Their low ratios of surface area to volume and relatively poor thermal conductivity assure that they will burn slowly, performing better under many fire conditions than unprotected noncombustible construction. It should not

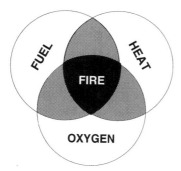

4.5 The three elements needed to start a fire.

be surprising, therefore, that the fire resistance of a timber frame is lowest where that ratio is locally at its highest. Such conditions typically occur at the edges, ends, and penetrations of members.[42]

Simply put, the relative fire resistance of a heavy timber is inversely proportional to sa/v, where sa is its surface area and v is its volume. Therefore, to improve the fire resistance of timber, either decrease its surface area or increase its volume.

The shortest line enclosing a given area is a circle. By extension, the smallest area enclosing a given linear volume is a cylinder. In a subsequent chapter, we will see that this shape optimizes the structural performance of timber columns having the same slenderness ratio in all directions. Now we can see that this shape also maximizes the fire resistance of timber columns having the same degree of fire exposure in all directions. If round columns are unavailable, impractical, or uneconomical, the next best column shape is the square, a rectangle with an aspect ratio of 1. As the aspect ratio increases, the surface area increases, reducing fire resistance.

Just as each knot or check can affect a timber's strength, each corner or penetration affects its fire resistance. In the vicinity of edges or holes, there is very little volume but enormous surface area. To the extent that they are aggregations of edges and holes, timber connections are particularly vulnerable in a fire. Conditions are even worse where increases in surface area coincide with substantial decreases in volume, such as occur in lapped and spliced joints. The overall *cross-sectional area* of such a joint—if tight—may be the same as the typical area of each member beyond the joint. The *fire resistance* of such a joint, however, will be far less. The joint will fail first because its integrity is controlled by a member of less-than-full cross-section. In a lapped joint, for example, the area of each member at a joint is only half of its area beyond the joint. The joint is therefore more vulnerable in a fire. Mortising a timber so that a steel connector sits flush with its surfaces may look good, but will result in similar vulnerability.

Because the overall fire resistance of a timber structure is governed by its weakest link, the designer must identify and eliminate or minimize vulnerable aspects. Some of the more common strategies involve:

- Rounding or chamfering corners
- Covering or blocking the spaces between spaced members[43]
- Placing steel connectors on the inside rather than the outside of timber joints
- Coating exposed steel connectors with intumescent paint
- Adding wood trim rather than subtracting timber cross-section to achieve architectural relief
- Minimizing the surface area of holes by using a few large bolts rather than many small ones

- Avoiding joints that require substantial reductions in the areas of the joined members

- Minimizing the ratios of surface area to volume among the timber members

Interfaces with Other Materials and Systems

With very few exceptions, timber structures typically support and/or are supported by other building systems. In mill construction, for example, masonry exterior walls provide vertical support for timber floor framing, while being laterally braced by them. In many buildings otherwise framed exclusively with steel or concrete, timber roof construction is permitted. Timber columns may be built into walls framed with wood or steel studs. Regardless of the particular combination, the interfaces between timber and other construction materials have important implications in a fire.

Timber may be more or less fire resistant than the other construction materials in a given building. In general, it will perform better than unprotected steel or lumber, but not as well as concrete or masonry. After a fire, the barn—stripped of siding—illustrates the former, while the farmhouse—survived only by the chimney—illustrates the latter.

Fire is a progressive phenomenon; assuming adequate supplies of oxygen, fuel, and heat, the longer it lasts, the more pronounced and widespread the damage. Depending on the manner in which the building systems are connected to one another, the failure of one can precipitate the failure of another. To minimize the impact of failure of the timber structure on the integrity of other buildings systems, and vice versa, one must become familiar with a building's construction details and the relative fire resistances of their components. Then, answer these questions:

- Which components are likely to fail first?
- What will happen to the components that fail, as they fail?
- Which components are thereby put at risk of consequential damage?
- What is the source of the risk?
  The impact of falling debris
  The geometric relationships between the systems
  The connections between the systems

The force of gravity guarantees that failed components will *want* to fall. Whether they *actually* fall depends on how they are anchored and to what. At one extreme are single fire walls; these are structurally independent of the buildings on either side of them specifically to assure that they will remain standing should either of those buildings collapse. At the other extreme are double fire walls; each layer is dependent on the adjacent building for lateral support. Collapse of that building nearly assures that the single wall attached to it will collapse with it. In most cases, building construction falls somewhere between these two extremes; in most cases, different materials are connected to one another because structurally they have to be; but in a fire, you would rather that they weren't.

An interesting and common condition occurs where a timber floor frames into a masonry wall. If the wall is pocketed to receive the joist ends, the pockets create areas of weakness in the wall. Under extended fire exposure, the joists will snap and their ends will rotate. To prevent the rotating joists from imparting an eccentric upward load on the tops of the pockets, the ends of the joists should be tapered, producing *fire cuts*. Unfortunately, this may be counterproductive, for it "lessens the inherent resistance of the joist to fire and can precipitate floor collapse."[44] While the fire cuts may decrease the consequences of timber failure, they may actually increase its likelihood. An alternative approach involves framing the floor joists into or onto a continuous timber ledger bolted to the wall. Wall pockets and their attendant problems are thereby avoided. If the floor framing fails, the joists will rotate off or away from the ledger (see Fig. 4.6).

In general, joist and beam hangers are designed exclusively to resist vertical loads. In plan view, the nails that secure each hanger to the face of the ledger are perpendicular to those that secure that hanger to the sides of the joist end, but all are in single shear under vertical loads. In the event of joist failure, the nails in the sides of the joist will still be in single shear, but those in the ledger will be loaded primarily in withdrawal, a much weaker orientation. As long as the total withdrawal resistance of the nails is less than the overturning resistance of the wall, failure of the floor framing should not precipitate collapse of the wall.

Pin connections, such as those at the bases and peaks of three-hinged arches, are called that because of their inability to resist rotation. Thus, timber failure is less likely to produce damaging stresses in supports of this type than in those previously discussed. In a fire, the most vulnerable parts of a three-hinged arch are the steel hinge at the peak and the timber of relatively small cross section permitted in its vicinity. Failure of this connection could drop both half-arches to the floor or could leave them leaning precariously against one another.

The fact that wood's strength is highest under short-term loads is certainly helpful in a fire, especially if the structure is subjected to the onslaught of falling debris and the weight of water from fire-fighting operations. The problem is that fires can change both the distribution of loads and the distribution of the structure's capacity to accommodate them.[45]

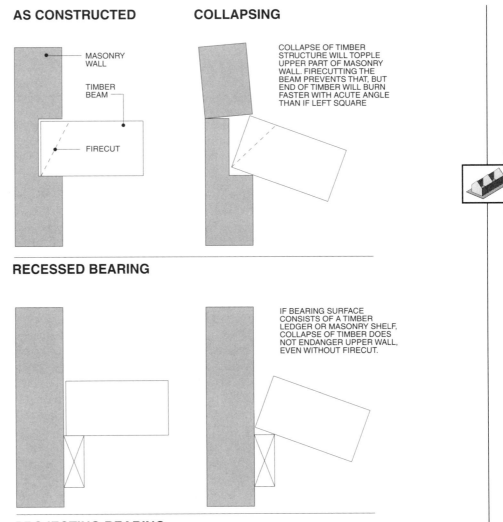

AS CONSTRUCTED

MASONRY WALL

TIMBER BEAM

FIRECUT

COLLAPSING

COLLAPSE OF TIMBER STRUCTURE WILL TOPPLE UPPER PART OF MASONRY WALL. FIRECUTTING THE BEAM PREVENTS THAT, BUT END OF TIMBER WILL BURN FASTER WITH ACUTE ANGLE THAN IF LEFT SQUARE

4

Code Issues

RECESSED BEARING

IF BEARING SURFACE CONSISTS OF A TIMBER LEDGER OR MASONRY SHELF, COLLAPSE OF TIMBER DOES NOT ENDANGER UPPER WALL, EVEN WITHOUT FIRECUT.

PROJECTING BEARING

4.6 Interface of timber and masonry bearing wall.

Summary

Fire safety is a function of a number of factors, of which structural performance is but one. The earlier fire is detected and the occupants notified, the more time they will have to evacuate the building. During its incipient stages, a building fire involves a limited area. The longer it burns, the more difficult and hazardous egress becomes. In a sense, there is a race between the fire and those trying to escape it. The goal of building fire safety, then, is to make sure that detection, alarm, and evacuation can be completed before conditions within the building and along its exit routes become intolerable.[46]

In his book, *Why Things Bite Back,* Edward Tenner distinguishes between loosely and tightly coupled systems:

> In human terms, even thousands of people on a crowded beach form a loosely coupled system. . . . There is risk to each swimmer . . . but little chance that one swimmer's mishap will spread to dozens of others.
> . . . Now imagine the same crowd packed in a stadium, surrounded by gates, turnstiles, wire mesh, and other control devices. . . . in installing these new barriers, the management has turned the place into a much more tightly coupled system. The barriers serve to keep troublemakers off the field. Unfortunately, they also make it more likely that a single problem will be tragically amplified.[47]

Buildings are inherently tightly coupled systems. Their walls and partitions are barriers no matter what, but they may be helpful or harmful depending on the occupants' relationship to them. That is why the speed with which the occupants can get to the exits and with which the exit openings can be protected is so important.

The responsible use of timber in construction starts with the recognition that it is both structure and fuel. Member sizes, shapes, and connections are acceptable when and if that fuel is consumed slowly. The major determinant is the ratio of a member's surface area to its volume; the lower it is, the greater the relative fire resistance. Measures that reduce the availability of fuel, such as fire retardants and intumescent coatings, or that reduce the available heat, such as sprinklers, permit larger timber buildings.

Although both fire walls and fire separation walls can permit the construction of larger heavy-timber buildings, they work in different ways. Fire walls enable oversize timber buildings to be subdivided into compartments of allowable size. Single fire walls are independent of the buildings around them, while each of the halves of a double fire wall is secured to one of those buildings. Fire walls bound the structure, but are outside of it. They minimize the extent of rated partitions, but require that the construction type be adequate for the most hazardous use within it. Fire separation walls, by contrast, are used to separate different uses, protect exit components, or enclose shafts. They are bounded by structure and supported by rated structure. They maximize the utilization of the available floor area, but increase the extent of rated partitions. Some may be required regardless of occupancy.

Early in the design process, the architect must weigh the costs and benefits of each of the allowable construction types, taking account of their implications with respect to separation, compartmentation, fire protection, and firefighting access. Heavy timber, carefully detailed and properly specified, may be left exposed to view in completed buildings. The costs of suspended ceilings can thus be avoided, as can the costs of the additional sprinkler heads that would be required in the concealed combustible spaces they would produce. Even where the structure is exposed to view, sprinklers are a good idea, to prevent fire from spreading along timber surfaces.

Notes

1. CWC, *Wood Reference Handbook*, p. 67.
2. AITC *Tech Note #7*, p. 1.
3. BOCA 1996, p. 56.
4. BOCA 1996, p. 56.
5. BOCA 1996, p. 58.
6. BOCA 1996, p. 57–58.
7. BOCA 1996, p. 67.
8. NFPA, *National Fire Codes*, p. 221-5.
9. BOCA, 1996, p. 71.
10. BOCA, 1996, p. 79.
11. Lathrop, p. 206.
12. NFPA, *Fire Protection Handbook*, p. 5–23.
13. NFPA, *Fire Protection Handbook*, p. 17–5.
14. BOCA 1996, p. 65.
15. Brannigan, p. 107.
16. Buchanan, p. 43.
17. BOCA 1993, p. 54.
18. BOCA 1996, p. 27.
19. BOCA 1996, p. 27.
20. BOCA 1996, p. 18.
21. IBC, p. 3.21.
22. BOCA 1996, p. 72.
23. Lathrop, p. 203.
24. BOCA 1996, p. 78.
25. AITC *Tech Note 7*, p. 1.
26. BOCA 1996, p. 62.
27. BOCA 1996, p. 72.
28. BOCA 1996, p. 28.
29. Grosse, p. 20.
30. TCM, p. 1–21.
31. AITC, *Tech Note 7*, p. 7.
32. BOCA 1996, p. 61.
33. BOCA 1996, p. 62.
34. NFPA, *National Fire Codes*, 921–25.
35. Brannigan, p. 256.
36. TCM, p. 1–21.
37. NFPA, *Fire Protection Handbook*, p. 5–28.
38. BOCA 1993, p. 65.
39. BOCA 1993, p. 74.
40. BOCA 1993, p. 60.
41. Grosse, p. 2.
42. NFPA, *National Fire Codes*, 5–24.
43. NFPA, *National Fire Codes*, 5–24.
44. Brannigan, p. 83.
45. Brannigan, p. 22.
46. Parker, p. 20.
47. Tenner, p. 16.

4

Code Issues

PART

2

Timber Design
and Detailing

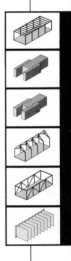

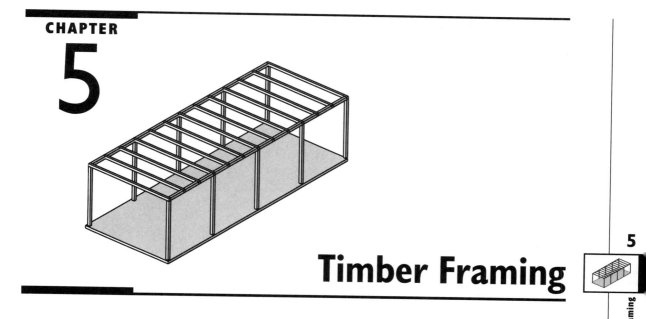

Timber Framing

"H-m-m-m . . . the first thing to do," murmured Horton,
"Let's see . . .
The first thing to do is to prop up this tree
And make it much stronger. That has to be done
Before I get on it. I must weigh a ton."

DR. SEUSS, *Horton Hatches an Egg*

Introduction

In a given building, the universe of practical framing arrangements is limited only by the architect's particular design objectives. The selected arrangement of the framing members determines, in turn, how the loads and forces acting on a building will be distributed among those members. Each member must be capable of bearing the loads and forces acting on it, and transmitting them to the members that support or brace it.

Design objectives that influence the framing concept include function, esthetics, economics, regulations, and service. Functional requirements determine the room sizes and ceiling heights and may affect the column spacing. Esthetic objectives may include symmetry, repetition, consistency, hierarchy, simplicity, visibility, order, drama, enclosure, lightness, restraint, and their complements. Economic objectives are not necessarily limited to construction cost; they may also include economy of means (i.e., doing the most with the least) and environmental impact. Regulatory requirements are predominantly those found in building codes, limiting the heights and areas of buildings according to their respective construction types. Service issues include durability and maintenance. When these objectives are in conflict, as they often are, the architect and structural engineer must find an appropriate balance.

Once the appropriate weight has been given to each of the design objectives, the expression of the final design concept is limited only by the designer's fluency with the properties of the structural materials.

Timber materials have properties fundamentally different from those of steel and concrete. Whether in the form of sawn lumber, glulams, or parallel strand lumber, timber properties are directional. In timber construction, there is no way of achieving structural continuity of the sort typical of steel and concrete construction. For that reason, solid timber-framing systems are exclusively one-way. To achieve two-way structural action in timber construction requires the use of open assemblies, such as trusses and space frames.

The directional nature of timber does not just affect the overall distribution of the framing members, but also their individual behaviors. This chapter will start with a look at the ways in which timber members react to external forces. Following that will be an overview of various framing systems, with a discussion of their relevant attributes.

The Evolution of Timber Materials

One of the aspects of timber which intrigues me is how it has evolved as a building material. When transportation costs were high, designers and builders were effectively limited to working with the stock of local timber. As transportation costs declined, demand for the best timber grew, leading eventually to near exhaustion of its supply. Glulam manufacturing processes, uncompetitive when large timbers were plentiful, penetrated the market for large timbers at that time, and enabled the production of members larger than any that came from nature. Today's explosive growth in demand for high-quality timber framing has again altered the economics of timber construction; the issue is now how to do more with less. (See Fig. 5.1.)

Up until about 50 years ago, the properties of each timber beam and column came from the tree from which it was cut. With its large cross-section, a sawn timber took a long time to dry and was likely to crack and twist in the process. And, although the quality of its inner fibers contributed to its structural performance, grading was based solely on the member's surface characteristics. As a result, there was considerable variation in the properties of sawn members within each grade.

With the advent of glulams, the production process changed from one of division to one of repetitive assembly. With glulams, the structural properties of the laminations could be tailored to the structural requirements of the job. The strongest pieces could be placed in the regions of highest stress—typically the top and bottom surfaces of beams. Thus, not only the structure, but the structural members themselves, could be engineered. The small cross-sections of the lami-

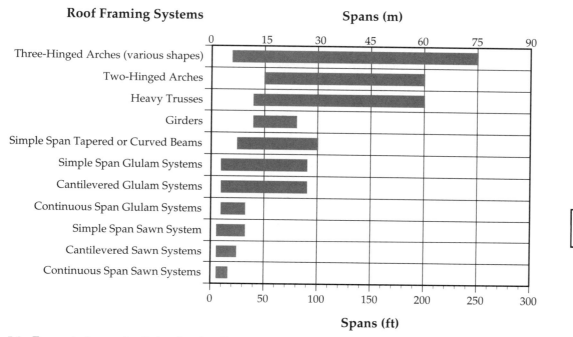

5.1 Economical spans for timber framing (derived from *Timber Construction Manual*).

nations enabled them to dry more quickly and with less checking, and permitted their stiffness to be machine tested, revealing essential information about the structural quality of each lamination. Lumber tested this way "has less variability in mechanical properties than visually graded lumber."[1] For this reason, the performance of glulams fabricated of machine stress-rated lumber is far more predictable than that of sawn members and of glulams composed of visually graded lumber.

In *reinforced* glulams, high-strength synthetic fibers supplant high-strength wood fibers in resisting the greatest stresses. The lumber laminations contribute relatively little to the strength of a reinforced glulam; instead, their role is more that of a filler, holding the tensile and compressive reinforcing as far apart as possible. These synthetic fibers are not only stronger than natural ones, but more consistent in their properties. Consequently, the performance of reinforced glulams is even more predictable than that of conventional glulams.

Loads

The goal of a structural system, and of each of its components, is to accommodate all of the applied loads without permanent deformation or unreasonable deflection. Loads can be vertical or horizontal, static or dynamic, concentrated or distributed, and of short or long duration.

- *Live loads*—those associated with occupancy—are typically regulated by the building code. These uniformly distributed loads may range from 40 psf (200 kg/m²) for apartments to 100 psf (500 kg/m²) for meeting rooms; from 60 psf (300 kg/m²) in the reading rooms of libraries to 150 psf (750 kg/m²) within their bookstacks; and from 50 psf (250 kg/m²) in private offices to 100 psf (500 kg/m²) in public lobbies.[2]

- *Dead loads*—those associated with the permanent building construction—can be calculated from the intended materials and assemblies. Where nonbearing interior partitions are anticipated, allowance for an additional uniform load of 20 psf (100 kg/m²) is required.[3]

- *Environmental loads* are caused by snow, wind, and earthquake. Snow will slide off steep roofs, especially if the roofing is smooth. Consequently, this can be said of snow loads: *The greater the incline, the greater they decline.* Snow will drift into areas of low pressure, as at discontinuities in roof planes. Snow will blow off roofs exposed to the wind. Therefore, snow loads are higher the more sheltered the environment. Lateral loads, such as those from wind and earthquakes, are addressed in Chapter 9.

The actual loads on a given member are a function of its *tributary area,* the building area it supports. At floor joists, the tributary area is simply the product of the spacing and the span. At columns, the load at a given elevation is a function of the tributary areas supported above. The total loads in the columns at the first floor of a three-story building will be the sum of the live, dead, and snow loads at the roof, plus the live and dead loads at the third and second floors. And, if the framing plans are different from floor to floor, the tributary areas may vary. Tributary areas follow from the layout of the framing. Therefore, if the architect intends to leave the framing exposed, and has some ideas about how it should be arranged, it is critical that these ideas be communicated to the structural engineer as early as possible in the design process.

Dormers can be used to illustrate the relationship between framing plan and tributary area. Two dormers of exactly the same shape and size may give rise to entirely different load distributions. It all depends on whether the dormer is supported by the roof or by the floor below. (Because it is tied into the skin of the primary structure, I consider a dormer framed out of the roof akin to a tight knot; because it can be structurally independent of the skin, a dormer supported on the floor, and framed through the roof, is, by contrast, like a loose knot.) The total load may very well be the same in each case, but the load paths from the roof to the foundations will be very different.

In residential wood-frame construction, joists and rafters are at close enough spacings that a significant amount of load sharing is assumed among adjacent members. Heavy timbers, even if regularly spaced, are typically too far apart to behave that way. For that reason, the design value increases associated with repetitive members only apply to dimension lumber.

The total load on a structure is not necessarily the sum of all of the individual loads that could occur; as with diversity factors in mechanical and electrical design, allowances are made for the probability that various combinations of structural loads will be at their maximum level simultaneously. It is highly unlikely, for example, that a major seismic event will coincide with a record wind gust, or that an entire bay, an entire floor, or all floors of a building will be fully loaded at the same time. In general, the larger the tributary area or the larger the design live load, the greater the allowable live load reduction.[4] Consult your applicable local code for the load combinations and reductions which apply to your project.

Forces and Stresses

Without realizing it, you probably already have an intuitive feel for the relationship between the orientation and distribution of forces and timber behavior. You know, for example, that a plank is bouncier at midspan than near its support, and that a narrow plank is somewhat bouncier than a wide one of the same thickness. You also know that a plank turned on edge is far stronger than the same plank lying flat, but that its top edge will buckle in compression if overloaded. In the next few pages, we will take a systematic look at the properties that come into play in reaction to various load orientations and combinations.

When the wind blows against a sailboat, you don't just *expect* it to move, you *want* it to move. When the wind blows against a building, you want it to stay where it is. That is why the study of the forces in buildings is within the branch of engineering called *statics;* a building must be capable of accommodating internal and external, horizontal and vertical service loads *without moving*. The structure, and each and every piece within it, must be in equilibrium. Every force must be resisted by an equal and opposite force. Otherwise, as with a house of cards, there would be progressive collapse.

Geometrically, the ways in which forces can act on a framing member fall into one of two broad categories: those in which the action and reaction are aligned with one another, and those in which they are not (see Fig. 5.2). When the forces are aligned, they can either push toward each other, causing *compression,* or pull away from each other, causing *tension*. When the forces are not aligned, the effect on the loaded member will depend on the degree of nonalignment.

When the forces pull or push just past one another, the result is *shear.* "A pair of scissors perfectly demonstrates shearing action; two forces slide by each other and destroy the material in between."[5] This

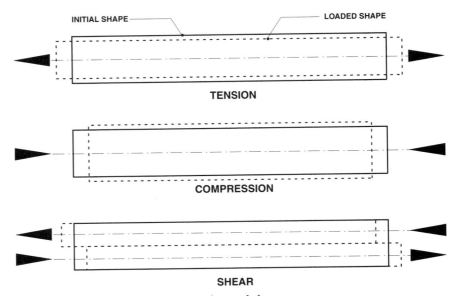

5.2 A comparison of tension, compression, and shear.

only works, however, if the surfaces of the opposing blades are essentially in the same plane. The greater the space between the blades, the harder it is to cut the material. Instead of shearing, it will bend. The same thing is true of timber framing members; the greater the distance between the lines of action and reaction, the greater the *bending*.

The orientation of the forces with respect to the member determines how it will behave. Tensile or compressive forces acting along the member's long axis are called, not surprisingly, *axial* forces. Resistance to axial tension is a function of a member's cross-sectional area and the strength of its wood fibers. Because long slender members are more likely to buckle under axial compression than short fat ones, resistance to axial compression is a function of a member's slenderness. All else being equal, a member loaded in axial compression will buckle in the direction of its smaller cross-sectional dimension or *thickness,* rather than its larger cross-sectional dimension or *width.*

A column is a common example of a member under axial compression. The length of a column divided by its thickness is called its *slenderness ratio.* The greater a column's slenderness ratio (i.e., the more slender the column), the greater its tendency to buckle. Reducing a column's slenderness ratio is a matter of reducing its length or increasing its thickness (see Fig. 5.3).

Let's take a moment to see how a column's thickness and lateral support conditions affect its slenderness ratio. Assume a column length of 120 in (3 m), a cross-sectional area of 50 in^2 (325 cm^2), and lateral support in both directions at the top of the column.

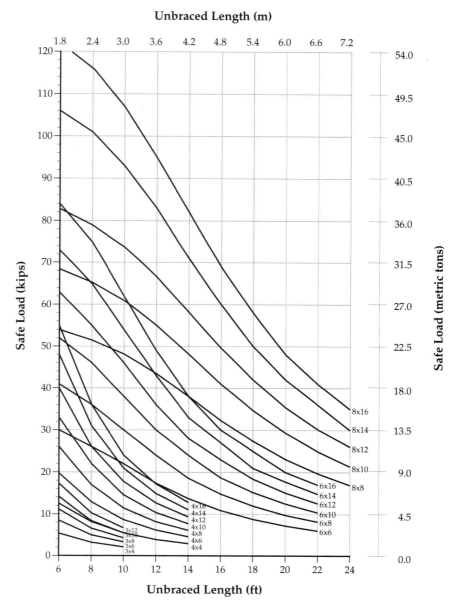

Unbraced Length (m)

Safe Load (kips)

Safe Load (metric tons)

Unbraced Length (ft)

5.3 Timber column capacities for sawn no. 1 Douglas fir–larch (based on Parker).

- A rectangular column with sides of 5 in (12.7 cm) and 10 in (25.4 cm) will have a slenderness ratio of 120/10 = 12 about its width, but 120/5 = 24 about its thickness. Therefore, the thickness will control.

- A square column of 50 in^2 (325 cm^2) will measure 7.10 in (18.0 cm) along both sides and have a slenderness ratio of 120/7.1 = 17 along both axes.

- A round 8-in-diameter column will have a slenderness ratio of 120/8 = 15 in all directions.

When these results are compiled in tabular form, it is easy to see how a column's thickness affects its slenderness ratio.

Column shape	Rectangle (1:2)	Square	Round
Slenderness ratios			
Thickness	24	17	15
Width	12	17	15
Controlling	24	17	15
Tendency to buckle	Greatest		Least

From the preceding, it is clear that a round column, besides having uniform resistance to buckling in all directions, has less tendency to buckle than any other shape of the same cross-sectional area when lateral bracing is provided at the same spacing along both axes. Obviously, buckling on the diagonal will never control in a rectangular column, since the diagonal of a rectangle is always longer than either of its two sides.

Interesting things happen when the lateral support conditions are not uniform in all directions. Again assume a column length of 120 in (3.0 m) and a cross-sectional area of 50 in² (325 cm²), but this time assume an additional lateral support parallel to the members thickness, at midheight (i.e., at 60 in [1.5 m]).

- The rectangular column will now have a slenderness ratio of 120/10 = 12 about its width, and 60/5 = 12 about its thickness.

- A square column with sides of 7.10 in (18.0 cm) will have a slenderness ratio of 120/7.10 = 17 along one axis and 60/7.10 = 8.5 along the other.

- A round column 8 in (20.3 cm) in diameter will have a slenderness ratio of 120/8 = 15 along one axis and 60/8 = 7.5 along the other.

When compiled in tabular form, it is easier to see how different lateral support conditions for member width than for member thickness affect the controlling slenderness ratios, assuming midheight bracing parallel to thickness.

Free length	Rectangular (1:2)	Square	Round
60 in (1.5 m) for wide face	12	8.5	7.5
120 in (3.0 m) for narrow face	12	17	15
Controlling	12	17	15
Tendency to buckle	Least	Greatest	

From the preceding, it is clear that the most appropriate *shape* for a column is a function of its specific lateral bracing conditions. For a given cross-sectional area, the most efficient column shape will have slenderness ratios that are equal along both the x and y axes. Given a free length l and a cross-sectional dimension d:

$$\frac{l_x}{d_x} = \frac{l_y}{d_y} \quad \text{or} \quad \frac{d_x}{d_y} = \frac{l_x}{l_y}$$

In other words, the column's *aspect ratio* (the ratio of its thickness to its width) should be equal to the ratio of the free lengths along the wide and narrow faces. In the rectangular example just given, those ratios were both 1/2.

Determining the appropriate thickness and width is easy, given the free lengths. Assume that $l_y = (2)(l_x)$. Based on the above equation, $d_y = (2)(d_x)$. The area of the column is given as: $A = (d_x)(d_y)$. Substituting $(2)(d_x)$ for d_y and 50 in² for A yields:

$$50 \text{ in}^2 = (2)(d_x)(d_x).$$

Solving the equation yields

$$d_x = 5 \text{ in.} \quad \text{And, since} \quad d_y = (2)(d_x), \quad d_y = 10 \text{ in.}$$

Column Stability Factors

The curve that correlates compression design value to slenderness ratios is nonlinear; its shape roughly corresponds to the behavior of columns of three classes. Short stout columns will fail in compression; long slender columns will fail by buckling; and intermediate columns may fail in compression or buckling. In recognition of the inherent instability of long slender columns, the slenderness ratio for columns is limited to 50, but is permitted to go as high as 75 under the lighter temporary loads of construction.[6]

For a given slenderness ratio, the *column stability factor* (C_p) is the degree to which the tabular compression design value (F_c) must be reduced to prevent buckling, and relates to the adjusted compression design value (F'_c) as follows:

$$F'_c = C_p F_c$$

For a given load P, the required cross-sectional area (A) of a column is given as:

$$A = \frac{P}{F'_c}$$

This can be rewritten as:

$$A = \frac{P}{C_p F_c}$$

Clearly, the lower the column stability factor, the greater the area needed to support a given load. While a constant area of 50 in² had been assumed for all the different slenderness ratios discussed in the examples earlier in this chapter, it is now evident that the required area will in fact vary with the slenderness ratio. If you want to know the tabular compressive design value required to support a given load, knowing a column's area and column stability factor, the above equation can be rewritten as:

$$F_c = \frac{P}{A C_p}$$

In this form, the inverse relationship between the column stability factor and tabular compressive strength is apparent. It assumes that you have tentatively selected a column size, which in turn has enabled you to calculate the applicable slenderness ratios.

Fixity of the ends of a column will reduce its tendency to buckle, thereby increasing its load capacity. The foregoing discussion assumed that the ends of the columns were relatively free to rotate, a safe assumption in most timber construction due to the nature of the material and the ways it can be connected.

Column Sizing

Here are the essential steps, then, in preliminarily sizing a timber column:

1. Compute the load that the column will support (P).

2. Establish the height(s) at which each column axis will be laterally braced.

3. Determine the free length along each of the column's faces.

4. Assume a column size with an aspect ratio similar to the ratio of the free lengths.

5. Calculate the slenderness ratios associated with the assumed column size.

6. Assume a species and grade, and calculate its ratio of E/F_c.

7. Project up from its larger slenderness ratio, in Fig. 5.4, to its E/F_c ratio, and across from there to its column stability factor.

8. Divide the column's load by the product of its area and stability

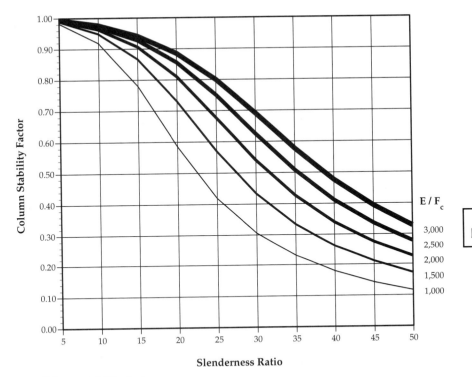

5.4 Column stability factors.

9. Compare the required compressive design value that results, to the compressive design value of the species assumed in step 6.

10. If the latter is greater than the former, the column works. Otherwise, decrease the controlling slenderness ratio or use wood with a higher E/F_c ratio and/or a higher compressive design value (F_c). Then test the column by repeating this procedure, starting with step 1.

Using this procedure, it is possible to determine any number of column properties given the others. Starting with the column length and an assumed column stability factor, you can approximate the slenderness ratio and, by extension, the column thickness. Or, starting with the column load and an assumed member size and properties, you can determine its maximum acceptable column length. Regardless of your specific approach, however, column selection involves enough variables that at least a little trial and error is inevitable. For preliminary design, here is a useful rule of thumb for *square timber columns, one-story high:*

Total tributary area, in square feet = 10 times the nominal column area, in square inches.[7]

Multistory Columns

All of the preceding examples have assumed that the load on the column is constant throughout its length. While this may be true in some simple one-story buildings, the columns in many buildings accumulate loads from intermediate levels. In that case, it is necessary not only to analyze separately the *lengths* of the various column segments, but also their *loads*.

Balloon-framed houses derived much of their strength from the fact that their studs ran unbroken from foundation to roof. Columns that are continuous from foundation to roof contribute more to a building's strength than one-story columns. Even though it would maximize structural efficiency, it is usually impractical to vary the cross-section of a continuous column in proportion to its loads. Assuming a constant area, the most heavily loaded portion of the column will control the design.

We saw earlier how a column's thickness, rather than its width, could control its slenderness ratio. There is an analogous situation when analyzing different column lengths with different loads, and it starts with this now familiar equation:

$$A = \frac{P}{(C_p)(F_c)}$$

If conditions were such that the column area was fully loaded at every story, the following would be true (referring to the stories with the subscripts 1 and 2):

$$\frac{P_1}{(C_{p1})(F_c)} = \frac{P_2}{(C_{p2})(F_c)}$$

If the column were not fully loaded at every story, we would no longer have an equation; one side would be greater than the other and would control the design. But assuming equality and that the species and grade of the column will also be constant throughout its length, we can eliminate F_c from both sides, yielding:

$$\frac{P_1}{C_{p1}} = \frac{P_2}{C_{p2}} \quad \text{or} \quad \frac{P_1}{P_2} = \frac{C_{p1}}{C_{p2}}$$

The column stability factor is a function of the physical properties of the member and its slenderness ratio. We've already determined that the physical properties, thickness, and width of a continuous column will be uniform throughout its length. The only variable left is its length. From Fig. 5.4, it is clear that the stability factor is inversely proportional to column length. For a continuous column, then, we can conclude that *those portions that do not control the design can be longer than those that do.*

You will be surprised how frequently this fact will come in handy. The most common situation involves the sloping roofs common on many timber buildings. To support these, interior columns often need to be taller on the second floor than the first, yet are more lightly loaded.

Notwithstanding the preceding, what often controls the column size are its connections rather than its unbraced lengths. This will be discussed in greater detail in Chapters 6 and 7. Whatever notching or bolting takes places at intermediate floors or framing intersections will reduce the net section of the member and must, therefore, be considered a critical condition.

Modeling a Timber Beam

There is a 15-min experiment that will give you an excellent feel for the behavior of a timber beam under load. All you need is the cardboard from the back of an 8.5 × 11 pad of paper, a paper cutter, some white glue, a black marker, and two stacks of books of equal height.

Cut the cardboard into 11 strips 1 in (2.5 cm) wide. (If you don't have a paper cutter, a scissors will work fine, as long as the width of the strips is consistent.) Making shallow cuts with the matte knife, score the bottom of the first strip about 1 in on center, edge to edge. Place the book stacks about 7.5 in (18.75 cm) apart. Affix each strip of cardboard to the previous one with a single longitudinal bead of glue run end to end. Complete the gluing as quickly as possible, since this experiment works best before the glue dries.

To align the surfaces, tilt the assembled stack on end, and press each strip against the table. (Avoid getting glue on the table by working on scrap paper or newsprint.) Then do the same thing with the stack on edge. Flip the stack over, so the "stacked" edge is now up. Using the marker, draw five parallel "registration lines" from the top of the stack to the bottom: one near each end and one each at the quarter points (see Fig. 5.5a).

Position the stack so that it spans between the book stacks, and the edge with the lines on it is facing you. Splay your fingers out along the length of the stack and apply uniform pressure to it. If the glue has begun to dry, it may be necessary to pick up the stack and bend it more forcefully, working outward from midspan (see Fig. 5.5b).

If you've done your work properly, and have bent the stack far enough (it is okay to bend the stack almost into a quarter circle), several things will begin to happen:

Buckling of the top surface

Widening of the scores in the bottom surface

Sliding of the ends of the strips past one another

Opening up of the end joints

Skewing of the end registration lines

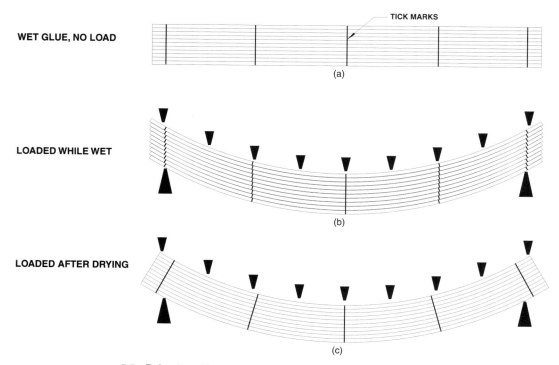

WET GLUE, NO LOAD

TICK MARKS

(a)

LOADED WHILE WET

(b)

LOADED AFTER DRYING

(c)

5.5 Behavior of beam model. (*a*) After lay up. (*b*) Loaded while glue is wet. (*c*) Loaded after glue has dried.

Beam Behavior

Like a piece of wood, your curved beam is composed of fibers (cardboard strips) and a substance holding them together (glue). Like a slender column, the top layer buckled because it was in compression. The scores widened because the bottom layer was in tension. The ends slid past one another because adjacent layers were compressed or stretched to different degrees. Because the shearing action is most pronounced at the ends of a beam, the joints were opened and the registration lines were skewed more there than at midspan.

Flatten the beam, let it dry, and then try bending it again (see Fig. 5.5*c*). Apply some pressure and bend it a little more. Observe what happens when you remove the pressure: It springs back to its dried shape. The dried beam's capacity to spring back to its original shape once you stop bending it is a measure of its *elasticity*. The load at which the beam starts to make funny noises (as the glue starts pulling away from the cardboard) is called the *elastic limit*, and can usually be sensed in the form of decreased resistance to bending. Beyond the elastic limit, the beam will not return completely to its original dried shape; its behavior is then characterized as *plastic*. For stresses below the elastic limit, deflections are proportional to loads.[8]

Stiffness

Although many construction materials exhibit elastic behavior—wood and steel being the most familiar—each has a different degree of elasticity or *stiffness*. Given two timber beams of identical cross-section carrying identical loads, the one that deflects less is considered stiffer. The stiffness of a given material is the ratio of its *stress* to *strain,* and is referred to as its modulus of elasticity or E. Stress is the load per unit area, while strain is the percentage change in the length of a member under a given stress.[9]

From experience, you know that it takes a heavy load to deflect a beam enough to be noticeable. It should not be surprising, then, that the elastic moduli of common structural building materials tend to be quite high: 30,000,000 psi (210,000 MN/m²) for steel, and generally between 1,000,000 psi (7000 MN/m²) and 2,000,000 psi (14,000 MN/m²) for various wood species, parallel to the grain.[10]

The modulus of elasticity of a structural material is important both in terms of its own performance and in terms of the performance of the materials attached to it. When you refer to a floor as "bouncy," you are really saying that it deflects too much as you walk on it. That's equivalent to inadequate stiffness. Such a floor makes you uncomfortable because it "gives" underfoot. If materials weak in tension, such as gypsum board, are attached to the undersides of such floor systems, they will crack. In most circumstances, bounciness and cracking of the finish can be avoided, and its serviceability thereby maintained, if deflections are limited to ⅟₃₆₀ of the span, or 1 in (2.5 cm) for a span of 30 ft (9 m). (To avoid cracking materials that are particularly weak in tension, such as masonry veneers, their stud back-up is often engineered to limit deflections to half that amount.) Where stiffness is less an issue, such as in roof structures without ceilings, the deflection criteria is much less severe: ⅟₂₄₀ of the span or more. The problem with stiffness is that it is a comfort rather than a life-safety issue. "Value engineering typically makes a structure as light as possible to meet strength requirements, and then clients complain."[11]

At first glance, the enormous difference in their respective moduli would seem to favor steel over timber as a structural material. Bear in mind, however, that steel weighs approximately 15 times as much as wood, per cubic foot. To really compare steel and wood systems, we need to look not only at the properties of the materials, but the geometric properties of the members incorporating them.

Moment of Inertia

When a spinning figure skater pulls in her outstretched arms, her spinning accelerates. As the distance from her center of rotation to her extremities diminishes, so does her resistance to spinning. In a similar manner, a beam's resistance to bending is a function of the distribution of its mass with respect to its centroid. This property is called its *moment of inertia,* or I. For a member of rectangular cross-section, as is typical of timber construction, and where b = member width, and d = member depth:

$$I = \frac{bd^3}{12}$$

The mathematical product that takes account of both the composition and shape of a beam is EI (elastic modulus × moment of inertia) and is called its *flexural rigidity.*[12] With this product, we can perform a fair comparison between timber and steel.

Rolled steel sections, such as wide-flange beams, come in a limited number of sizes. In a given situation, an engineer would customarily select the lightest shape having the required moment of inertia. If, for example, an I of 500 in⁴ (21,000 cm⁴) is required, the following sections will suffice: W14 × 53, W16 × 40, and W18 × 35. Weighing only 35 lbs/ft (52 kg/m), the W18 is the most efficient section. The flexural rigidity of this member is given as:

$$EI = (30 \times 10^6)\,(0.51 \times 10^3) = 15 \times 10^9 \text{ lb/in}^2 \qquad (1.0 \times 10^9 \text{ kg/cm}^2)$$

In contrast to rolled steel sections, glulams can be fabricated to just about any size. Southern pine glulams are available with an E of 1.6×10^6 psi (0.1×10^6 kg/cm²). This is just over 5 percent of the E of steel. Therefore, to achieve an equivalent flexural rigidity, we will need an I approximately 20 times that of the W18 × 35. A member 28.5 × 5 in (0.72×0.125 m) will have a cross-sectional area of approximately 1 ft² (0.09 m²), which will weigh about 35 lbs/ft (52 kg/m) for this species, and have an I of 9,645 in⁴ (405,000 cm⁴). For this member, then:

$$EI = (1.6 \times 10^6)\,(9.6 \times 10^3) = 15 \times 10^9 \text{ lb/in}^2 \qquad (1.0 \times 10^9 \text{ kg/cm}^2)$$

In other words, a southern pine glulam 28.5 × 5 in has the same flexural rigidity as a W18 × 35 rolled steel section. Therefore, under an equivalent load, it will deflect the same amount. The steel section is much shallower, of course, because its wide flanges distribute its mass more efficiently, *but it is no lighter.* The glulam weighs the same as the lightest steel section having the required moment of inertia. In general, I have found that glulams are equivalent to steel in this respect only for rolled sections with thin flanges. The structural benefit of thicker flanges cannot be matched by glulam's relatively inefficient rectangular cross-sections. (Now you can put your cardboard beam aside for awhile. Save it, for we will examine it again during our subsequent look at glulams.)

If the bottom layer is in tension and the top is in compression, then somewhere within the beam must be a layer that is in neither tension nor compression. That layer, called the *centroidal* or *neutral axis,* is the middle or sixth layer of your 11-layer sandwich. Just because it is neutral doesn't mean it has no role in the structural behavior of the beam. Without it, the beam would be shallower. From experience, you know that such a beam would bend more easily.

A beam's resistance to bending is a function of its depth. From our experiment, it is clear that the top and bottom layers, the *extreme fibers,* do the bulk of the work. With that it mind, the great strength of I-beams is easy to understand: The bulk of their material is as far as possible from the neutral axis. "If the cross-section is symmetrical with respect to its centroidal axis . . . the extreme fiber stresses in tension and compression are equal."[13]

Shear, Bending Moments, and Deflection

For a given set of spanning and loading conditions, a beam must be able to resist the induced shear and bending stresses and be able to limit deflection to a tolerable amount. Its capacity to do so is a function of its material and its geometric properties. Material properties vary widely among species and cannot be predicted from their visible characteristics. Through a comprehensive testing program, the American Forest and Paper Association has determined appropriate design values for common species, by grade and use. This information has been compiled in its publication *Design Values for Wood Construction.* Geometric properties, on the other hand, are easy to calculate given the width and depth of a member (see Fig. 5.6).

When I am in the midst of preliminary design of a timber building, one of the first things I want to know about the structural system is its depth. With this information I can establish tentative floor-to-floor heights, which in turn determine allowances for stairs. If ductwork is to be run between floor or roof beams, structural depth may also be important to the mechanical engineer. The ratio of depth to width affects a beam's tendency to buckle and, therefore, the spacing of bridging and blocking. The capacity of a beam, like that of a column, can be increased by reducing its slenderness ratio. In a beam or column, this is achieved by bracing the vertical face at intermediate locations between its ends.

Knowing a relevant structural formula can help you understand how a change in a particular variable will affect the result. To understand the case at hand, the formuli for all three conditions—shear, bending, and deflection—must be arranged in a manner that allows them to be compared. Let's start by defining terms:

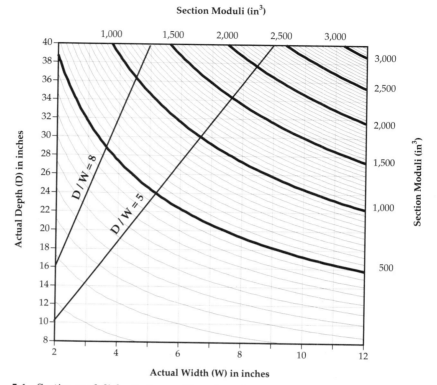

5.6 Section moduli for timbers with rectangular cross sections. To determine the required section modulus for a given bending member, divide the maximum bending moment it will experience by the design value in bending of the intended species and grade, modified by applicable adjustment factors. The greater the slenderness ratio of the member's cross section (D/W), the greater its tendency to buckle. Solid timber blocking is often appropriate when the ratio is between 5 and 8. Higher ratios should be avoided.

l = span

w = load per unit length

b = width of member in inches

h = depth of member in inches

f_v = actual stress in shear

f_b = actual stress in bending

E = modulus of elasticity

d = deflection of member under load

D = deflection as a fraction of span

V = shear stress

M = bending moment

F_b = design value in bending

F_t = design value in tension

For shear and bending in a simply supported uniformly loaded timber beam, it is easy to compare stresses to the capacity of the member to resist them, and then to equate them to one another and solve for member depth:

	Stress	Resistance	Combined form	Solved for depth
Shear:	$V = \dfrac{wl}{2}$	$f_v = \dfrac{3V}{2bh}$	$f_v = \dfrac{0.75wl}{bh}$	$h = \left(\dfrac{0.75wl}{bf_v}\right)^1$
Bending:	$M = \dfrac{wl^2}{8}$	$f_b = \dfrac{6M}{bh^2}$	$f_b = \dfrac{0.75wl^2}{bh^2}$	$h = \left(\dfrac{0.75wl^2}{bf_b}\right)^{1/2}$

Deflection is usually handled a little differently, at least partially because what is "allowable" is a function of the role and location of the member—span/360 for floors versus span/240 for roofs—not its material or geometry:

$$d = \frac{5wl^4}{384EI}$$

where

$$I = \frac{bh^3}{12}$$

but since we know that the tolerable amount of deflection d is a fraction of the span, we can substitute (span × fraction of span), or $(l \times D)$ for d, yielding:

$$ID = \frac{5wl^4}{384E\left(\dfrac{bh^3}{12}\right)}$$

Solving for ED and h (depth) and putting in the same form as previously used for shear and bending yields deflection:

Combined form $\quad ED = \dfrac{0.16wl^3}{bh^3}$

Solved for depth $\quad h = \left(\dfrac{0.16wl^3}{bED}\right)^{1/3}$

I find the product ED—elastic modulus × allowable deflection as fraction of span—to be of particular interest. Take, for example, a floor beam where $E = 1.8 \times 10^6$ and $D = \frac{1}{360}$. (The units are not important.) In this instance,

$$ED = \frac{(1.8)(10^6)}{(0.36)(10^3)} = 5000$$

Now, take a roof rafter directly overhead, with the same width, span, and load, and where $D = \frac{1}{240}$. If you wanted to use rafters of the same depth (h) as the floor beams below them, you could meet your deflection criteria using a material with a proportionately lower E:

$$ED = \frac{(1.2)(10^6)}{(0.24)(10^3)} = 5000$$

In other words, just as f_v relates shear to shear resistance and f_b relates bending to bending resistance, ED relates deflection to deflection resistance. For that reason, I refer to it as the *deflection resistance factor*, and label it f_d. With that, let us summarize the above by listing all three formuli, solved for member depth:

Shear $\qquad h = \left(\dfrac{0.75wl}{bf_v}\right)^{1/1}$

Bending $\qquad h = \left(\dfrac{0.75wl^2}{bf_b}\right)^{1/2}$

Deflection $\qquad h = \left(\dfrac{0.16wl^3}{bf_d}\right)^{1/3}$

Let us now take the indicated root of the span, so we can extract it from the parenthetical portion of each equation. Let us also rearrange what is left inside, to simplify comparisons between the equations:

Shear $\qquad h = 0.75l\left[\left(\dfrac{w}{b}\right)\left(\dfrac{l}{f_v}\right)\right]^{1/1}$

Bending $\qquad h = 0.87l\left[\left(\dfrac{w}{b}\right)\left(\dfrac{1}{f_b}\right)\right]^{1/2}$

Deflection $\qquad h = 0.54l\left[\left(\dfrac{w}{b}\right)\left(\dfrac{1}{f_d}\right)\right]^{1/3}$

While there are many similarities between these formuli, there are also important differences. The similarities include the ratio of load per unit length (w) to member width (b) in each equation. The load per unit length (w) is the product of the load per unit area (L) and the spacing or *loaded width* of the framing members (S):

$$w = LS$$

and therefore:

$$\frac{w}{b} = L\left(\frac{S}{b}\right)$$

Earlier in this chapter, we saw that the bending resistance of a beam varies as the first power of its width, but as the second power of its depth. From the above equation, we see that the load per unit length (w) varies as the first power of the member spacing (S). *Therefore, a simple way to match the bending capacity of a timber beam to its load is to make its width proportional to its spacing.* For the purposes of this discussion, assume that that strategy is employed, and that S/b—the ratio of member spacing (in feet or meters) to member width (in inches or tenths of a meter)—equals 1. Notice how simple the equations become:

$$\text{Shear} \qquad h \propto 0.75l\left(\frac{L}{f_v}\right)^{1/1}$$

$$\text{Bending} \qquad h \propto 0.87l\left(\frac{L}{f_b}\right)^{1/2}$$

$$\text{Deflection} \qquad h \propto 0.54l\left(\frac{L}{f_d}\right)^{1/3}$$

In all three equations, the required height of the member is a function of its span (l) and of the ratio of its load per unit area to its resistance (L/f_x). The selected member must satisfy all three of the equations. In simple terms, the shallowest acceptable member will be that with the greatest height given by these equations. Although the required depth for shear is directly proportional to the span, the required depths for bending and deflection are not. As shown in Fig. 5.7a and b the factor that controls under lighter loads may be different from the factor that controls under heavier loads.

As discussed in Chapter 2, structural properties vary by grade. Furthermore, the properties vary independently of one another. Design values in shear, for example, vary much less than those for bending. Therefore, the controlling factors may be different for one grade than for another. Compare Fig. 5.7a and b for two different grades of Douglas fir–larch beams.

An examination of the design values assigned to each species and grade reveals no correlation whatsoever between those for shear, bending, and modulus or between the various grades within a given species. Consequently, it is impossible to predict which will control. The only universal rule is that the selected member depth must satisfy these three criteria: shear, bending, and deflection (plus a fourth, bearing, which will be discussed in Chapter 6). Which formula will control in a specific case depends on the *combination* of load, member properties, allowable deflection, and—although it was ignored for the purposes of this discussion—member width. Nevertheless, the graphs do suggest that shear is the controlling factor under heavy loads.

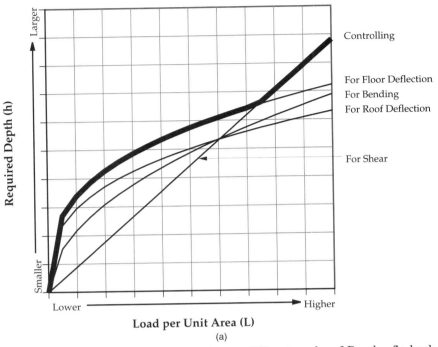

5.7 Timber beam depth determinants for two different grades of Douglas fir–larch. For any load, the greatest depth controls the design, as indicated by the dark line. (*a*) Select structural. (Design values from *Design Values for Wood Construction.*)

Loaded Width

In sizing a beam, what matters is its span and the load per unit length, not the *loaded width,* the width over which the load is distributed. If the spacing of the beams is doubled and the load per unit area is halved, *w*—the load per unit length—will stay the same. The beam doesn't know or care about the width over which loads on it were accumulated. The deck cares, however, since the loaded width of the beam is equivalent to the span of the decking.

There are actually several types of load to be considered: dead, live, snow, wind, etc. It is appropriate here to consider the relationship between dead load—the weight of the building including its structure—and the span.

The linear relationship between span and member depth means that doubling the span will double the required depth. All else being equal, this means that the weight of the beams per unit area will vary in proportion to their span. For short spans, the beams are shallow, so their weight constitutes a small percentage of the load and has little impact on beam design. For longer spans, the self-weight of the beams is a much more significant factor. Take for example a floor with a live load of 30 psf (155 kg/m²), a loaded width of 40 in (1 m), and a span of 10 ft (3 m). Assume a beam of 36 pcf (600 kg/m³) of size 4 × 12 in (0.1 × 0.3

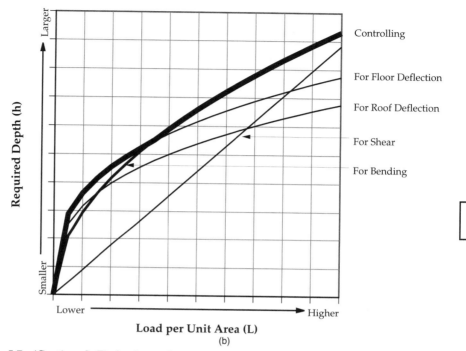

5.7 (*Continued*) Timber beam depth determinants for two different grades of Douglas fir–larch. For any load, the greatest depth controls the design, as indicated by the dark line. (*b*) No. 2. (Design values from *Design Values for Wood Construction.*)

m) with a cross-sectional area of 0.33 ft² (0.03 m²). The live load per unit length will be 100 plf (150 kg/m), and the weight of the beam per unit length will be 4 plf (6 kg/m), or 4 percent of the live load. If the span is doubled and the loaded width halved, the weight of this beam (which may now be undersized) will equal 8 percent of the unchanged live load. If, on the other hand, the width of the beam is increased in proportion to increases in beam spacing, the weight of the *beam* per unit area will remain the same, but the depth and therefore the weight per unit area of the *decking* will need to increase. For preliminary design, the rule of thumb for deck sizing is:

Decking span, ft = 3 × nominal decking thickness, in[14]

To summarize:

- Increases in beam span usually result in increases both in beam depth and in the dead load of the floor beams as a percentage of the floor's live load.

- Increases in beam spacing usually result in increases both in the depth of decking and in the dead load of the floor decking as a percentage of the floor's live load.

- Increases in beam width in proportion to increases in live load do not usually affect the ratio of the dead load of the floor structure to its live load.

Since the depth of the beam affects the total load it can carry, the problem of setting beam depth is obviously indeterminate. For that reason, a structural weight of between 5 psf (25 kg/m^2) and 10 psf (50 kg/m^2) is usually assumed during preliminary design, regardless of span. The problem with this approach is that it misrepresents the relationship between span and structural weight, making it difficult to compare the relative efficiencies of various framing concepts. Nevertheless, for preliminary design, the following rules of thumb may be useful:

Span of sawn beam, ft = 1.0 × nominal depth of sawn beam, in[15]

Span of glulam beam, ft = 1.5 × actual depth of glulam beam, in[16]

Equivalent Thickness

In timber structures, one of the measures of structural efficiency is *equivalent thickness*. It is the average thickness of the structure or a designated portion thereof. To visualize it, imagine that you could melt a timber structure, and that there was a dam around the perimeter of the area being studied to confine the melted structure; the equivalent thickness would be its depth. To calculate it, take the total volume of wood in the assembly of interest, and divide it by the area covered. For example, in a floor platform, the equivalent thickness would be determined by dividing the sum of the volumes of each piece of framing and decking by the floor area. The volume of a particular framing member is the product of its actual cross-sectional area and its length. The same approach can be used with decking, but since it is typically of uniform thickness over the entire area, its *equivalent* thickness is equal to its *actual* thickness.

If the price per unit volume of all timber products were the same, and if erection costs were a function solely of timber volume, then the cost of a timber structure would be proportional to its equivalent thickness. Unfortunately, things are not that simple. The price of timber is a function not only of its volume, but also of its grade, length, and cross-sectional area; in general, the better, the longer, or the larger the material, the higher its material cost per unit volume. The cost of erection is a function not only of timber volume, but also of the types of connections, the piece count, and the average piece weight; in general, the greater the complexity of the connections, the number of pieces, or the weight of the pieces, the higher the erection cost per unit volume. For the foregoing reasons, equivalent thickness is a better measure of relative cost when comparing alternatives that are moderately different from one another than ones that are radically different.

For an equivalent thickness comparison to be useful, it must be fair. Assume, for example, that the bay size in each system is slightly different, and that one system fills the entire floor area with a whole number of bays, while the other requires special bays along the perimeter to fill out the area. In this example, the equivalent thickness of a typical bay would not necessarily be representative of the equivalent thickness of the entire floor. The greater the number of typical bays, the lower the impact of such exceptions. Another prerequisite of a fair comparison is that the assemblies studied include all affected components. Analyzed solely in terms of their horizontal elements, small bays appear far more efficient than larger ones. Differences are reduced dramatically when columns are included in the analysis. And finally, make sure that the members in competing systems are stressed to approximately the same degree. Members that are understressed may be oversized. This can skew the results of a comparison.

If timbers were available in an infinite range of sizes, equivalent thickness would vary continuously with changes in span and load. In the real world, however, timbers are available only in discrete sizes, so equivalent thickness varies in steps. Complicating matters further is the fact that fire resistance, not structural resistance, sets the minimum sizes of heavy timbers. Short or lightly loaded members may have to be larger than structurally necessary to retain their fire resistance. Thus, the validity of equivalent thickness comparisons may be questionable for smaller member sizes.

From the discussion earlier in this chapter, it is clear that in comparing two bending members of identical span and cross-sectional area, the deeper one will have a much greater load capacity. Similarly, in comparing two bending members of identical span and load capacity, the deeper one will have a smaller cross-sectional area. This reinforces the importance of aspect ratio in equivalent thickness comparisons. Therefore, for a fair comparison, make sure the ratio of member depth to width is approximately the same in each of the systems being considered. In any comparison, it is helpful to draw a diagram of the various framing conditions (see Fig. 5.8) and to graph the results (see Figs. 5.9a, b, and c).

The presence of concentrated loads or *combined members*—those subject to axial and bending loads simultaneously—can affect equivalent thickness calculations. Girders supporting beams at their third points, for example, will have lower maximum moments that those loaded uniformly for their entire spans. Similarly, a roof truss supporting purlins only *at* its panel points can have a much lighter top chord than one where the purlins land *between* them. Finally, a floor deck subject solely to gravity loads can often be thinner than, and avoid the blocking required for, one also intended as a diaphragm against lateral loads. Because of their potential affect on the outcome of an equivalent thickness comparison, it is a good idea to include in the comparison a list of the structural assumptions associated with each system.

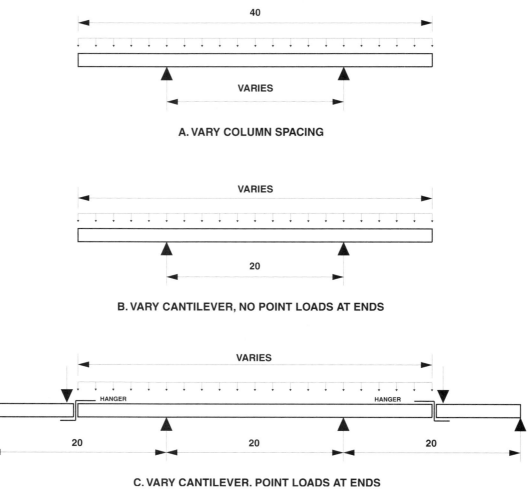

5.8   Framing and loading for three different cantilevering strategies.

Since the goal is usually to maximize economy of the entire project, not just the timber construction, comparisons between different framing systems need not be limited to the arrangement and volume of timber required. Other aspects worth considering include implications at the foundation and at the roof, and the potential benefit, if any, of *composite action*. Although timbers typically have rectangular cross-sections, wide flanges can be approximated in the field by gluing plywood decking to the tops of beams and joists.[17] The stiffness of this assembly is greater than that of the framing members acting alone. Furthermore, because the glue eliminates contact between the underside of the decking and the top of the framing, squeaking is reduced. As with wood I-beams, glued floor systems increase structural efficiency.

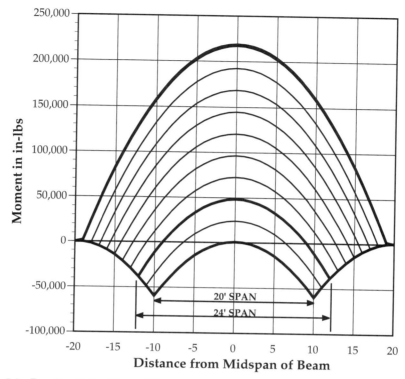

5.9 Results of the three different cantilevering strategies. (*a*) Vary column spacing. The illustrated trends apply regardless of specific spans and loads.

The percentage of the total load on a beam represented by its dead load increases with increasing span (see Fig. 5.10). This is due to the fact that if beam B is twice the span of beam A, it not only is twice the length, but also requires a somewhat greater depth. Assuming the beam spacing and load per unit area is the same in both cases, the live load will increase at the same rate as the span, while the dead load will increase at a higher rate than the span. At some span, the dead load represents such a large portion of the total load that the use of beams becomes uneconomical. Chapter 8 discusses trusses, an economical alternative. Chapter 14 includes an example of an equivalent thickness comparison from an actual project.

Framing Plans

For economy, the framing plans should be as simple as possible, consistent with the design concept (see Fig. 5.11). Don't change framing direction, framing depth, bay size, member size, column orientation, species, grade, timber type (sawn, glulam, etc.), bearing details, decking, or any other aspect unless you have a very good reason. Recognize

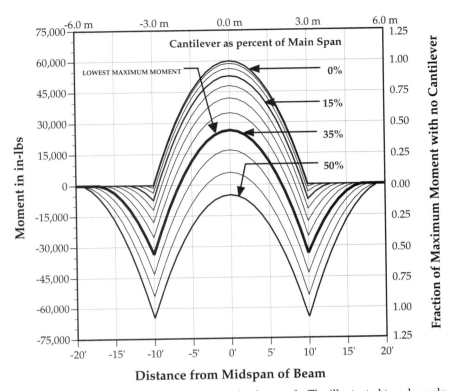

5.9 (*Continued*) (*b*) Vary cantilever, no point loads at ends. The illustrated trends apply regardless of specific spans and loads.

that adherence to the foregoing guidelines will expedite and improve the quality of construction even if it results in some minor structural inefficiencies. It will also reduce the effort required to coordinate your drawings and details and the likelihood of errors and inconsistencies. There is a learning curve in any fabrication process. The simpler and more repetitive the fabrication and erection of your timber frame, the greater the contractor's opportunities to economize.

Understand the industry's conventions and use them to advantage in the context of your design. Think of the big picture: If the framing in all but one bay must be of the highest grade, the savings in material from using a lower grade in that bay are likely to be more than offset by the extra effort required of the fabricator and erector in keeping track of the exceptions and installing them in the appropriate places. On the other hand, if differentiating between pieces will eliminate unnecessary fabrication steps, by all means do it, particularly if the differences between the resulting members will be obvious to the erector. An example of this would be the decision to factory stain only those pieces that will be exposed to view in the completed building. In gen-

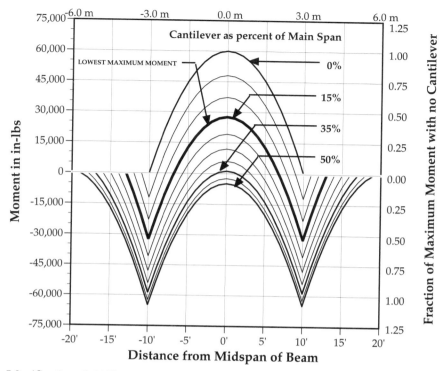

5.9 (*Continued*) (*c*) Vary cantilever, point loads at ends. The illustrated trends apply regardless of specific spans and loads.

eral, *distinctions are worth making when they are evident to the contractor and when they help organize the work.*

Understand the relationship between span and member size: Bending moments and bending resistance vary as the square of the span and member depth, respectively. Bending resistance varies linearly with member width. Thus, increasing member width is an inefficient way to accommodate a longer span, but a reasonably efficient way to accommodate a larger running load. Although pound-for-pound timber has the same flexural rigidity as steel over a whole range of spans, timber's rectangular cross-section is inherently inefficient; for longer spans, therefore, consider using timber trusses.

Combined Members

Members subject to combinations of bending and axial forces are referred to as *combined members*. The combinations may involve bending and axial compression or bending and axial tension. Although these conditions might sound unusual, they are much more common than you think:

Span (m)

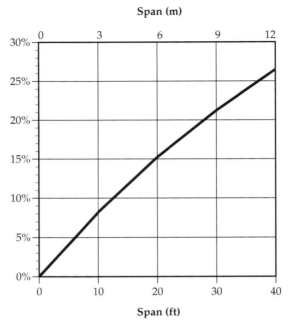

Span (ft)

5.10   Dead load of framing as a percent of total load, for different spans. Bending is assumed to control. Although shown for a *specific* species, grade, and load, this *general* trend applies over a wide range of spans and loads.

- The collar ties in an attic used for storage
- The top chord of a timber truss to which roof decking has been directly applied
- The bottom chord of a timber truss sagging under its own weight
- The perimeter columns of a building during high winds

We have already seen that the most efficient prismatic shape for a column uniformly braced in all directions is a square, but that the most efficient shape for a beam is a rectangle of greater depth than width. How do we reconcile these conflicting geometric requirements, when a member is *both* a beam and a column or hanger?

While the precise mathematics are beyond the scope of this book, the following simplification is sufficient for most preliminary design purposes. When designing for bending or tension alone, the actual stress in the member may not be more than its design value. In other words:

$$\frac{f_t}{F_t} \leq 1.0 \qquad \text{and} \qquad \frac{f_b}{F_b} \leq 1.0$$

5.11   Roof framing, Wolf Trap Farm, near Washington, D.C.

Under combined tension and bending, whatever capacity is not used to resist bending is assumed available to resist tension, and vice versa. As an equation, this can be written:

$$\frac{f_t}{F_t} + \frac{f_b}{F_b} \le 1.0 \quad [18]$$

Adjustment Factors

Timber's properties are dependent on its use conditions, both internal and external. Internal conditions are those inherent to the member, and include its moisture content and geometry; external conditions are those outside the member, and include loads and their durations.

The structural performance we expect from a timber is fundamentally different from that which we expect from a tree. The ability of a tree to bend in the wind is one of its most remarkable attributes. Absorbing that energy in its trunk minimizes the loads acting at its base. If trees were unyielding, they could be uprooted by gentle breezes. In a building, by contrast, stiffness is desirable. Excessive flexibility in the structure can lead to distress in the finish materials attached to it.

Up to this point, this chapter has assumed, among other things, that

- Live loads are of reasonably long (10-year) duration

- Members are reasonably dry

- Members will be used under ordinary temperature ranges and not heated above 150°F (65°C)[19]

- Bending members are oriented so that loads are applied to their narrow faces

- Bending members are braced sufficiently to keep them from rotating or being displaced[20]

In general, the tabulated design values in *Design Values for Wood Construction* must be multiplied by appropriate adjustment factors, to account for deviations from the above conditions. The adjusted numbers are known as *allowable design values.* "Some of the adjustment factors will cause the tabulated stress to decrease, and others will cause the stress to increase. When factors that reduce strength are considered, a larger member size will be required to support a given load."[21] Underdesign is never an option. For that reason, consideration is required for factors that increase member size, but is not required for factors having the opposite effect.[22] While each factor should be considered separately, their combined result is what matters. In the following paragraphs, we will briefly examine the adjustment factors most relevant to preliminary design.

The *load duration factor* takes account of the fact that wood's capacity to accommodate stress increases in inverse proportion to the duration of load. For that reason, this adjustment factor is greater than 1 for load durations of less than 10 years. The durations most commonly encountered include snow loads with a factor of 1.15; wind and seismic loads with a factor of 1.6; and impact loads, with a factor of 2.0. Presumed to be far longer than 10 years, permanent loads such as dead loads have a factor of 0.9.[23] The load duration factor is the only factor that can be more or less than 1, depending on specific circumstances.

The conditions under which the other adjustment factors apply vary. Those relevant to timber construction include:

- The *wet service factor,* which reflects the reduction in timber properties resulting from moisture content in excess of 19 percent

- The *temperature factor,* which reflects the loss of timber strength associated with prolonged exposure to temperatures in excess of 150°F (65°C)

- The *beam stability factor,* which addresses the problem of buckling in deep unbraced beams

- The *size factor,* which reduces bending design values in members with depths greater than 12 in (0.3 m)

- The *treatment factors,* which require that the impact of pressure preservatives and fire retardants on timber properties be considered

- The *form factor,* which equilibrates the moment capacity of beams with round or diamond-shaped cross-sections to that of square beams of equivalent area[24]

Structural Glued-Laminated Timbers (Glulams)

At the time when many of New England's timber bridges and barns were built, the timbers needed for their construction were readily available nearby. The use of timber in these utilitarian structures was a natural consequence of its local availability.

Explosive growth in the use of wood products has led to a dramatic decline in both the quality and quantity of the remaining timber resources. The shortage of high-quality timber has led to large price increases, creating opportunities for the manufacturers of alternative structural products to expand their shares of the market. At the same time, changes in the relative costs of labor, materials, energy, and capital have forced designers and builders to think more broadly about buildings and the materials and systems of which they are composed.

More than 100 years ago, piano makers realized that they could enhance the strength of pin blocks by cross-laminating multiple wood layers. The advent of stronger adhesives eventually led to the development of structural glued-laminated timber—glulam—revolutionizing the field of timber construction. No longer were the properties of a member predetermined; the manufacturer now had the capacity to tailor those properties to the demands of each job. (See Fig. 5.12.)

Glulams consist of thin layers of dry lumber—laminations—glued together in a controlled environment[25] to produce a member with overall properties superior to those of its laminations. The composite performance of a glulam, the fact that the whole is greater than the sum of its parts, is primarily due to the distribution of lumber defects among the layers. In general, the larger the member cut from a tree, the greater the likely quantity and size of knots and other defects within it. Whereas a two-by-eight has relatively little material hidden from view behind its wide faces, a six-by-eight has quite a bit. For these two reasons, chances are that a sawn timber will be weaker than and will be graded lower (i.e., with less confidence) than a group of pieces of dimension lumber of equivalent overall cross-section.

As illustrated during fabrication and testing of our cardboard beam, the radius to which laminations can be comfortably curved is primarily a function of their thickness. From personal experience gained during the design and construction of a small-diameter wood-framed cylinder sheathed with three layers of ¼-in plywood[26], I can say with confidence that the thinner the laminations, the tighter the possible radius. While ¾-in (18-mm) laminations are typical for tightly curved glulams, "these thinner laminations are not used for straight or slightly curved glulams because cost is heavily influenced by the number of glues lines in a member."[27] For economy, 1.5 in (36-mm) laminations are the norm.

The depth, as we have seen, has more influence than the width in the bending resistance of a rectangular beam. The most important factor,

5.12   Straight and curved glulams in New Public Library, Maumelle, AR. Fennell & Purifoy, architects. (Fennell & Purifoy.)

in other words, is the *distribution* of mass with respect to the neutral axis. Wide-flange steel beams derive their efficiency from their unique shape. Indeed, a W12 × 40 has two thirds of its mass within ½ in (12 mm) of its top and bottom surfaces, yielding a moment of inertia of over 300 in⁴ (12,600 cm⁴) in a cross-sectional area of only 12 in² (77 cm²). Contrast this with a glulam of equivalent depth and moment of inertia. In a section roughly 2 in (5 cm) wide, this member has twice the area of the steel beam, but is only one fourth the width. Its mass is distributed uniformly over its depth, so that it acts the same as if it were concentrated halfway between its centroid and its top and bottom surfaces.

The process by which glulams are manufactured or laid up helps to utilize lumber more efficiently. Recall that axial stresses range from zero at the neutral axis to a maximum at the top and bottom edges of a beam. In the case of a steel beam, the distribution of mass roughly corresponds to the distribution of stress—more at the extremes, less near the neutral axis—but the physical properties of the material are uniform throughout. In a sawn timber, like a steel beam, the physical properties of the material are still relatively uniform throughout, but the rectangular cross-section prevents any correspondence between the distribution of mass and stress. A glulam, like a sawn timber, has a rectangular cross-section, but can be fabricated of laminations whose properties correspond to the stresses associated with their locations within the beam. Those at the extremes need be much stronger than those near the core. In the case of a wide flange, where a beam's physical properties are constant throughout its depth, efficiency derives from optimizing its shape. Where the properties can be customized, as in the case of a glulam, efficiency derives from *maximizing the correspondence between actual and allowable stresses*. It should not be surprising, therefore, that it is only a slight exaggeration to say that the lamination at the centroid need only be slightly stronger than sawdust.

If wood's strength in tension and compression were equal, the top and bottom laminations of a glulam beam could be the same. However, a quick glance at a table of timber design values will reveal for each species and size classification that those for tension are typically far lower than those for compression.[28] Therefore, if the extreme laminations were identical, the one in tension would be fully stressed at a lighter load than the one in compression, and would therefore control the design. To achieve *balanced construction,* glulam fabricators save their best lamstock for outer *tension* laminations; to make sure each of these ends up in the proper orientation, they clearly mark the top or bottom of each beam. To accommodate cantilevers or load reversals, however, both outer laminations may have to be of the same high quality.

There are other important differences between sawn timbers and glulams. A timber's length and depth are limited by the dimensions of the tree from which it is cut; glulams can be fabricated to virtually any length by finger jointing the ends, and to virtually any depth by increasing the number of laminations. Timber sizes are given in nominal dimensions, which are essentially those the member had before it was seasoned and surfaced; glulam dimensions are those of the finished member. Because the moisture content differential between the core of a large timber and its faces is typically much greater than that between the dry laminations for a glulam, checking tends to be a much greater problem with timbers than with glulams.[29]

Reinforced Glulams

The strength of a glulam derives entirely from the properties of its lumber laminations. The properties of these are a function of their species and grades. As far as I know, no one has yet figured out how to grow trees consistently stronger than those we already have. Therefore, it would be fair to say that, up to now, the strength of a glulam has also been *limited by* the properties of its lumber laminations. In other words, the only way to increase the strength of a glulam beam of the highest structural grade is to increase its size.

In reinforced glulams, plastic fibers of extremely high strength are located strategically within the cross-section of the member, in areas of highest stress (see Fig. 5.13). "The composite reinforcement is so strong that a 1.8 mm thick layer of it as strong as six 38 mm high grade Douglas fir laminations. Thus, only small amounts of reinforcement are required."[30] In fact, the reinforcement quantities are so small that the member's strength can be increased substantially without noticeably increasing member size.

As noted earlier, wood's tensile strength tends to be much lower than its compressive strength. For that reason, the design of conventional glulams is controlled by tension. Unlike conventional glulams, but like reinforced concrete, the enormous tensile strength of its reinforcement enables the reinforced glulams to be controlled by compression. Other comparisons with reinforced concrete are also instructive, as shown in Table 5.1.

Whereas the modulus of elasticity of glulam laminations gets no higher than about 2,500,000 psi (17,500 MN/m²), that of the reinforcement panels used in reinforced glulams is between 3 and 6 times higher.[31]

5.13 Cross section of FiRP® glulam beam. Compression and tension reinforcements are visible near the top and bottom of the beam, respectively. Each area of reinforcement is protected by a single wood lamination, referred to as a *bumper.* (Wood Science & Technology Institute [N.S.], Ltd., Corvallis, OR.)

TABLE 5.1 A Comparison of Reinforced Glulam and Reinforced Concrete

Aspect	Reinforced glulam*	Reinforced concrete
Reinforcement material	High-strength plastic fiber plates	Deformed steel rods
Reinforcement units	0.09 in (2.3 mm) × width of member	⅜ in (9 mm) to 1⅜ in (54 mm) OD
Reinforcement yield strength	143,000 psi (aramid)	40,000 psi (grade 40 steel)
Reinforcement method	Adhered to lumber laminations	Embedded in concrete
Reinforcement installation	Laid up with other laminations	Wired in place prior to pour
Specific gravity of member	0.50 (Douglas fir)	2.35 (stone concrete)
Reinforcement quantity	A function of its thickness	A function of its size and spacing
Reinforcement protection	Lumber bumpers 1.50 in (36 mm) thick	Concrete cover 1.50 in (36 mm) thick[†]
Length of reinforcement	Partial or full length of member	Partial or full length of member
Ratio of elastic moduli	$E_{aramid}/E_{Douglas\,fir} = 10$	$E_{steel}/E_{concrete} = 15$[‡]
Member depth increments	Lamination thickness	Depths may vary continuously
Member width increments	A function of available lam stock	Widths may vary continuously

*Derived from information furnished by the Wood Science and Technology Institute, Corvallis, OR.
[†]Timoshenko, p. 151.
[‡]Timoshenko, p. 150.

Reinforced glulams use high-strength fiber-reinforced plastics, rather than costly high-grade lamstock, to maximize structural performance. As a result, they have the potential to turn the economics of framing selection on its head. Earlier in this chapter, I showed that a conventional glulam weighs roughly the same as a steel section of equivalent flexural rigidity. Reinforced glulams, with their strength concentrated near their top and bottom surfaces, are more efficient than conventional glulams and should, therefore, weigh less than a steel section of equivalent flexural rigidity.

Timber Framing as an Architectural Concept

Although it must ultimately satisfy a number of structural criteria, the design agenda for an exposed timber frame, I think, should be set initially by the architect. Even if it is not possible to generate it all at once, that agenda should eventually include an overall framing concept, as well as strategies for dealing with details *among* timber components and *between* timber components and those of other systems. From my experience, the most attractive timber frames are those that are most completely integrated into the fabric of their buildings.

Integration begins with the identification of the timber frame's boundaries and geometry. Ask yourself questions like these:

- What will fill in the vertical openings between widely spaced columns?
- Is lateral bracing part of the frame or part of another system?
- If the structure will have to be sprinklered, how will the pipes be arranged?

- Where will ducts and conduits be run?
- Will there be any secondary structures such as dormers or bays?
- Will secondary framing rest on top of or nest into the primary framing?
- Will the locations of partitions coincide with those of overhead structural members?
- Will the rhythm of the framing be interrupted to accommodate openings and shafts?
- Will the framing be exposed inside or outside the building, or both?
- Will the framing pass through the weatherskin?
- How and where will other materials be fastened to the frame?
- Will adjacent materials be able to accommodate in-service deflection of the frame?
- Are the proposed details constructable within a conventional construction sequence?

5.14 Long-span glulam arches, Back Bay Railroad and Rapid Transit Station, Boston, MA. Kallmann McKinnell & Wood, architects. (Roger N. Goldstein.)

Answers to these sorts of questions will narrow considerably the range of possible design solutions. To go further, put yourself inside the building using a real or virtual architectural model. Study the visual implications of various combinations of member size and spacing. For longer spans, compare the spatial consequences of trusses to those of deep beams and arches (see Fig. 5.14). Determine whether the layout of the nonstructural elements—especially interior partitions—enhances or detracts from the legibility of the timber frame.

The space inside a timber-framed barn is clearly very different from that within a covered bridge. I believe that much of the character of a timber structure derives from the interplay between its rhythms and its hierarchies (see Figures 5.15 (a) and (b). Visualize what happens as the spacing of timber joists in a floor changes. When they are only a few feet apart, thin plywood decking will do. As the gap between them increases, the depth of the joists and the thickness of the decking must also increase. At some point, deck panels may have to be replaced with deckboards. At large enough spacings, even they get so heavy that imposition of a secondary framing system is

(a)   (b)

5.15 Two examples of the interplay between rhythm and hierarchy: (a) Delaware Aqueduct, Lackawaxen, PA, John A. Roebling, engineer. (b) Lath House, Phoenix, AZ.

justified. What were joists are now beams, and the issue of spacing starts over with the new joists.

Like it or not, your frame and each of its components will fall somewhere along this structural continuum. The key is deciding where. The limits of the continuum may be set by other concerns: The floor-to-floor height may be so tight that deep beams would be unacceptable, or the clear space between the joists may have to accommodate recessed ductwork or lighting fixtures. Codes may also limit your options: If sprinkler heads are required in every joist space, but cannot be more than 12 ft (3.6 m) on center, any spacing other than that would necessitate more than the minimum number of heads. The goal should not be to optimize every system—it is nearly impossible to do so in any building—but to be aware of the constraints and their relationships to one another. In other words, understand how a change in one building system affects the others.

The vocabulary of timber structures is inherently different from that of cast-in-place concrete and steel. Timber and steel framing members are linear and can therefore span in only a single direction (see Fig. 5.16). As evidenced by waffle slabs, a concrete floor can span in both directions at once. Regardless of the number of conceptual levels in the framing hierarchy (joist, beam, girder, etc.), the tops of all of the members in a steel-framed platform, including cantilevers, lie essentially in one plane. The same can be said for cast-in-place concrete. Of all the conventional framing systems, only precast concrete

5.16 Exposed timber framing in gallery of Denmark's Louisiana Museum.

shares with heavy timber the option of framing *into* or *over* other members. Despite this similarity, the differences between the sizes, weights, and spanning capabilities of precast concrete and timber elements are profound.

As previously discussed, the range of framing options is essentially continuous in the horizontal plane. In the vertical plane, by contrast, the choices are limited to two: flush versus layered. Cantilevering can be achieved by layering the framing throughout the floor or roof or simply by lowering selected primary members. Other advantages of layering are that it allows multiple-span secondary members (resulting in greater stiffness) and that it enables the creation of continuous channels within the structure (for running mechanical and electrical systems). The major disadvantages of layering are that the capacity of the primary members may have to be reduced to reflect the loss of bracing by the secondary framing and the deck, and that the basis of cross-grain shrinkage will now be the sum of the depths of all layers, rather than the depths of the deepest members in the only layer.

The fact that timber is exclusively a one-way framing system has not prevented some designers from trying to make it look otherwise. When blocking is installed at the same spacing as the members being blocked, the overall appearance is that of a nonhierarchical grid, but a closer examination of the intersections reveals the deception. A more straightforward way to flatten the structural hierarchy or expand beyond the linear limitations of individual timber elements is to use them to create structures, like domes, which are *form resistant* (see Figures 5.17 (*a*) and (*b*)—whose stability derives from their curved, continuous shape[32]—or trusses. Here, however, the issue is no longer to define the point where the typical bay falls on the span continuum; instead, the spans within each bay vary widely. To allow for this variation, the decking throughout could be sized for the longest span, or its thickness could increase with the span. If the tops of the radial timbers were stepped, the changes in deck thickness would be visible inside the dome. If not, the changes would be visible on the outside. Another option would be to insert a secondary framing system only where it was justified by the span.

Summary

Framing members are sized to accommodate a variety of factors, some of which are under the complete control of the designer and some of which are not. Those in the first category include their structural properties, spacing, spans, and cross-sectional proportions. Those in the second include the various horizontal and vertical loads acting on the structure and their corresponding durations. These aspects are mandated by building codes, but are a direct consequence of the overall form of the building and of the distribution of its various uses.

(a)

(b)

5.17   Two different form-resistant timber structures in one building: (*a*) Glulam half-arches, acoustic metal deck, and steel compression ring comprise faceted dome. (*b*) Glulam half-arches, timber decking, and custom steel connector comprise faceted half-dome. Police & Justice Building, West Orange, NJ. The Goldstein Partnership, architects.

Several design considerations apply to every member type: its structural properties, its degree of continuity, and whether it is loaded solely in tension, compression, or bending, or in a combination of modes. For each specific type—column, hanger, diagonal brace, girder, beam, joist, and rafter—there are also specific design considerations: length, cross-sectional proportions, and support conditions for compressive members; cross-sectional area for tensile members; span and depth for bending members.

The bending moment in a beam increases as the square of its span; its bending resistance increases as the square of its depth. Various preliminary framing schemes can therefore be compared by adjusting the depths of bending members in proportion to their spans. To identify the smallest acceptable bending member for a given condition in later stages of design, determine the minimum sizes needed for shear, bending strength, and stiffness, and select the largest of the three.

The design values assigned to different timber-framing materials relate to their quality and consistency. Sawn timbers are graded on the basis of their visible characteristics; little confidence is placed in the bulk of each member hidden from view, so performance expectations are relatively low. Glulams are composed of a multitude of small members, each of which can be individually tested; the lay-up process controls the quantity and distribution of strength-reducing characteristics, enabling the material quality to be matched to stresses, and resulting in a higher level of performance. Fiber-reinforced glulams, like reinforced concrete, derive their structural properties primarily from the strength and reliability of their reinforcement; because the wood within their cross-sections is used primarily as a spacer, the influence of its defects on the structural performance of the member is reduced, resulting in even higher performance.

Member sizes are limited by their method of manufacture. Sawn timbers can only be as large as the trees from which they are milled. The widths of glulams—whether or not reinforced—are limited by the available lamstock, but their depths are limited only by the imaginations of their designers, the capabilities of their fabricators and erectors, and the practical difficulties associated with shipping and handling. Although timbers of all types have relatively inefficient cross-sections compared to those of structural steel, the structural capacities of timber, pound for pound, are roughly equivalent to those of steel, over similar spans.

During preliminary design, a timber frame can be thought of as a hierarchical assemblage in which the depth of the hierarchy and the geometric relationships among its members are still in flux. Varying the spans or spacings of the members at any level of the hierarchy may affect the sizes and proportions of the members at all levels. Relative costs among different timber materials and among the respective sizes under consideration are important determinants in finalizing the design.

Do not consider the timber frame only in isolation; instead, think of it also in terms of its role as an armature supporting the rest of the building. Make each decision about the frame in anticipation of the work of the following trades. Whatever care is taken to design a beautiful frame will be compromised unless appropriate provisions are made for whatever mechanical and electrical systems will be required. Recognize that sprinkler systems are a common requirement in non-residential heavy-timber buildings and that, unlike ductwork and wiring, sprinkler pipes usually cannot be hidden behind the exposed structure. Finally, study the possible geometric relationships between the timber frame and each of the other building systems, and choose those that reinforce your design intentions.

Notes

1. Breyer, p. 134.
2. BOCA 1993, p. 151.
3. BOCA 1993, p. 150.
4. BOCA 1993, p. 151.
5. Benson 1988, p. 27.
6. NDS, p. 14.
7. Derived from Allen, *Companion,* p. 49.
8. Gordon, p. 40.
9. Gordon, p. 52.
10. Gordon, p. 54.
11. "Too Flexible Floors . . .", p. 94.
12. Timoshenko, p. 114.
13. Timoshenko, p. 115.
14. Derived from Allen, *Companion,* p. 53.
15. Derived from Allen, *Companion,* p. 59.
16. Derived from Allen, *Companion,* p. 61.
17. Faherty, p. 7.30.
18. NDS, p. 14.
19. NDS, p. 6.
20. NDS, p. 19.
21. Breyer, p. 165.
22. Breyer, p. 165.
23. NDS, p. 6.
24. King, p. 10–19.
25. TCM, p. 1–5.
26. Goldstein, "Geometry . . .", p. 79.
27. Breyer, p. 191.
28. AFPA *Design Values for Wood Construction,* p. 20.
29. TCM, p. 1–10.
30. Canadian Society, p. 1.
31. ICBO 5100, Table 3, p. 22.
32. Levy, p. 33.

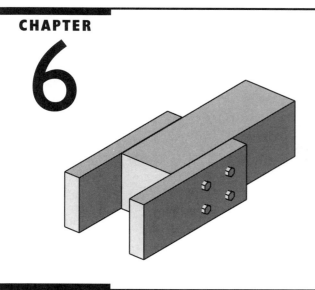

CHAPTER

6

Fasteners and Connectors

Union gives strength.
AESOP, *The Bundle of Sticks*

Introduction

Every timber structure is a framework, a "structure composed of parts fitted and joined together."[1] In this chapter, we shall be concerned with how the parts of a timber frame, its *members,* are united, or *connected.* Although dozens of cultures, over hundreds or thousands of years, have developed an enormous variety of timber connections, we shall be concerned primarily with first principles, with their common structural underpinnings.

A timber structure must be able to accommodate the loads acting on it, without collapse, displacement, or permanent deformation. If downward (gravity) loads were the only ones buildings had to accommodate, you could build simply by stacking materials on top of one another; joists on beams, beams on columns, and columns on foundations. The wooden block towers we erected as children were like this; they consisted of unconnected members. The capacity to transmit loads from one member to another was a function of their contact area and compressive strength. In the real world, of course, things are much more complicated than that. The certainty of gravity loads, combined with the high probability of wind, earthquake, and snow loads means that a tower of wooden blocks is no more stable than a house of cards (a point driven home each time a child gleefully removes one of its pieces and watches it collapse).

Structural integrity is closely related to structural geometry. A square wooden frame with hinges at the corners is a framework, but an unstable one, easily turned into a parallelogram. A triangular frame similarly connected cannot be similarly transformed. Its shape is fixed by its geometry, irrespective of the fixity of its joints. Picture what happens if one of the hinged connections in a square frame is replaced with a rigid brace. Suddenly, one connection has stabilized an entire structure. Finally, just for fun, try building a pentagonal frame with hinged connections. Replacing one of its hinges with a rigid brace reduces but does not eliminate its deformability. Thus, rigidity in a single connection does not necessarily yield integrity in an entire structure.

There are several important lessons here:

- Structural integrity involves both individual connections and the overall assembly to which they belong.
- Bracing can be effective, even if not provided at every joint.
- Geometry is of crucial importance to the stability of a structure.

Timber Connection Categories

A timber frame is composed of discrete members. You can't weld them like steel, or meld them like concrete. Conceptually, your options are quite limited: Shape the members so that they interlock, join them with nails or bolts, or use some type of timber connector (usually metal). This chapter covers fasteners and connectors, while Chapter 7 covers traditional joinery.

Because each of these connection types has such distinctive design and construction implications, the architect and engineer should select an approach to connection design as early as possible in the design process. That approach need not be limited to a single category; indeed, the beauty and economy of a timber frame can be enhanced by tailoring the type of connection to the specifics of each joint. There is nothing inherently wrong with mixing connection types on a single project, as long as both the fabricator and erector have the proper range of skills and equipment. There may even be times when more than one type should be combined in a single joint. (See Fig. 6.1).

Regardless of the number of members in a timber connection, each must be individually checked for adequacy. Since the strength of a connection is controlled by the strength of its weakest link, it is a good idea to begin your analysis with the member you believe will control. If you're right, you'll avoid some backtracking; but even if you're wrong, no time will have been lost, because you have to check every member anyway.

Modeling the Connection

A timber connection, like the structure of which it is a part, is a three-dimensional construct. Its strength is closely related to its geometry.

(a)   (b)

6

6.1 Two views of an encyclopedic example of timber connections in Roebling's Delaware Aqueduct, an early iron suspension bridge. (*a*) Elevation of end of timber truss. At lower left, truss diagonal is mortised into battered post. At bottom right, horizontal thrust is resisted by iron plate and threaded rod behind tapered shim. At center right, tapered shim for hanger plate is let into bottom of beam to prevent transverse displacement. At center left, bolts connect spaced beam to battered post. At top right, U-bolt projects above beam. (*b*) Truss suspended by U-bolt from catenary. The lengths of the U-bolts varies to accommodate the varying elevation of the catenary. These connections are shown in context in Fig. 8.16.

(See Fig. 6.2). Because minor changes in its geometry can have a major impact on its strength, each connection should be studied through the use of hand-drawn or computer-generated 3-D drawings—isometric or perspective—or physical models. And, since the role of the background members is at least as important as those in the foreground, make sure to "disassemble" each connection for the purposes of analysis, and to project to its surfaces any critical information that would otherwise be hidden.

Types of Fasteners

There are many types of fasteners used in timber construction, but each falls into one of two broad categories: those for which holes are not required or are only provided to ease driving and minimize splitting,

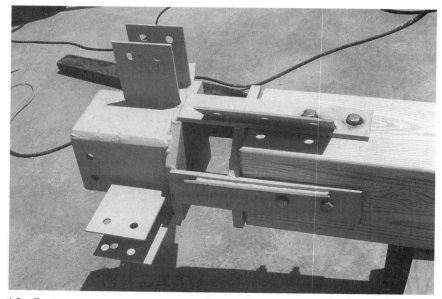

6.2 Custom connector to accommodate multiple members from multiple directions.
This connection is shown in context in Fig. 14.8.

and those for which holes are required. Nails and spikes are among the
fasteners that do not require holes. Bolts, lag screws, split rings, and
shear plates are among the fasteners that cannot be installed without
them. In general, a given fastener's category is a function of its diameter; the smaller a fastener's diameter the less likely a hole will be
needed to install it. Nails are driven with hammers and held in place
by friction with and pressure from surrounding wood fibers. Even
though a pilot hole may be needed to prevent splitting, the diameter of
the hole is always at least slightly smaller than that of the nail shank,
to prevent the nail from falling out of the hole. This is not the case with
fasteners *requiring* holes; "Bolts are installed in holes drilled slightly
(1.0 to 2.0 mm [$\frac{1}{32}$ to $\frac{1}{16}$ in]) larger than the bolt diameter to prevent any
splitting and stress development that could be caused by installation
or subsequent wood shrinkage."[2] Consequently, they can and should be
installed without a hammer. But regardless of the type of fastener, it
must be able "to carry and transfer loads over a large enough area of
the wood so that the wood fibre in contact with the fastener is not
deformed."[3]

Due to the directional properties of wood, fasteners in timber connections are most effective when they are loaded laterally *and* oriented
with their axes perpendicular to the grain of the connected wood members. For given loading conditions, member thicknesses, and species,
several generalizations can be made:

- The *design value* of a fastener is roughly proportional to the area of the fastener that bears on wood; for nails and bolts this area is proportional to fastener diameter d.

- The *number* of fasteners (n) required to resist a given load is, therefore, inversely proportional to the bearing area of each one.

- Therefore, for fasteners of a given length to resist a given load, a number of fastener combinations will suffice, as long as the *product* of the fastener diameter and the number of fasteners—$d \times n$—remains roughly constant.

There is an interesting mathematical relationship between the ranges of common fastener diameters. From $6d$ to $60d,$ nail shank diameters (in round numbers) run from ⅛ (3 mm) to ¼ in (6 mm). Common bolt diameters run from ½ (12 mm) to 1 in (24 mm). Split ring and shear plate diameters run from 2½ (63 mm) to 4 in (100 mm). Roughly speaking, then, each regime extends from diameter d to diameter $2d,$ and the diameters in each subsequent regime are 4 times larger than those of the previous regime. Therefore, comparisons between regimes are easy to make: If it takes a single 1-in bolt to resist a given load, it will take roughly 4 times as many $60d$ nails. Put another way: $(1)(1$ in$) = (4)(¼$ in$)$.

Unfortunately, comparisons of this type are of limited value. First of all, different fastener types have different sizing systems. Whereas a bolt of a given length is available in a variety of diameters, a common wire nail of a given length is available in only a single diameter. As a result, to work in a particular connection, nails must be selected by length, while bolts may be selected by length and diameter. Second, different fastener types have different bearing geometries. Whereas nails and bolts extend through the entire thickness of each member in a connection, split rings and shear plates are housed only ⅜ (9 mm) to ⅝ in (15 mm) into the members. As a result, their bearing lengths are considerably less than those of nails and bolts.

Notwithstanding these limitations, one conclusion is inescapable: To accommodate the large loads typical of nonresidential timber construction, nails and spikes are usually impractical. So many would be needed that splitting of the wood members would be extremely likely. Consequently, the rest of this chapter will focus on the larger fasteners (i.e., those that require holes) and some of the typical issues that influence the engineering of connections that employ them. (See Fig. 6.3).

The type of fastener is not the only decision to make when developing an approach to connection design; you must also decide whether the connections need to be engineered. Homebuilders often use nailing schedules "which give the number, size and type of nail, and the direction of driving (e.g., face nailing, toenailing) to be used for different connections."[4] This is practical for light-frame buildings because such

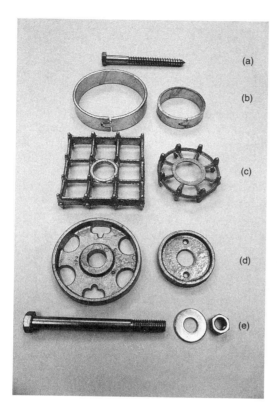

6.3 Timber connection hardware. From top to bottom: (*a*) Lag screw. The length of the unthreaded portion should approximate the thickness of the side member. Consistent with their different purposes, the threads of lag screws are different from those of bolts. (*b*) Split rings in two different sizes. Rings are inserted into matching grooves precut with special tools into meeting faces of overlapping timbers. To ease installation, their cross-sections are slightly tapered. The smaller size is used in members of 2-in (5-cm) nominal thickness, the larger in members of 3-in (7.5-cm) or greater nominal thickness. (*c*) Spike grids in two different shapes. Spike grids are commonly used in pole, dock, piling, bridge, and trestle construction. Square units can be flat on both faces, or flat on one and curved on the other, for interfaces like that shown in Fig. 6.21. (*d*) Shear plates in two different sizes. After insertion in a precut dap of matching thickness, a shear plate ends up flush with the face of the timber. Shear plates are used in pairs where sideplates are timber, or individually where sideplates are steel. (*e*) Bolt with nut and washer. The length of the unthreaded portion should approximate the total thickness of *all* members. Connections with shear plates, split rings, or spike grids must be bolted together. For split rings and shear plates, use ½-in (12-mm) bolts for small units and ¾-in (18-mm) bolts for large units. For spike grids, use bolts up to 1 in (25 mm) in diameter. (Connectors manufactured by Cleveland Steel Specialty Co.)

buildings employ a fairly common and limited set of framing conditions and usually have so much structural redundancy that the integrity of the frame is not dependent on the adequacy of any one connection. By contrast, timber buildings tend to have unique designs and details and relatively little structural redundancy. The integrity of the whole structure may very well be dependent on the integrity of each of a number of connections. Consequently, as a general rule, *connections must be specifically engineered for every timber construction project.*

The spans and live loads typical of heavy-timber buildings are generally much greater than those in wood-framed houses, requiring substantially stronger connections. The design values of bolts are so much higher than those of nails and spikes that they are the fastener of choice in contemporary timber construction, except at decking and sheathing, where nails are the norm. Since splitting is always a concern with nails, it is best to use the smallest diameter nail having the requisite design value, and to install it as far as possible from the ends or edges of the members being connected. Always install nails in side grain, never in end grain. And, when installed in side grain, recognize that nails loaded in withdrawal have lower design values than those loaded in shear.

Nomenclature of Bolted Connections

There are several conventions associated with bolted connections. If you were to travel along the surface of the connection parallel to the applied loads, the bolts encountered at any point occupy its *rows*. If you subtract the area occupied by those bolts from the member's total cross-sectional area, what remains is its *net section*. Every member has *edges* and *ends* beyond the bolt or group of bolts. The distance between the bolts nearest an edge or end and the edge or end itself is called the *edge distance* or *end distance*. " 'Edge distance' is the distance from the edge of a member to the center of the nearest bolt, measured perpendicular to grain. . . . End distance is the distance measured parallel to grain from the square-cut end of a member to the center of the nearest bolt."[5] Further subdivisions distinguish between bolts bearing toward or away from the edge or end of a member: *loaded edge distance, loaded end distance, unloaded edge distance,* and *unloaded end distance.* Collectively, I refer to end and edge distances as *timber setbacks.* (See Fig. 6.4)

In terms of bolt diameters, the required spacing *between rows of bolts* is substantially less than that required *between bolts in a row.* The spacing between rows is intended to prevent splitting and to maintain a reasonable net section. The spacing between bolts in a row is intended to accommodate nonuniform shear stresses resulting from bending of the bolts under load.[6]

The fact that end and edge distances are proportional to bolt diameter and are measured to the *center* of the nearest bolt, rather than to its *edge,* is critical to understanding the geometry of bolted connections. If end and edge distances were measured to bolt centerlines but given as fixed dimensions, then the holes for large bolts would be much closer to the ends and edges of the members than those for small bolts. To prevent such a potentially dangerous situation, end and edge distances are given as multiples of bolt diameter. For end distances, the multiple increases with the bolt load and is higher toward the loaded end than toward the unloaded end.[7] Similarly, for edge distances: The multiple is higher toward the loaded edge than toward the unloaded edge. Measuring to the centers rather than the edges of the bolts eases layout and simplifies calculations.

To create a bolted connection, the faces of two or more members must overlap sufficiently to accommodate the bolt(s) and all required edge and end distances. In a connection consisting of two members and one bolt, there is one interface between the members. This *shear plane* slices across the shaft of the bolt. It is here that the bolt must resist the

PARALLEL-TO-GRAIN LOADING

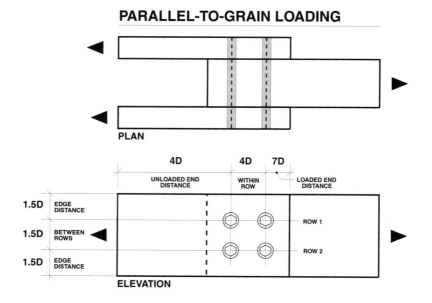

PLAN

4D	4D	7D
UNLOADED END DISTANCE	WITHIN ROW	LOADED END DISTANCE

1.5D EDGE DISTANCE

1.5D BETWEEN ROWS

1.5D EDGE DISTANCE

ROW 1

ROW 2

ELEVATION

PERPENDICULAR-TO-GRAIN LOADING

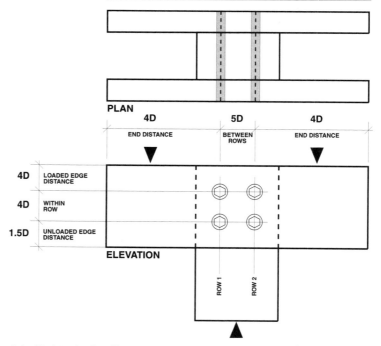

PLAN

4D	5D	4D
END DISTANCE	BETWEEN ROWS	END DISTANCE

4D LOADED EDGE DISTANCE

4D WITHIN ROW

1.5D UNLOADED EDGE DISTANCE

ELEVATION

ROW 1   ROW 2

6.4 End and edge distances and bolt spacings for simple joints, for bolts of diameter D. Bolt rows are always aligned with direction of load. Distances and spacings are shown for side members. Joint design must also account for distances and spacings of main members (see Fig. 6.6). For hardwoods, loaded end distance under parallel-to-grain loading can be reduced to 5 D. Spacing between rows for perpendicular-to-grain loading can be reduced when bolt length in smaller of main or side members, divided by D, is less than 6. (Derived from *National Design Specification for Wood Construction.*)

shearing action of the two wood members acting in opposite directions. The bolt is said to be in *single shear*. In a connection with three members, there are two shear planes; hence, *double shear*. Generally speaking, then, the total number of shear planes in a bolted connection equals 1 less than the number of members.

If you look at a connection from the side, you can see the arrangement and thicknesses of its members. Imagine what happens to the bolts once loads are applied. In single shear, the axes of the bolts rotate in concert with the applied loads. In double shear, the outer members pull the axes one way, and the inner one pulls them the other. The nonsymmetry of single shear is an example of *eccentricity*. Because it tends to produce twisting of the joint, eccentricity should be avoided whenever possible. One simple way to achieve this is to use an odd number of members in every bolted connection, distributed symmetrically about the middle member. The simplest noneccentric bolted connection has three members. The center one is the *main member,* the outer ones are the *side members.*

To install a bolt, a hole must be drilled through each of the members it will join. Where there is a hole, there is no wood. Creating a bolted connection reduces the amount of wood in each member, in the vicinity of the connection. For this reason, the connections are often the weakest links in a bolted timber structure. If there are multiple rows of bolts within the width of a member, the net section may be incapable of resisting the applied loads. When referring to timber structures, architects and engineers often say "the connections control the design." Now you know what they mean.

Each member in a timber connection must be sized to accommodate the stresses induced in it by the applied loads. Depending on the arrangements of the loads, the stresses along the length of the member may vary or remain constant. Where a member is loaded exclusively in tension or compression—as in a timber truss (see Chap. 8)—the stresses within the member will be relatively constant throughout its length. To offset the material removed for the bolts, the overall cross-sectional dimensions of the member will have to be larger in the vicinity of the connection than elsewhere along its length. Because timbers—whether sawn or glue-laminated—tend to have constant cross-sections, much of the wood away from the connections will be structurally underutilized. From this perspective, it is easy to understand why rounding the entire cross-section of or chamfering the corners of that portion of a truss member which is away from its connections does not reduce the member's overall capacity. These operations simply produce elegant dogbones.[8]

Due to timber's directional structure, its design value is generally greater along, or *parallel to grain,* than it is across, or *perpendicular to grain.* It should not be surprising, therefore, that the design value of a bolt will be greater in a connection that loads all of the members parallel to grain than in one that loads some members perpendicular to grain.

Group Action Factors

The design value of a group of bolts is a function of the number of bolts in each row and of the controlling area A_c—the lesser of the cross-sectional areas of the aggregate of the side members and of the main member. As the number of bolts in a row increases, the design value of each bolt decreases. As the controlling area increases, the design value of each bolt increases. For simplicity, think of the ratio of bolts-in-a-row to cross-sectional area as the *bolt density:*

$$\text{bolt density} = \frac{\text{bolts per row}}{A_c}$$

The greater the bolt density, the lower the design value of each bolt. The impact is minimal, however, until the number of bolts in a row exceeds 3. The multiplier that incorporates these effects is called the *group action factor* (GAF).

The cost of fabricating bolted connections is usually minimized when the required capacity is achieved with the fewest bolts. To compare the capacities of various bolt layouts, you must first know how to calculate the capacity of each one. The capacity of the bolts in a given connection is approximated by the following equation:

$$\text{Capacity of bolts} = \text{design value per bolt} \times \text{bolt rows} \\ \times \text{bolts per row} \times \text{GAF} \times \text{spacing factor}$$

To simplify things, let's call the product of the last three variables—bolts per row, GAF (group action factor), and spacing factor—the *row capacity factor.* This factor tells you how much (or little) benefit there is to increasing the number of bolts in a row. As will be explained later in this chapter, the spacing factor ranges between 0.75 and 1.00, and relates to the ratio of the minimum allowable spacing to the minimum spacing necessary for full design value. Obviously, if the timber length available for bolting is fixed, the number of bolts in a row can only be increased if the spacing between those bolts decreases. If the product of the three variables—the row capacity factor—increases either marginally or not at all with the addition of a bolt to each row, then it will be necessary to consider other strategies if you wish to increase the overall capacity of the bolts in a given connection. As shown in Fig. 6.5, the point of diminishing returns is more pronounced and is reached at shorter bolt runs for small member cross sections than for larger ones.

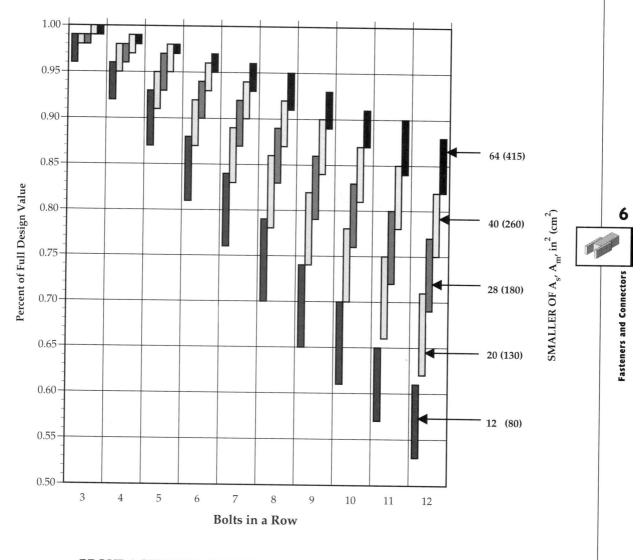

GROUP ACTION FACTORS WITH WOOD SIDE MEMBERS

BASED ON 2 AREAS:

A_m = GROSS CROSS-SECTIONAL AREA OF MAIN MEMBER, AND

A_s = SUM OF GROSS CROSS-SECTIONAL AREAS OF SIDE MEMBERS.

AT TOPS OF BARS, THE AREAS ARE EQUAL; AT BOTTOMS, ONE IS HALF THE OTHER

6.5 Group Action Factor (GAF) graph. For "area ratios" between 0.5 and 1.0, interpolate between the values at the top and bottom of the applicable bar. For areas between those which are graphed, interpolate between corresponding locations on the adjacent bars, for the applicable number of bolts in a row. (Based on values in *National Design Specification for Wood Construction*.)

Dimensional Constraints in Bolted Connections

In a given connection, the controlling end and edge distances are a function of the geometry of the connection and the directions of the loads. To keep track of these limitations, I recommend that you create what I call a *constraint map*.

The role of a constraint map in the design of a timber connection is analogous to that of a survey showing required setbacks and other zoning constraints, in the development of a site plan for a subdivision. Both the zoning setbacks in the survey and the timber setbacks in the constraint map establish the limits of construction. In the survey, the area bounded by the setback lines is called the *buildable area;* the number of units permitted within that area is often a function of their size. In a timber connection, the area bounded by the setback lines could be called the *boltable area;* the number of bolts permitted within that area is also a function of their size. For property, setbacks are relative to the street, and are measured at the front, rear, and side yards; for timber connections, they are relative to the forces, and are measured at the loaded and unloaded ends and edges. Just as there are special zoning requirements for lots with streets on two sides, and there are special connection requirements for joints subject to forces in two directions. The monetary value of a piece of real estate is a function of the quality of the surrounding neighborhood; the structural value of a bolt in a timber connection is a function of the quality of the surrounding wood, as measured by its specific gravity. To limit housing density, zoning setbacks may be proportional to unit size. To limit bolt density (and prevent splitting), timber setbacks are proportional to bolt diameter. *One important difference:* Zoning setbacks are measured to the faces of the nearest units, while timber setbacks are measured to the centers of the nearest bolts.

To create a constraint map follow these 10 steps (see Fig. 6.6):

1. Sketch the connection in multiple views or in three dimensions, drawing the joint at large scale, or, if using CAD, zooming in on the appropriate portion of the joint, making sure that the edges of all members are visible.

2. Indicate the width and thickness of each member, and compute its cross-sectional area.

3. Calculate the ratio of the sum of the areas of the side members to the area of the main member.

4. Produce a composite drawing in two dimensions, using a different color or line weight for each member, and showing the outlines of the connected members at a typical shear plane.

5. Using the sense of the force in each member, establish its applicable end and edge distances for an assumed bolt size. (Where mem-

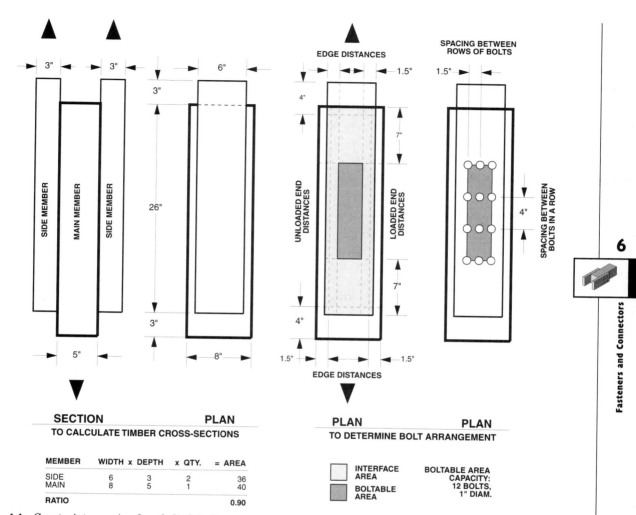

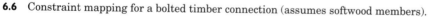

6.6   Constraint mapping for a bolted timber connection (assumes softwood members).

bers are subject to sense reversals, the controlling distances will be those that are furthest away from the ends or edges.)

6. Indicate the edges and ends of the connected members with solid lines, timber setbacks with dashed lines, and loads with arrows.

7. Outline the area within which all members overlap, label it *interface area,* and fill it with a light tone.

8. Within the interface area, fill the zone bounded by the innermost sets of setback lines with a darker tone. This represents the *boltable area* of your connection, the maximum extent of the centers of its outermost bolts.

9. On a clean plan, lay out the bolts within the boltable area, using the applicable spacing requirements for bolts in a row and between rows.

10. Compare the total available bolt capacity (accounting for the group action factor) to the required capacity; if the former is greater than or equal to the latter, the connection works.

Whereas timber setbacks at any one shear plane are derived from the geometries of the individual members there, the *controlling timber setbacks* can only be determined by overlaying the individual setbacks of all the members in the connection and identifying the most extreme cases. Remember that timber setbacks are a function of the direction of the load with respect to the axis of each member—parallel or perpendicular to grain—and of the direction of the load with respect to the perimeter of each member—loaded versus unloaded edges and ends. In a connection composed of two members of different width, the setbacks associated with the narrowest will usually control in connections where both members are loaded parallel to grain.

Determining the Number of Bolt Rows

Determining the number of bolts that can fit within the boltable area involves the use of *modular arithmetic*. Think back to grade school arithmetic: When one number is divided by a smaller one, the answer consists of a whole number and a remainder. (If the larger number is a multiple of the smaller one, the remainder is 0.) Dividing 14 by 5, for example, yields the whole number 2 and a remainder of 4. The conventional answer would be 2.8, which would round off to 3. In modular arithmetic, however, the answer is either 2 (the result of dividing and truncating) or 4 (the remainder). As a mathematical expression, *div* is an "arithmetic operator that divides a number to its left by a number to its right, ignoring any remainder, resulting in just the whole part"[9] and *mod* is an "arithmetic operator that divides the number to its left by the number to its right, ignoring the whole part, resulting in just the remainder."[10] Using our example, 14 div 5 = 2, and 14 mod 5 = 4.

The reason for using modular arithmetic in analyzing timber connections is that bolts only come in whole numbers; you can't buy or install fractions of a bolt. For example, for parallel-to-grain loading using timbers of common width, the maximum number of rows of bolts is given by the expression:

$$(w_b) \text{ div } (1.5d) + 1$$

where w_b is the boltable width—the width of the boltable area from the constraint map—and d is the diameter of the bolts in inches. Assuming a boltable width of 4.5 in (a 4 × 8 with 1½-in edge distances), here are the results for different bolt diameters:

Bolt diameter, in	1.000	0.875	0.750	0.625	0.500
Number of rows of bolts	4	4	5	5	7

As previously discussed, the boltable width will vary with bolt diameter. For parallel-to-grain loading, it just so happens that the spacing required between the edge of the members and the first row of bolts—the edge distance—is the same as the spacing required between each row of bolts. Therefore, for parallel-to-grain loading using members of common width, the maximum number of rows of bolts is given by the expression: (w_m) div $(1.5d) - 1$, where w_m is the overall width of the member. For a member width of 7.5 in (as would be the case with 4 × 8s), here are the results for different bolt diameters:

Bolt diameter, in	1.000	0.875	0.750	0.625	0.500
Number of rows of bolts	4	4	5	7	9

As previously discussed, the area occupied by the bolts themselves is deducted from the cross-sectional area of the member. The area remaining, the *net section,* represents the weak link in the member. The net section for a member of a given thickness and width is given as:

Net section = (thickness × width)
    − (thickness × number of rows of bolts × bolt diameter)

Dividing both sides by the thickness of the members yields the *net width:*

Net width = width − (number of rows of bolts × bolt diameter)

Plugging the maximum number of bolt rows derived above, into this formula, yields the following results, assuming Douglas fir–larch in double shear, with a main member 7.5 in (18.75 cm) wide and 3.5 in (8.75 cm) thick, and with side members of the same width but only 1.5 in (3.75 cm) thick:[11]

Bolt diameter, in	1.000	0.875	0.750	0.625	0.500
Number of rows of bolts	4	4	5	7	9
Net width, in	3.500	4.000	3.750	3.125	3.000
Net width/member width, percent	47	53	50	42	40
Capacity per ⅛ in of diameter, compared to ½ in	1.66	1.48	1.30	1.14	1.00
Capacity of all bolts, compared to ½ in	1.48	1.15	1.08	1.11	1.00

The fact that bolts come in only a few discrete sizes, in combination with the fact that fractional bolts do not exist, produces nonlinear results. However, if we limit our analysis of this table to only those two bolt diameters that result in an actual spacing of $1.5d$ with no remainder—1 in (2.54 cm) and ½ in (1.27 cm)—and extend this table to ¼ in diameter, a general trend is clear: the smaller the bolt diameter, the smaller the net width and net section. This is so because the boltable width increases as the bolt diameter decreases—the smaller the bolt, the closer it can be installed to the edge of the member.

To understand the significance of this trend, it is necessary to refer to tables of bolt design values, such as those in the *National Design Specification for Wood Construction*. After analyzing the information found there, we can draw a simple but profound generalization for parallel-to-grain bolted connections, for all of the specific gravities and thickness combinations covered, and for both single and double shear connections. Letting dv refer to design value, d refer to bolt diameter, and the subscripts $_l$ and $_s$ refer to large and small bolts:

$$\frac{dv_l}{dv_s} \geq \frac{d_l}{d_s}$$

To put it simply, the rate at which bolt design values increase is the same or greater than the rate at which their diameters increase. Our previous finding was that larger bolts result in greater net sections than smaller bolts, resulting in more efficient timber utilization. Our new finding demonstrates that the capacity of a larger bolt is at least as much as that of a smaller bolt per increment of bolt diameter. Together, these findings mean that the capacity of a parallel-to-grain connection is maximized through the use of a small number of large bolts, rather than a large number of small bolts. (As mentioned earlier, this is also likely to minimize the cost of the connection.)

This conclusion should be balanced against the desirability of diversity, of not putting all of one's eggs in one basket. Spreading the load over more bolts means less reliance on each and, therefore, less likelihood that the failure of one will lead to progressive failure of all. In the context of these contradictions—a few larger bolts for economy or many smaller ones for diversity—final determination of bolt size and quantity should be left to engineers experienced in timber construction.

Determining the Number and Spacing of Bolts in Each Row

To calculate the maximum number of bolts in each row, use the same rules of modularity as illustrated previously for calculating the number of rows of bolts. Keep in mind, however, that:

- End distances are greater than edge distances.
- Spacing between bolts in a row is greater than spacing between bolt rows.

- The ratio of end distance to spacing between bolts in a row is greater than the ratio of edge distance to spacing between bolt rows.

- The smaller the bolt diameter for a given boltable length, the more bolts will fit in each row.

- The more bolts in each row, the lower the design value of each, because of the *group action factor.*

These factors explain why the interface areas of connections where all members are loaded parallel to grain tend to allow the most bolts when they are rectangular and have their long dimensions parallel to grain.

In a bolted connection, both the bolts and the wood remaining around them (the net section) must have sufficient capacity to resist the applied loads. Since the connection is only as strong as its weakest link, what is needed is a design in which these two aspects are in balance. For wood members to qualify as heavy timbers, most building codes mandate that they not be less than a certain thickness and width. In projects where such provisions apply and where structural efficiency is the goal, I recommend that the spacing of the framing be large enough that the member sizes mandated for fire resistance are also justified for structural resistance. After all, that portion of the cross-section of each member that is not performing a structural role is essentially going to waste.

When a bolted timber connection is under axial load (tension or compression), specific stresses arise within each member. The wood behind the shank of each bolt in a row is pushed in one direction, while the wood between rows is pushed in the other (see Fig. 6.7). The capacity of the member to accommodate these equal and opposite forces is a function of its shear strength and the area over which the shearing forces are distributed. The maximum bolt load is that which the wood in contact with the bolt shaft can tolerate without crushing. The relationship between the spacing of bolts in a row and their design values consists of five regimes, shown graphically in Fig. 6.8, and numbered in accordance with the following descriptions.

1. When the spacing between bolts in a row is too small (less than 3 times the bolt diameter), the shear planes are so small that they will fail under extremely light loads, clearly an unsafe condition.

2. When the spacing is large enough to assure safe and predictable performance, the design value of the bolts (i.e., the maximum loads they can carry) will be proportional to their spacing.

3. When the shear planes are just large enough for the maximum bolt load, the spacing is *optimal;* the load and the load capacity of the shear planes are in balance.

4. When the bolts are loaded too heavily, the wood in contact with them will crush, at which point the size of the shear planes—a function of bolt spacing—is irrelevant.

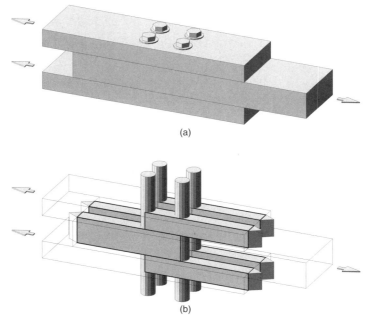

(a)

(b)

6.7 Three-member bolted timber connection under axial load. (*a*) Appearance of the outside of the assembly. (*b*) Simplified diagram of what happens on the inside of the members. The bolt shafts act like dowels, loading the exposed end grain within half of each hole. Vertical shear planes resist the forces induced when the wood *behind* the bolts tries to move one way and the wood *between* them tries to move the other. The shear planes extend to the loaded end of each member.

5. When the bolt spacing is larger than optimal, the risk of crushing prevents any further increase in design value. There is no structural advantage to having oversize shear planes.

For bolts loaded parallel to grain, the relationship of the spacing between bolts in a row s and bolt diameter d can be summarized in a simple and elegant *spacing equation:*

$$s = 4d$$

The load of each bolt in a connection conforming to this equation can be thought of as being resisted by two shear planes, each with an area 4 times as long as the bolt diameter. With design values in shear so much lower than those in compression, a balance between the allowable load in shear—design value in shear (F_v) × shear area—and the allowable load in compression parallel to grain—design value in compression (F_c) × compression area—can only be achieved if the area in shear, as shown, is approximately 8 times the area in compression:

$$(F_v)(t)(4d)(2 \text{ shear planes}) \cong (F_c)(t)(d)$$

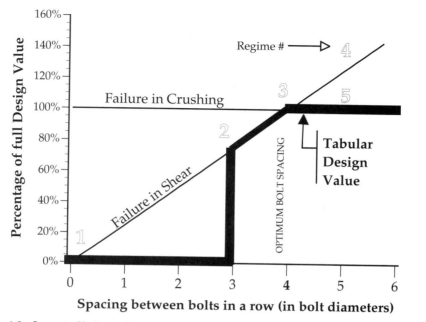

6.8 Impact of bolt spacing on design value, showing five regimes.

As discussed earlier, doubling the bolt diameter at least doubles the allowable load per bolt, while distributing the larger compressive forces over a proportionally larger area inside the bolt hole. As illustrated in the foregoing spacing equation, doubling the bolt diameter also doubles the bolt spacing which, for a given member thickness, doubles the shear area. Doubling the shear area doubles the shear capacity of the bolt. Thus, a balance is maintained between compression and shear, regardless of the bolt diameter.

Reduced Design Value: An Example

Although tightening the bolt spacing might enable you to increase the number of bolts in each row, the result is actually a net *decrease* in the row capacity factor. To understand why, consider the following example, where b_l is the boltable length, b_r is the number of bolts in a row, and d is the bolt diameter. For full design value, the number of bolts in a row is given as:

$$b_r = (b_l) \text{ div } (4d) + 1$$

while for the design value at the minimum spacing, the number of bolts in a row is given as:

$$b_r = (b_l) \text{ div } (3d) + 1$$

Thus, a boltable length of 12 in (30 cm), can accommodate a row of 4 bolts spaced 4 in (10 cm) apart at full value or a row of 5 bolts spaced 3 in (7.5 cm) apart at reduced value. Because the design value of each bolt is reduced by the ratio of the actual spacing to the full value spacing, the reduced value in this case is three quarters of the full value. If we assign a full value of 1.0, the reduced value is 0.75. Thus, the combined value of the 4-bolt row is 4×1.0, or 4.0, while the combined value of the 5-bolt row is 5×0.75, or 3.75. And, because the group action factor is inversely proportional to the number of bolts in a row, the negative impact of decreasing the bolt spacing is even more pronounced.

The forces acting on the main member and on the side members must be in equilibrium. The bolts are sized and spaced to resist these opposing forces. Assuming all members are of the same species, the bearing area of each bolt in the main member must approximate the total bearing area in the side member(s). In general, bolt design values are higher when the thickness of the main member is roughly equivalent to the sum of the thickness of the side members and when the specific gravity of the wood is higher (see Fig. 6.9). So, just as there is no increase in bolt design value when the spacing of bolts in a row is more than 4 times the bolt diameter, there is generally no appreciable increase in bolt design value when the main member is more than twice the thickness of each side member. As previously discussed, however, bolt design values parallel to grain are substantially higher than those perpendicular to grain. (See Figs. 6.10a and b and 6.11a and b).

Perpendicular-to-Grain Loading

To understand the difficulty of designing connections where one or both members is loaded perpendicular to grain, consider three connections where bolts could be employed, each taken from the second floor of a timber-framed building, and each containing three members with the same combination of thicknesses:

- *Column:* A connection consisting of a two-story column where both members of a spaced beam are bolted to its sides at the second floor. The gravity load is parallel to the grain of the column (the main member) and perpendicular to the grain of the beams (the side members).

- *Hanger:* A connection consisting of a timber hanger the top end of which is sandwiched between the members of the same spaced beam. The gravity load is parallel to the grain of the hanger (the main member) and perpendicular to the grain of the beams (the side members).

- *Ledger:* A connection consisting of two ledgers, one bolted to each face of an interior second-floor beam. The gravity load is perpendicular to the grain of the floor beam (the main member) and of the pair of ledgers (the side members).

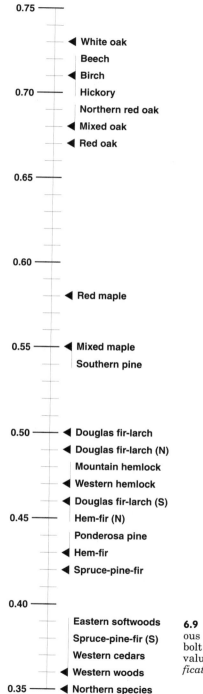

0.75

◀ White oak

Beech

◀ Birch

0.70 — Hickory

Northern red oak

◀ Mixed oak

◀ Red oak

0.65

0.60

◀ Red maple

0.55 — ◀ Mixed maple

Southern pine

0.50 — ◀ Douglas fir-larch

◀ Douglas fir-larch (N)

Mountain hemlock

◀ Western hemlock

◀ Douglas fir-larch (S)

0.45 — Hem-fir (N)

Ponderosa pine

◀ Hem-fir

◀ Spruce-pine-fir

0.40

Eastern softwoods

Spruce-pine-fir (S)

Western cedars

◀ Western woods

0.35 — ◀ Northern species

6.9 Specific gravities for various species. Use with graphs of bolt design values. (Based on values in *National Design Specification for Wood Construction.*)

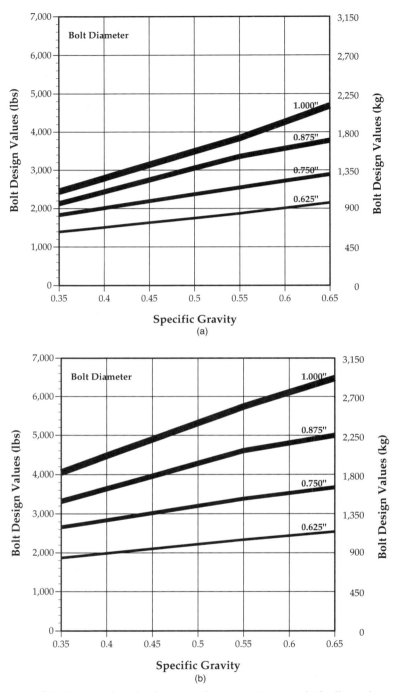

6.10 Bolt design values for three-member connections in which all members are loaded *parallel* to grain; double shear connections. (*a*) Side members of 1.5-in (3.75-cm) actual thickness; main member of at least 3-in (7.50-cm) thickness. (*b*) Side members of 3.5-in (8.75-cm) actual thickness; main member of at least 4.5-in (11.25-cm) thickness. Where side members are ¼-in (6-mm) ASTM A36 steel side plates, the applicable design values are roughly the same as those shown for connections with a main member of the same thickness, but with *wood* side members equivalent to main member thicknesses. (Based on values in *National Design Specification for Wood Construction*.)

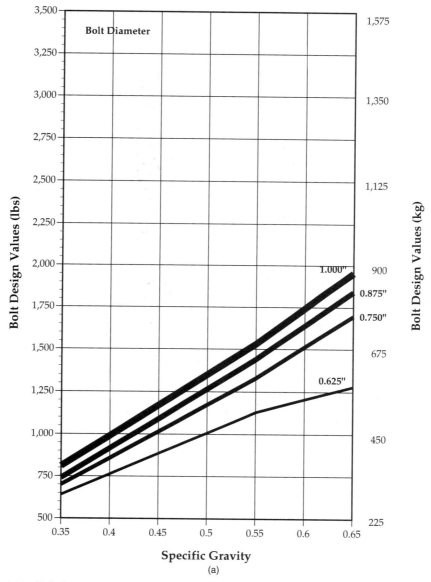

6.11 Bolt design values for three-member connections in which the main member is loaded *parallel* to grain and the side members are loaded *perpendicular* to grain; double shear connections. Where the main member is loaded *perpendicular* to grain, and the side members are loaded *parallel* to grain, design values are typically equal to or greater than those shown in these graphs. (*a*) Side members of 1.5-in (3.75-cm) actual thickness; main member of at least 3-in (7.50-cm). (Based on values in *National Design Specification for Wood Construction.*)

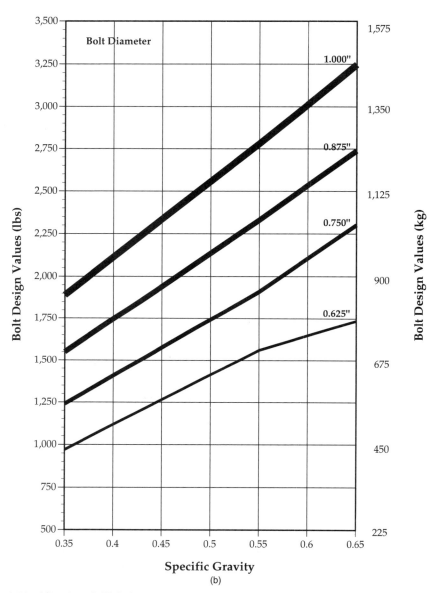

6.11 (*Continued*) (*b*) Side members of 3.5-in (8.75-cm) actual thickness; main member of at least 4.5-in (11.25-cm) thickness. (Based on values in *National Design Specification for Wood Construction.*)

Now consider the differences between these details. In the column and hanger connections, perpendicular-to-grain loading is found only in the side members; in the ledger connection, it is the case in all members. The fact that the ends of the column are a floor above and below the connection means that the column's end distances will not be a factor in establishing the boltable area. The loaded end of the hanger, by contrast, is flush with the tops of the beams. Because the loaded end distance parallel to grain in the hanger must be larger than the unloaded edge distance perpendicular to grain in the beams, the hanger's loaded end distance will control. Ledgers, by definition, are shelves for supporting secondary members' framing into primary ones; therefore, they are customarily fastened to the lower portion of a beam or girder.

From experience, you already know that wood splits easily along the grain; if it were not so, splitting firewood would require more than an ax. A saw blade narrows away from its leading edge to prevent binding (see Chap. 2). An ax blade, by contrast, widens away from its leading edge to pry wood fibers apart as it is driven. Up to now, we have focused on the fact that wood is weaker in compression across the grain than along the grain. Now it is evident that it is weaker still in *tension across the grain*. Therefore, avoid connections like our ledger, where tension is induced perpendicular to grain by a significant load suspended below the centroid of a beam.[12] (See Fig. 6.12.)

The stresses in bolted connections can originate either outside or inside their members. The discussion so far has involved the former; now let's examine the latter. Even if seasoned somewhat prior to erection, timbers continue to shrink—particularly across the grain—long afterward. Where all the members in a bolted connection are the same

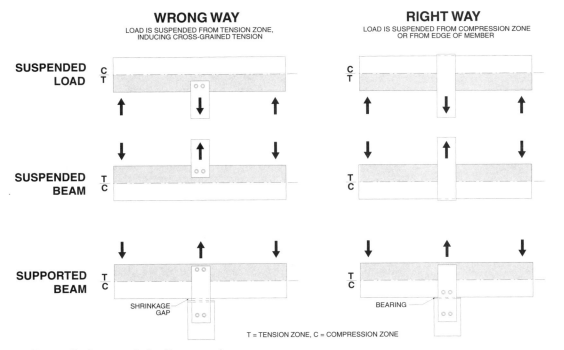

6.12 Perpendicular-to-grain loading scenarios.

width and are loaded parallel to grain, all will tend to shrink in the same perpendicular direction, and by roughly the same amount. The bolts will end up slightly closer together than where they started, but their shafts will still be parallel to one another, because all the members are pulling in the same direction.

Problems can arise when one member is loaded perpendicular to another or where the shrinkage of one member is restrained by another of a different size or material. In the former, bolt displacement resulting from shrinkage across the grain of one member could induce stress along the grain of another, and vice versa. Not only will the bolts be pulled closer together, but their shafts will end up tilted. The latter could arise at a splice in the bottom chord of a truss. Bolt displacement during shrinkage of the main member could be restrained by the smaller and consequently drier side members. When the irresistible force of shrinkage encounters the essentially immovable bolts, the result will be a split. The ends of members are particularly susceptible to splitting because they have so little area over which to distribute perpendicular-to-grain tensile forces and because splits may already have begun there at the time of felling. The best way to minimize these stresses is to limit the distance between the outer bolt rows on a single splice plate to 5 in (12.5 cm).[13] Where the distance between the outer bolt rows on the main member is greater than that, use multiple splice plates to minimize the accumulation of shrinkage stresses.

Bolt Bearing Conditions

The installed capacity of any fastener is a function not only of the design of the connection of which it is a part, but also of the quality of construction. Bolt design values assume uniform bearing of the bolt against the loaded portion of the bolt hole. Achieving uniform bearing requires holes of the proper size and texture:

- *If the hole is too large,* the bolt will rotate along its axis, loading the wood nonuniformly, resulting in localized crushing and in a consequent reduction in the capacity of the bolt.

- *If the hole is too small,* the member may split during installation of the bolt or after shrinkage of the wood around it.[14]

- *If the hole is too rough,* the area of wood in contact with the bolt will be reduced, resulting in localized crushing. The crushing will continue until the bearing area is sufficient, resulting in an oversized hole that could prevent the joint from behaving as predicted.

- *If the threaded portion of the bolt bears on wood,* the effective bearing area will be reduced, with the same consequence described for rough bolt holes.

If bolted connections are to perform as intended, control of the size and quality of the bolt holes is imperative, as is use of properly proportioned bolts. To achieve smooth bolt hole surfaces, the contractor must use sharp bits, and must *run* them at high speed and *feed* them at low speed.[15] To assure uniform bearing, the contractor must use bolts that are unthreaded in the portion that will be in contact with wood.

The above preparations are necessary, but not sufficient, to insure proper bolt performance. They must be supplemented by specific steps during installation and after occupancy:

- To prevent localized crushing when the bolt is tightened, use washers or steel plates under the heads of all bolts and under all nuts.

- To accommodate the shrinkage that typically occurs after the building is enclosed and during the first heating season, tighten any loose bolts, one year after installation.[16]

- To enable the members to continue to expand and contract as their moisture content changes in service, make all connections snug tight, but no tighter.

Split Rings and Shear Plates

Split rings and shear plates represent two types of timber connector (see Fig. 6.3). Although they transmit loads somewhat like bolts, they differ from them in several significant ways.

In a bolted connection, each bolt can act in two ways: in shear to resist loads in the plane of the members it connects, and in tension to prevent the connection from coming apart normal to that plane. Each bolt penetrates the full thickness of each of the connected members and acts in shear over that entire distance. To maintain adequate edge distances, bolt diameters cannot be larger than one third the width of the narrowest member in the connection, and tend to be quite a bit smaller. Because the substantial loads typical of timber connections cannot usually be resisted by a single bolt, multiple bolts are common. End and edge distances limit the boltable area.

Split rings and shear plates have several important similarities and differences. Split rings come in *one* piece, and are available in two sizes: 2½ in (6.25 cm) diameter and ¾ in (1.9 cm) deep, or 4 in (10 cm) diameter and 1 in (2.5 cm) deep. Shear is transmitted through the connector; a bolt is needed simply to keep the connection from coming apart. Half of the depth of a split ring is housed in each member, so shear is transmitted through the ring. Shear plates come in *two* pieces, in diameters and with embedment depths similar to those of split rings. As with split rings, a bolt is needed to keep the connection from coming apart, but because shear plates are flush with the surfaces of the members in which they are embedded, the bolt also acts as the sole means of transmitting shear from one shear plate to another, or from one shear plate into a steel side plate. Because design values are primarily a function of connector bearing area, it should not be surprising that the design values and net sections of similarly sized split rings and shear plates are nearly identical, all else being equal. (See Fig. 6.13.)

Unless the wood has already shrunk, housing a large steel cylinder within a wood member would sound like a recipe for disaster. Not to worry; split rings incorporate provisions for shrinkage. The circumference of the groove for a split ring is slightly larger than the circumference of the ring itself. To fit in the groove, the "split" in the

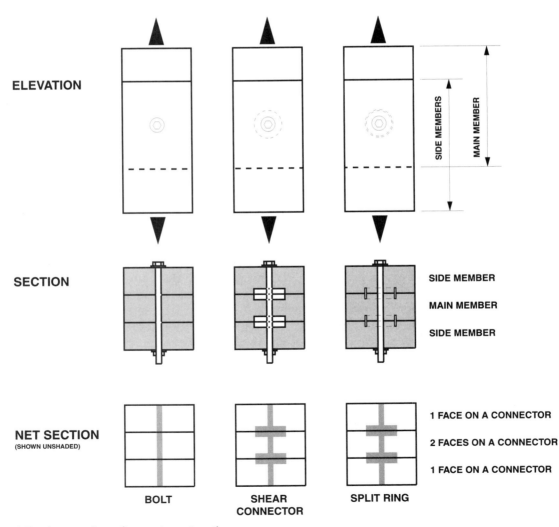

6.13   A comparison of connector net sections.

ring is opened slightly. As the member shrinks, the circumference of the groove shrinks along with the size of the gap at the split end of the ring.

Connection Geometries

Although there are a large number of possible combinations of member widths and orientations in timber connections, most fit within a relatively small number of categories: Members are or are not from the same species, are or are not of a common size, and are or are not all loaded parallel to grain. It is not customary to mix species within a building, so the first issue is usually moot. With respect to the other issues, however, one can conclude the following:

> When all members are loaded parallel to grain, the *orientation* of all ends and edges is the same:
>
> - If the widths of all members are the same, then their actual edge *distances* are the same and are coincident. Thus, the controlling edge distance is the same as the edge distance of each of the members.
> - If the widths of the members vary, then the controlling edge distance is that of the narrowest member.
>
> When some members are loaded perpendicular to grain, the loaded and unloaded edge and end distances of each member must be checked to determine which ones control.

Staggered Fasteners

The discussion so far has assumed that the bolts are arranged in regular rows and columns, but what if the bolts in each row are staggered with respect to those in adjacent rows? Recognizing the vital role of the net section—the wood between and beyond the bolt rows—in resisting the loads imparted by the bolts, you should not be surprised to find that this material is somewhat sacred. Under the NDS, it is permitted to wander somewhat as long as its overall width is maintained.

For parallel-to-grain loading, if the spacing between rows is at least 1.5 times the bolt diameter, it doesn't matter whether the bolts in adjacent rows are aligned as long as the spacing of bolts within each row is adequate. Due to end distance requirements, however, the interface area required for a given number of bolts of a given size will be larger if they are staggered than if they are aligned. If the spacing between adjacent rows is small—less than one quarter the stagger—then those two rows are considered one[17] and are subject to the associated spacing requirements and group action factor. Staggering of bolts in connections involving perpendicular-to-grain loading should be avoided.

Fastener Coordination

When designing bolted timber connections, establish general esthetic objectives before addressing specific conditions. Are you trying to minimize the number of fasteners? Where multiple fasteners are required, would you prefer a few long rows or a larger number of shorter rows? Given a choice, would you prefer a large or small number of bolts, split rings, or shear plates?

Consideration must also be given to issues of coordination and conflict avoidance. Particularly in trusses and other assemblies where multiple members converge, there is always the possibility that the connections that have been designed cannot be readily constructed. Sometimes, the puzzle is solved by rethinking the sequence of assembly. At other times, the only answer is redesign. In all cases, there is the risk of delay and added cost. Therefore, coordination is imperative. (See Fig. 6.14.)

6.14 Multiple glulams converging at a custom steel connector. The ends of the four intermediate members were mortised to receive the connector's plates. The connector's center is open to allow passage of the pendant of a lighting fixture. This connection is shown in context in Fig. 5.17*b*.

Wherever a member has fasteners in multiple faces, check that they will not interfere with one another. Remember: Two fasteners cannot occupy the same space at the same time. This type of conflict can occur where bearing blocks are fastened at the same elevation to two adjacent faces of a column, where framing changes direction, or at the spring point where several braces radiate from a member. To avoid a conflict between bolts driven from adjacent or opposite sides, offset them vertically (see Fig. 14.8) or use lag screws instead of bolts. Regardless of the condition, make sure your calculations of net section account for *all* of the fasteners in *all* of the faces, and that your calculations of end and edge distances account for all offsets.

In connections where nuts are not desired, where there is no room for them, or where access to either the nut or the head of the bolt will be so limited that one cannot be prevented from turning during tightening of the other, conventional bolting may not be feasible. There are two common alternatives: carriage bolts and lag screws. After installation, a portion of each projects from only one side of a connection, the nut in the case of the carriage bolt, the head in the case of the lag screw. One hammer blow to the head of a carriage bolt sets its shoulders slightly into the wood surface; subsequent turning of the nut draws the shoulders further in.

Because a carriage bolt can only be installed if there is enough room on one side of the joint to insert it and set the head and, on the other, to hold the nut, it, like a bolt, requires two-sided access for installation. But, like a lag screw, it only requires one-sided access for tightening. Therefore, one way to select a fastener is to classify the connection in terms of the type of access required for its installation and tightening.

Fastener type	For installation	For tightening
Bolts	2-sided	2-sided
Carriage bolts	2-sided	1-sided
Lag screws	1-sided	1-sided

Loads at Angles to Grain

Design values have been established for bolts, lag screws, split rings, and shear plates for different species, member thicknesses, and fastener diameters, and for parallel-to-grain (0°) and perpendicular-to-grain (90°) loads. Unfortunately, fasteners can be loaded at any of an infinite number of angles in between, and often are, particularly in trussed assemblies. What to do?

If the design values parallel and perpendicular to grain *and* the angle between the direction of load and the direction of grain are known, the Hankinson formula enables you to determine the design value of a fastener loaded at that angle.[18] Because working with the formula can be time consuming, most people prefer to consult a graph, like the one in Fig. 6.15. Here is how to use it:

1. Divide the parallel-to-grain design value by the perpendicular-to-grain design value for the intended species and grade.
2. Find this ratio along the vertical axis.
3. Find the applicable angle to grain along the horizontal axis.
4. The value of the curve at which these two lines intersect is their Hankinson formula multiplier.
5. If the intersection is between two curves, interpolate between their values.
6. Multiply the multiplier by the parallel-to-grain design value to yield the design value at the given angle.

Direct Bearing Connections

Many joints in timber structures involve one member bearing directly on another. In a direct bearing connection, the stress in each member (s) is defined as the load (P) divided by the loaded area (A) or:

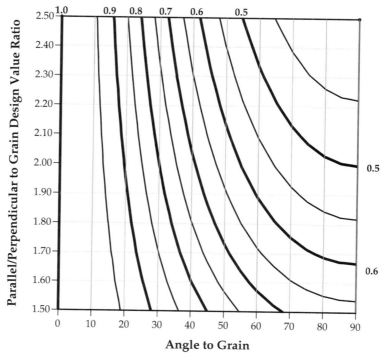

6.15   Hankinson formula multiplier.

$$s = \frac{P}{A}$$

In an isotropic material such as steel, the resistance of the member to crushing is independent of its orientation. In an anisotropic material such as timber, however, the resistance of the material is very much dependent on its orientation.

In timber structures, common locations of direct bearing connections include those where joists bear on beams and where beams bear on columns. In terms of member orientation, there is a big difference between these two examples: In the former, the side grain of each member is loaded; in the latter, the side grain of the beam bears on the end grain of the column.

Since timber is structured somewhat like a bundle of thick-walled straws, it should not be surprising that, in most species and grades, it will crush more easily when loaded from the side than when loaded from the end. The capacity of a direct bearing connection (i.e., wood against wood), like that of a bolted connection (i.e., wood against metal), is usually controlled by the wood member loaded at the greatest angle to grain. Therefore, in the case of a timber beam bearing directly on a timber column, the bearing area of the beam will control that aspect of the design of the connection.

In a bearing connection where one member is compressed perpendicular to grain and the other is compressed parallel to grain, the stresses at the bearing surfaces of the members can be equalized in several ways:

■ By utilizing a *combination of grades* where the compression design value of the beam perpendicular to grain is approximately equal to the compression design value of the column parallel to grain. Within Douglas fir–larch posts and timbers, an example would be the combination of Dense No. 1 for the beam and No. 2 for the column.[19]

■ By utilizing a *single grade* where the difference between the design values is insignificant. Within mixed oak posts and timbers, an example would be mixed oak No. 1.[20]

■ By utilizing a *timber connector between the members,* where the area of each of its bearing surfaces is inversely proportional to the compressive design value of the timber surface bearing on it. Within the standard product line of any connector manufacturer, an example would be a column cap where the beam seat is substantially larger than the column's cap plate.

These recommendation are meant for conditions where the bearing surface of the weaker member would otherwise be overstressed. From my own experience, however, I have rarely found this condition to control the design of the connection or the sizing of the members. Most of the time, the capacity of a column is limited by its slenderness ratio, and that of a beam, by its strength or stiffness. Even so, the structural engineer must always verify that the bearing conditions are satisfactory.

Standard Connectors

Anyone who has been around a residential construction site is familiar with the wide range of joist and beam hangers available, and their extremely low unit costs. Their manufacturers have responded to the proliferation of wood I-beams and laminated veneer lumber with entirely new connector series sized specifically to the widths and depths of these members. Like their predecessors, these connectors are usually stamped out of thin-gauge steel and provided with nail holes. The result is a product perfectly suited to the short spans and light loads typical of houses.

Many of the same manufacturers also make connectors to accommodate analogous conditions in timber buildings. Timber connectors use steel 2 to 4 times thicker, and have allowable loads 2 to 4 times greater than their residential counterparts. Welded rather than stamped, they do not lend themselves to the same degree of mass production. On the one hand, this makes them much more costly; on the other, it enables the manufacturer to customize its timber connectors to some degree, at nominal additional cost (see Fig. 6.16). To accommodate larger loads and larger members, timber connectors are often furnished with holes large enough for bolts instead of nails. Finally, timber connectors—when painted—often look good enough to leave exposed to view in a building, while lumber connectors do not. Due to manufacturing efficiencies, the cost per pound of capacity tends to be much smaller for lumber connectors than for timber connectors, particularly custom ones. Therefore, for economy in timber construction, use standard lumber connectors when neither strength nor appearance is critical.

Where one heavily loaded timber frames into the side of another, it is customary for the beam hanger to include a top flange for direct bearing on the carrying member. Where in-line beams or purlins frame in from opposite sides, their seats are welded to a central saddle. For particularly heavy loads, this flange may so thick—¼ in (6 mm) or more—that it will cause a bump in the floor or roof deck passing over it. While mortising the top of the carrying member is not advisable, as it reduces its structural properties and throws the top of the supported member out of alignment with the top of the carrying member, mortising the underside of the decking is acceptable. Slight shaving of the edges of the timbers may also be required to accommodate the inside radii between pairs of adjacent connector surfaces. Allowances must also be made for differential shrinkage between beams and girders (see Fig. 6.17).

Standard timber connectors are sized for the widths of standard sawn timbers and glulams. But if the goal is to mortise the members so that the connectors end up flush with their faces, several issues must be addressed. First, the dimensions of the mortised portion of the member must meet the code-mandated minimum dimensions for heavy timber. Second, the edge distances for any bolts must be measured from the mortised surfaces. Third, the depths of the mortises must relate to their locations: at seats, to match the connector's thickness; at sides, to match the connector's thickness plus room to ease installation and accommodate expansion. (See Fig. 6.18). Fourth, a detailed sketch of what is now a custom connector must be transmitted to the manufacturer for pricing, and the manufacturer must furnish a shop drawing of it for review, prior to fabrication. (For more on shop drawings, see Chap. 12.)

Drilling and Notching

A number of common building details require that the framing be drilled or notched. These include "bird's mouths" where rafters bear on columns and beams, seats for ledgers, and provisions for suspended loads. Most of these conditions involve connections of one sort or another, so it is appropriate that they be addressed here. (See Fig. 6.19.)

Different parts of a beam play different structural roles. In a simply supported beam under uniform load, bending induces the highest compressive and tensile stresses in the top and bottom of the beam, respectively. The highest shear stresses arise near the ends of the beam, and the highest bearing stresses, in the vicinity of the bearings. Just as holes for bolts reduce a member's effective cross section, notches in the top or bottom portions reduce its effective depth. Notches on the tension side are more problematic than those on the compression side, because they induce tension perpendicular to grain, which can lead to splitting starting from the corner of the notch.[21] For this reason, notches in glulams, as a percentage of member depth, can be as high as 40 percent on the compression side, but are limited to 10 percent on the tension side. But regardless of whether the notch is in the top or bottom of the beam, it should be near the ends, because fiber stresses are lower there than at midspan.

The zones of highest shear are near the ends of a member, at mid-depth. Therefore, holes for mechanical and electrical components should avoid these areas. Elsewhere along the beam, the AITC recommends that the diameter of horizontal holes in glulam beams be limited to the lesser of 1.5 in (3.75 cm) and 10 percent of the member depth, that there not be more than one hole per 5 ft (1.5 m) of member length, and that their nearest edges be no less than 8 bolt diameters apart.[22]

Another condition that can induce tension perpendicular to grain— and that should therefore be avoided—occurs when a heavy load is suspended from the lower (tensile) portion of a beam. Suspension from the compressive portion might be okay, but if there is room, the best place from which to hang is the top of the beam.[23] (See Fig. 6.12.)

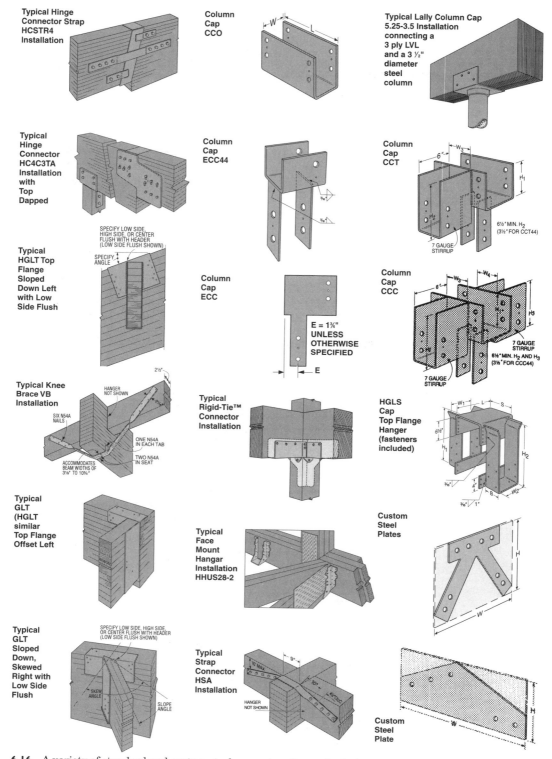

Typical Hinge
Connector Strap
HCSTR4
Installation

Column
Cap
CCO

Typical Lally Column Cap
5.25-3.5 Installation
connecting a
3 ply LVL
and a 3 ½"
diameter
steel
column

Typical
Hinge
Connector
HC4C3TA
Installation
with
Top
Dapped

Column
Cap
ECC44

Column
Cap
CCT

6½" MIN. H₂
(3½" FOR CCT44)

7 GAUGE
STIRRUP

Typical
HGLT Top
Flange
Sloped
Down Left
with Low
Side Flush

SPECIFY LOW SIDE,
HIGH SIDE, OR CENTER
FLUSH WITH HEADER
(LOW SIDE FLUSH SHOWN)

SPECIFY
ANGLE

Column
Cap
ECC

$E = 1¾"$
UNLESS
OTHERWISE
SPECIFIED

Column
Cap
CCC

7 GAUGE
STIRRUP

6½" MIN. H₂ AND H₃
(3½" FOR CCC44)

7 GAUGE
STIRRUP

Typical Knee
Brace VB
Installation

2½"

HANGER
NOT SHOWN

SIX N54A
NAILS

ONE N54A
IN EACH TAB

TWO N54A
IN SEAT

ACCOMMODATES
BEAM WIDTHS OF
3⅛" TO 10¾"

Typical
Rigid-Tie™
Connector
Installation

HGLS
Cap
Top Flange
Hanger
(fasteners
included)

Typical
GLT
(HGLT
similar
Top Flange
Offset Left

Typical
Face
Mount
Hanger
Installation
HHUS28-2

Custom
Steel
Plates

Typical
GLT
Sloped
Down,
Skewed
Right with
Low Side
Flush

SPECIFY LOW SIDE, HIGH SIDE,
OR CENTER FLUSH WITH HEADER
(LOW SIDE FLUSH SHOWN)

SKEW
ANGLE

SLOPE
ANGLE

Typical
Strap
Connector
HSA
Installation

W MAX

9"

10°

HANGER
NOT SHOWN

Custom
Steel
Plate

6.16 A variety of standard and custom steel connectors for use in timber construction. (Connectors manufactured by Simpson Strong-Tie Company, Inc. Printed with permission.)

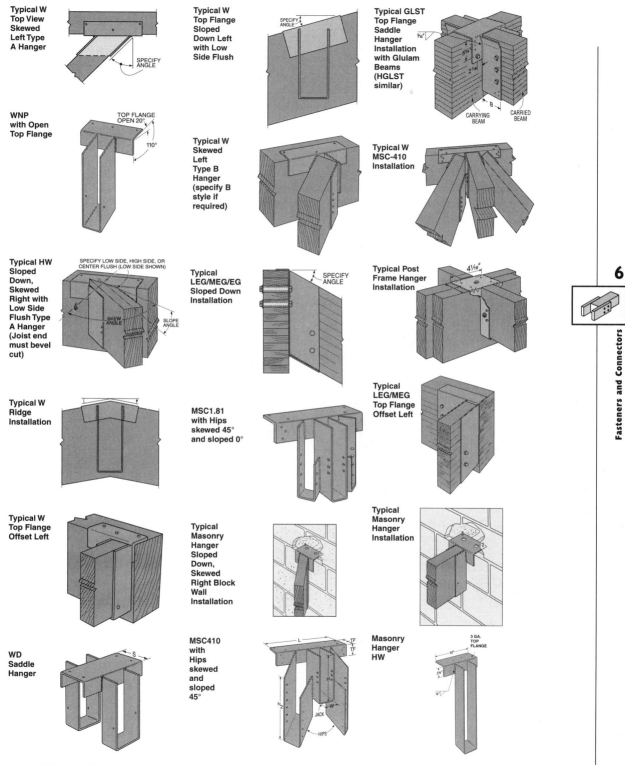

Typical W Top View Skewed Left Type A Hanger

SPECIFY ANGLE

Typical W Top Flange Sloped Down Left with Low Side Flush

SPECIFY ANGLE

Typical GLST Top Flange Saddle Hanger Installation with Glulam Beams (HGLST similar)

CARRYING BEAM

CARRIED BEAM

WNP with Open Top Flange

TOP FLANGE OPEN 20°

110°

Typical W Skewed Left Type B Hanger (specify B style if required)

Typical W MSC-410 Installation

Typical HW Sloped Down, Skewed Right with Low Side Flush Type A Hanger (Joist end must bevel cut)

SPECIFY LOW SIDE, HIGH SIDE, OR CENTER FLUSH (LOW SIDE SHOWN)

SKEW ANGLE

SLOPE ANGLE

Typical LEG/MEG/EG Sloped Down Installation

SPECIFY ANGLE

Typical Post Frame Hanger Installation

4 1/16"

Typical W Ridge Installation

MSC1.81 with Hips skewed 45° and sloped 0°

Typical LEG/MEG Top Flange Offset Left

Typical W Top Flange Offset Left

Typical Masonry Hanger Sloped Down, Skewed Right Block Wall Installation

Typical Masonry Hanger Installation

WD Saddle Hanger

S

MSC410 with Hips skewed and sloped 45°

L

TF

H

H₂

W

JACK

HIPS

Masonry Hanger HW

3 GA. TOP FLANGE

10"

2 3/4"

6

Fasteners and Connectors

6.16 (Continued)

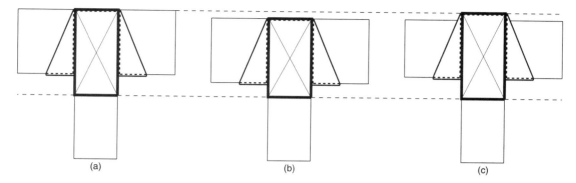

(a)   (b)   (c)

6.17   The impact of cross-grain shrinkage on member alignment at saddle hanger. (*a*) Before shrinkage, tops are flush. (*b*) Girder shrinkage leaves tops flush. (*c*) Beam shrinkage leaves girder proud. To avoid such problems, use well-seasoned material.

6.18   Pressure-treated glulam arch mortised for flush fit within custom steel shoe. The gap between the top edge of the shoe and the return of the mortise prevents bearing there. Diagonal rod bracing attaches to each side of shoe. This connection is shown in context in Fig. 9.10.

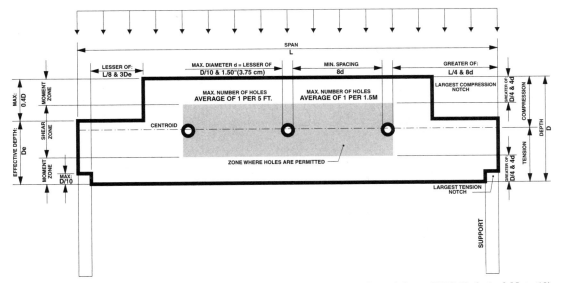

6.19 Allowable hole and notch locations and sizes in glulam beams (adapted from *AITC Technical Note 19*). Holes shown are in addition to those for connections. All holes and notches are subject to engineering review.

Connectors with deep pockets, such as beam and joist hangers, and column caps, usually have bolt holes in their side plates. The primary purpose of the bolts is to provide a positive connection between the member and the connector. Even though the seat of the connector is intended to carry the service load, shrinkage of the beam may leave it hanging by the bolts. This scenario—heavily loaded bolts trying to pull out of the top of a beam—is structurally equivalent to the scenario discussed in the foregoing paragraph, just turned upside down. To avoid a problem, the effects of shrinkage must be minimized. The simplest approach is to locate the bolts down near the seat, so shrinkage of only a small portion of the member depth is involved. If the bolts must be higher on the beam, then provide vertically slotted holes for them in the side plates, so that they are not restrained as the beam shrinks.

Summary

The loads applied to the members in a timber frame must eventually find their way to the foundations. Connections enable loads to be transmitted between timbers, either directly, or indirectly through intermediary devices. The most common intermediaries are timber connectors—split rings and shear plate connectors—and fasteners—bolts, carriage bolts, and lag screws. (Each type is subject to a set of adjustment factors similar to those applicable to framing and discussed in Chap. 5.)

6.20 Detail of three-dimensional trussing under church cupola. These connections were made with a large number of small bolts. St. Francis of Assisi Church, Raleigh, NC (Jon A. Condoret, architect.)

Bolted connections may include one or more bolts and two or more members. Bolts may be arranged in one or more rows, with one or more bolts per row. The total area available for bolting is a function of the diameters of the bolts and of the end and edge distances of each of the members. Bolt design values are a function of the spacing between bolts in a row and between rows, and of the number of bolts in a row. The higher the number of bolts in a row, the lower the design value of each. For economy and structural efficiency, it is usually better to maximize bolt diameter and minimize bolt quantity; for diversity, there is safety in numbers. (See Fig. 6.20.)

The design of a timber connection starts with a determination of the magnitudes and directions of the forces to be accommodated, with special attention to load reversals. The capacity of the connection will be a function of the species, orientations, cross-sectional dimensions, and arrangements of the members, and of the types, sizes, preparations for, and arrangements of the fasteners and connectors. (See Fig. 6.21.) The

6.21 Connection between round and rectangular members. The vertical plates only contact the column tangentially. Squaring the top end of the column would have enabled a more secure connection. This connection is shown in context in Fig. 5.15*b*.

appropriate fastener or connector for a particular joint will depend not only on the joint's required capacity, but also on its accessibility for installation. (Assembly of an especially complex connection may mandate a particular construction sequence.)

Most timber connections are subtractive—they require that material be removed—usually through drilling or mortising. In those cases, the structural adequacy of the remaining material—the net section—must be checked. Since the net sections of the members are usually lower at the connections than anywhere else, the design of the connections often controls the sizing of the members. While the arrangement of the framing determines the required capacity of each connection, the arrangement of the connection determines the available capacity of the framing.

There is an old saying to the effect that there are two ways to deal with a bridge that is too low: Raise the bridge or lower the water. Analogous reasoning applies to timber connections: Size the members to accommodate the reduced sections at the connections, or add enough

6.22 Bearing blocks bolted to timber column. Steel sideplates prevent lateral displacement of diagonal braces. This connection is shown in context in Fig. 9.3.

meat to the connections so that they don't control the design. If the net section at a connection does control member size—more likely under tension and compression than under bending—it can be thought of as an opportunity rather than a constraint; the designer has the option of reducing the net section of the rest of the piece to match that at the connections. Classic examples are the cylindrical midsections of square-ended truss bottom chords.

Where bolting is a subtractive process, enlarging the connections is additive and is particularly appropriate at heavily loaded columns. Rather than notching a column to create a shoulder for floor beams, consider enlarging the column with wood bearing blocks. (See Fig. 6.22.) In connections, where perpendicular-to-grain bearing would otherwise control the design, consider employing steel bearing plates or connectors sized to spread the load and thereby reduce or the eliminate the problem.

Notes

1. Flexner, p. 760.
2. CWC, *Wood Reference Handbook,* p. 244.
3. CWC, *Wood Reference Handbook,* p. 229.
4. King, p. 140.
5. NDS, p. 37.
6. King, p. 112.
7. NDS, p. 39.
8. Eyebars in early iron suspension bridges were developed for similar reasons.
9. Apple, p. 115.
10. Apple, p. 116.
11. NDS, p. 48.
12. NDS, p. 38.
13. NDS, p. 39.
14. Faherty, p. 5.63.
15. Breyer, p. 627.
16. CWC, *Wood Reference Handbook,* p. 244.
17. NDS, p. 30.
18. NDS, p. 120.
19. NDS *Supplement,* p. 32.
20. NDS *Supplement,* p. 34.
21. *AITC, Technical Note 19,* p. 4.
22. *AITC, Technical Note 19,* p. 3.
23. TCM, p. 7–696.

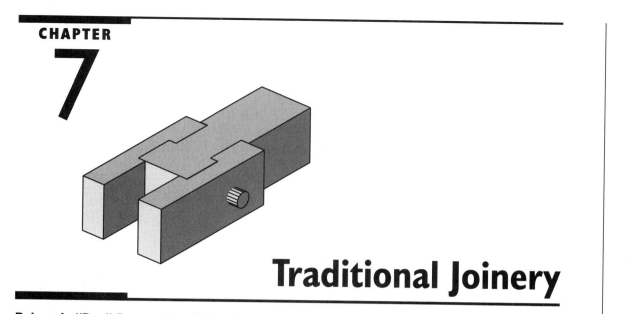

Traditional Joinery

Robert L. "Ben" Brungraber, Ph.D., PE, Engineer and Operations Director
Benson Woodworking Co., Inc., Alstead Center, NH

*The simplest form of joint is, as a rule, the strongest; complicated joints
are to be admired more for the ingenuity and skill of the carpenter in
contriving and fitting than for their strength of construction.*
FRED T. HODGSON, F.A.I.C., 1909

Introduction

The two timber connection methods that are currently most popular
with designers and fabricators are adhesives (in glulam members) and
metal mechanical connectors (bolts, lags, nails, shear plates, and split
ring connectors). There are other ways to connect timbers to one
another; some of these methods are thousands of years old; others are
being developed today. The older methods are termed *traditional joinery,* and are characterized by carefully crafted connections within the
members themselves, using mostly wood products in the process.

The pegged mortise and tenon joint is certainly the most common
and well known of the traditional joinery techniques. (See Fig. 7.1.)
That simple but elegant connection has been used in the construction
of thousands of remarkable structures, some of which would be a challenge to connect today with "more modern methods." The pegged mortise and tenon is still holding together a great many structures, some
of which we need to repair or to strengthen, or at least to evaluate for
structural capacity. Pegged mortise and tenons are also still being used
to connect a growing number of buildings, both residential and com-

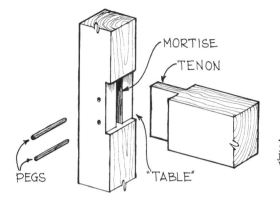

7.1 A simple mortise and tenon joint. (Brian J. Smeltz, Benson Woodworking Co., Inc.)

mercial. While it is the most recognized version of the traditional joinery techniques, it is far from being the only rendition. The family of "all-wood-joinery" schemes developed by dedicated and talented builders all over the world, over more than 2000 years, is even more varied and rich than one might suppose.

This chapter briefly covers the history of traditional joinery methods, various groupings of connections that use these methods, methods of evaluation, peg capacities and the future of traditional joinery methods in modern timber structures.

History of Traditional Joinery

We have been using wood to build structures for a very long time. The evolution of our timber connection methods was closely linked to the available tools and raw materials. The first primitive shelters were built by tipping or leaning branches against one another. A logical next step in the evolution of timber structures relied on "posts" embedded in holes dug with other branches. If all you had in your tool chest was a sharp stone, your first "timber connection" probably relied on vine lashing. Once you could chop wood, the more primitive forms of notching found at the intersections of log walls would be feasible. Those structures had a long run, until we had refined saws, chisels, and drills (or at the very least the brutally simple and amazingly effective spoon auger; see Fig. 7.2). Those powerful hand tools helped to make possible the incredible range of connection methods called *traditional joinery,* which dominated timber structures for a thousand years.

European

Europe has been described by at least one ugly American tourist as "big, old piles of stones you have to climb up in; and bells they insist on ringing at all hours." A great many of those stone buildings and bell

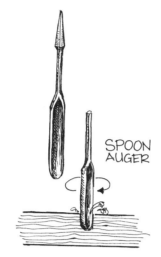

SPOON AUGER

7.2 For hundreds of years, until metal crafting skills could reasonably create a twist auger, the spoon auger was used to drill holes in timber. (Brian J. Smeltz, Benson Woodworking Co., Inc.)

towers have roofs that are framed with heavy timbers. Whether they were working with English brown oak to build cathedral roofs, or firs and spruces to build Norwegian stave churches (see Fig. 1.1), Europeans had a rich supply of timber, a pressing need for shelter from their harsh environment, and a teeming population of talented builders. Starting in about 500, Europeans were crafting permanent structures with traditional joinery methods. Their snow loads were significant, and their use of quite massive timbers reflects the conservatism inspired by uncertainty; at least the survivors often display that "overbuilt look." Their use of massive timbers (and usually hardwoods) in relatively small structures meant that there was sufficient reserve capacity so that the earliest joiners could rely on the simplest mortise and tenon joints. By the time Europeans were building structures large enough, with more carefully sized members, and for which simple mortise and tenon joints were not always adequate, they had developed iron to use in specific ways to reinforce critical joints. Heavy-timber framing, however, never lost its dominant position in Europe. They still use relatively heavy timbers in the roofs of nearly all their homes. The Europeans use amazing glulam techniques to build beautiful structures, ones that would almost certainly be built with steel or concrete in North America. With some exceptions—the English obsessed over some joints; the French delighted in incorporating chaotic timbers, and the Germans pushed the limits of complexity—Europeans did not need, nor did they develop, the more complex and sophisticated types of traditional joinery that involved splines, concealed necks, and keys. Those elegant expressions of the timber crafting art were developed even earlier, but on the other side of the world.

Asian

The Japanese were using fine woodworking techniques to build remarkable structures 2000 years ago. And they had learned those techniques from the even older Chinese traditions. (Even now, after working in traditional timber structures for 20 years, I look to thousand-year-old Asian work for inspiration. I have learned that if I choose to ignore them and just try to invent my own "new joint," I stand a very good chance of later finding that I have only just caught up with some thousand-year-dead Japanese craftsman.) Tradition has it that some of the older pagodas are 900 years old. It has long been the practice for the Japanese to totally rebuild some of their sacred structures every few decades, if only to pass on the requisite skills to another generation of craftsmen. While these buildings are in their original form and location, they are like the old family ax, great grandpa's— the one that's had six new handles and two new heads since he bought it.

The Japanese forests contained firs and cedars in abundance. The Japanese carefully managed their forests to reliably and sustainably yield prime material for their spectacular religious shrines and monarchical palaces. Their homes and trees tended to be smaller, and their taste perhaps a bit more refined than in Europe. Smaller timbers required more care to connect, and the immense number of sophisticated joinery designs they developed range from the sublime to the nearly outrageous. (See Fig. 7.3.) The timber framing industry also continued to flourish, uninterrupted, to the present. Annually, 45,000 companies in Japan cut the timbers for a million new homes.

North American

Given the primeval forests they found here, it is no surprise that heavy timber was the material of choice for the first North American colonists. Given our history, it should not surprise that the colonial building methods were based in European technologies. There remain a great many older North American heavy-timber bridges, barns, homes, and churches that are joined with pegged mortise and tenon connections. Timber framing, as it is often called, remained the dominant building method in the more populous areas of North America until the late nineteenth century. Unlike timber framing in Europe and the Orient, which has seen continuous use, American timber framing underwent a significant lull. The incredible pressure to build ever more housing and cities for the influx and continental spread of our immigrant population, as well as to do it faster with less material, and with less-skilled labor, absolutely demanded a new method. Industrialized sawmills and wire nail machines

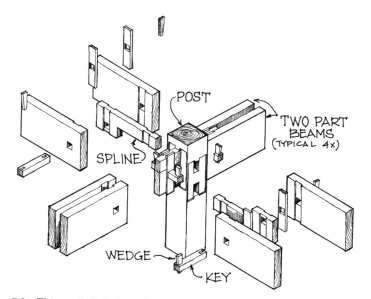

POST

TWO PART
BEAMS
(TYPICAL 4x)

SPLINE

WEDGE

KEY

7.3 This exploded view illustrates a Japanese connection between four beams and a post. It is not even the most complex way to build this joint. (Brian J. Smeltz, Benson Woodworking Co., Inc.)

made the balloon and platform framing techniques, developed in the late nineteenth century, feasible and much faster and cheaper than timber framing.

Timber framing in North America lay dormant for nearly a century, until an apparently unlikely revival of the methods and the craft began in the early 1970s. This rebirth of North American timber framing was inspired by a young generation's search for quality in the process and in the product of its work. That search took them to precedents found in older New England buildings. Since those structures were based in European traditions, their "new timber-framed structures" tended to be similar to those built in Europe 200 years earlier. The new timber framing has changed, however. The small businesses have gotten larger; the small do-it-yourself homesteading client has given way to the multimillion dollar commercial builder; the average framer has gotten older; the Asian traditions have informed the techniques, and the tools have gotten bigger and more powerful. The modern North American timber-framing industry numbers 400 companies, builds 4000 buildings a year, and is growing at least as fast as the rest of the construction industry. Its buildings express the massive and reassuring structure and warmth of timber, and the joy of exposed and obvious craftsmanship. The newest forms of timber-framed homes, with their separated, accessible, and unviolated structure, insulation, and services promise to be better suited to the twenty-first century than do entwined and aggravating stud-built homes. Timber framing is back in North America again, perhaps this time to stay.

Problems, Compromises, and Benefits with Traditional Joinery

We have already mentioned the skilled labor costs associated with crafting traditional joinery connections. There can be added material costs associated, as well. The very nature of traditional joinery is to remove material from the receiving or mortised member. Cutting a housing and a mortise and drilling some peg holes, all at the same cross-section, is a bit like trying to saw the member off in an inefficient, if elegant, way. Some members are sized simply to allow the timber to survive cutting the joinery with sufficient net capacity, rather than to handle the forces the basic analysis indicates are present. Even when the connections determine the member size, they can still control the structure's capacity. It is nearly impossible to cut a traditional connection that is as strong as the members connected.

There are other offsetting features associated with traditional connections. They are often lovely; much more attractive than galvanized hangers, for instance. (See Fig. 7.4.) They can be strong, stiff, and

7.4 Some traditionally connected timber structures serve solely as sculpture. This 40-ft (12-m) octagonal battered tower sits in the middle of the fifth floor of the Western Reading Room of the recent addition to the Denver Public Library. The "Denver Derrick" represents the western imagery of an oil rig, supporting a wagon wheel on the ceiling of the round room. Architect: Michael Graves. (Benson Woodworking Co., Inc.)

quite good at absorbing deflection under load—a great benefit in resisting seismic forces. Traditional joinery only rarely uses steel parts. In untempered environments, moisture can condense on steel surfaces, causing rust on the steel and decay in the contacting wood. The thermal expansion compatibility of similar materials is one reason traditional joinery can last so long, in some cases for many centuries. And, while cutting joints can be labor intensive, traditional joinery in the hands of skilled and well-equipped craftsmen, can compete economically with more modern and "sophisticated" connection methods.

Shrinkage

A carefully crafted joint can look like furniture when it is first assembled, and something quite primitive after the connected timbers have dried, shrunk, and twisted. Abutting faces can be separated by the timbers' shrinking, producing an unsightly result. This movement can have more than aesthetic impact on the joint's behavior.

A tenoned girt into a post mortise can lift up off the bearing surface and hang on the pegs, if it is lightly loaded and the pegs are placed high on the tenon. (See Fig. 7.5.) This problem is similar to that described in Chap. 6, where bolts are installed too high on a beam in a hanger. One way to minimize this preloading of pegs through shrinkage, is to hold the pegs as low as possible on the tenon, close to the bearing surface.

Because wood shrinks a lot across the grain, but very little along the grain, bearing surfaces cut at an angle to the grain will distort to new angles as the wood shrinks. This concentrates loading that was meant to be distributed, and can induce splitting when members are supported from their weaker points. The Germans advocate leaving a deliberate gap at rafter feet, so that the rafter does not bear at its point

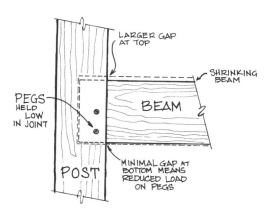

7.5 When pegging green timbers to a post, the pegs should be held as close as allowable to the lower edge of the beam. This minimizes the shifting of shear loads onto the pegs as the beams dry and shrink. (Brian J. Smeltz, Benson Woodworking Co., Inc.)

after it shrinks, causing the bearing surface to become more acute. (See Fig. 7.6.)

One way to mitigate the visual and structural impact of shrinking timbers is to fully house the joints. The mortised member also receives a housing, the size of the tenoned member. (See Fig. 7.7.) The tenoned member fits into this housing. A full housing requires that the mortised member be wider than the tenoned member. Having this "relief" in the parallel side faces of the mortised and tenoned members is much more forgiving than having the surfaces meet flush with one another. The housing can be designed to transfer shear forces from the tenoned member into the mortised one. The housing helps to prevent rotation about the longitudinal axis of the tenoned member as it dries in place. As the mortised member dries, a gap is not revealed as it would be at an unhoused, "butt fit" shoulder along the tenon. Fully housed joinery is an indicator of high-quality design. The costs of fully housing the connection include the increased labor and the fact that the tenons are shortened by the depth of the housing, all else being equal.

Tolerances

Specifying quality in a field that relies on craftsmanship as much as does traditional joinery can be a frustrating experience. Certainly, one quantitative measure of sound craftsmanship is cutting tolerances. What is feasible and reasonable; what is achievable, and what matters? To expect that joints will be cut to ¹⁄₁₆ in seems quite reasonable, but is overly simplistic. There are places where a ¼-in error will not show, nor matter. There are other places where the unforgiving human eye can discern a ¹⁄₆₄-in discrepancy in surface alignment. For aesthetic reasons, therefore, many framers try to avoid flush abutting faces. Not only can the eye see very tiny jumps in nearly flush surfaces, but very slight changes in alignment can be jarring. Try to bring narrower timbers into the sides of wider timbers, so that the faces do not align.

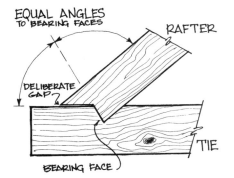

EQUAL ANGLES
TO BEARING FACES

RAFTER

DELIBERATE GAP?

TIE

BEARING FACE

7.6 The German building codes are quite clear about leaving a deliberate gap at rafter feet and similar angled cuts. The intention is to avoid inadvertently bearing point loads near the top of the rafter, as the timber dries and shrinks, making the cut angle more acute. (Brian J. Smeltz, Benson Woodworking Co., Inc.)

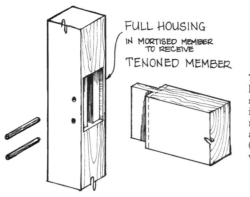

FULL HOUSING
IN MORTISED MEMBER
TO RECEIVE
TENONED MEMBER

7.7 A timber is said to be housed when its entire cross-section is let into a housing cut into the face of the receiving member. These housings have aesthetic and structural roles. (Brian J. Smeltz, Benson Woodworking Co., Inc.)

Member Topology Groupings

Members Parallel with One Another

Compression Splices

If two members actually abutted end to end and exclusively bore uniform compression forces relative to each other, then the required joint would be exceedingly simple and effective—two well cut cross-cuts bearing on each other. Even if one found a case where this simplest of joints would be adequate, the joint would be unnerving enough to passersby to require at least some capacity under reversed loading. But the basic principle still holds: End-to-end joints between timbers mostly in compression are the most straightforward of traditional joints. This is not to say that some very elaborate designs have not found their way into usage. (See Fig. 7.8.)

Most compression splices are derivatives of the simple lap joint. The joint is easy to cut and provides at least enough moment resistance to erect and to withstand incidental transverse loads. Old rules of thumb hold that the lap ought to be about 4 times the length of the larger of the two cross-sectional dimensions. The key to transferring compression forces is to get uniform bearing on the two pairs of faces. This means that pairs of faces ought all to be fairly perpendicular to the member axis and dead parallel to the matching face, so that they bear uniformly on one another. Getting these four bearing faces so accurately cut can be done with precise measuring and careful, machine-aided cutting.

Another, and time-honored, way to get all the faces to bear uniformly is through kerfing the mating surfaces at a splice. The two timbers are lapped and brought into bearing with one another; usually, at first, one pair of faces bears before the other and not uniformly. Next, the members are clamped and a hand saw is run down the gap at both bearing faces. Then, the two members are slid together by the width of this saw kerf, until they bear again, however imperfectly. The saw is run into the bearing surface gaps again. This slide and kerf sequence is repeated

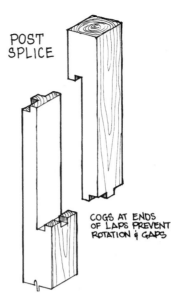

POST SPLICE

COGS AT ENDS OF LAPS PREVENT ROTATION & GAPS

7.8 This fairly elaborate splice between two timbers is designed to prevent relative movement between them as they dry and distort in place. (Brian J. Smeltz, Benson Woodworking Co., Inc.)

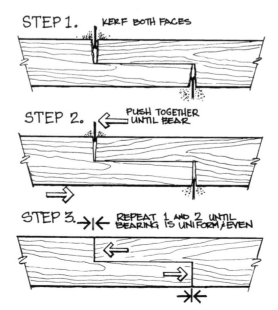

STEP 1. KERF BOTH FACES

STEP 2. PUSH TOGETHER UNTIL BEAR

STEP 3. REPEAT 1 AND 2 UNTIL BEARING IS UNIFORM/EVEN

7.9 When it is critical for the spliced members to bear completely and uniformly on each other, the simple lap splice is best. It gives maximum bearing area and allows the joint to be "kerfed" during cutting. This method of iterative fitting generates, reliably and simultaneously, two complete, uniform, and matched pairs of bearing faces. (Brian J. Smeltz, Benson Woodworking Co., Inc.)

until the saw leaves two uniform, kerfwide gaps that close perfectly. (See Fig. 7.9.) In very long members, with more than one of these splices, the splices are cut before the other joints are laid out, cut, and kerfed, so that the minor shortening as the result of kerfing does not affect the relative positions of the other connections.

Tension Splices

Most end-to-end member splices feel some tension forces, if only while they are being transported and erected. While the ability to bear tension parallel to the grain is among wood's strongest properties, making tension connections in timber has always been, and remains, the biggest challenge for builders using the material. As with compression splices, many tension splices are based on the simple lap joint. Interlocking and ornate versions of the lap joint number in the thousands and come from many areas and ages. Because the tension force is transmitted from one lap to the other, and not on the two bearing faces of the compressed lap, the intrinsic capacity of a lapped tension splice is limited to half that of the timbers themselves. And this is the theoretical maximum. The net section left after wood is removed for shear connectors between the two laps further reduces the splice capacity. The innate eccentricity in the load path between the laps can induce local bending failures that further reduce the joint capacity. (See Fig. 7.10.) The keying and interlocking found in the more elaborate lap joints are intended to counter the tendency for a lap joint to split and open up under tension forces.

There are various methods to transfer tension forces through shear connections between bypassing laps. One classic method is to neck the lap down even further, to less than half the cross-section, and to leave a "knob" on the end, which bears in compression on a similar knob left on the other lap. This effectively turns the tension connection into a compression bearing joint. Among the costs of this method is the increased necking, which further reduces the intrinsic capacity. Since the allow-

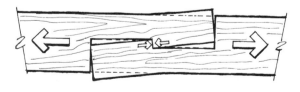

ECCENTRIC LOAD PATH

INDUCES BENDING, WHICH
CAN DISTORT AND SPLIT
MEMBERS

7.10 When loaded in tension, simple lap splices generate not only axial stresses, but also bending stresses, because of eccentricities in the load path between the lapped members. (Brian J. Smeltz, Benson Woodworking Co., Inc.)

(a)

(b)

7.11   Two examples of the many types of splice joints between parallel coaxial members. (*a*) Simple lap. (*b*) Complex lap. (Benson Woodworking Co., Inc.)

able bearing stresses are lower than the allowable tension stresses, the bearing area needs to be bigger than the net area of the necked lap. This means that the theoretical capacity of the members of this joint design family is even less than one third of the capacity of the members being spliced. Given the serious reduction this represents, and how much work it is to cut these joints, their great popularity with timber framers from many areas and centuries is a good indicator of just how difficult it is to tie timbers to one another, end to end. (See Fig. 7.11.)

Just about as soon as they could acquire it, timber framers everywhere used iron to selectively reinforce those joints that needed the most help. Tension splices were logical joints to reinforce. Hand-wrought pins were used in place of wooden pegs. (See Fig. 7.12.) Some really elegant forms were developed in the nineteenth century, when iron was still very expensive but the techniques to form it were already well established. Long span trusses, with their long tension chords and large member forces at the supports, saw a lot of innovation in the use of iron straps and rods. (See Fig. 7.13.)

Keyed Beams

Keyed beams are interesting examples of parallel timbers connected to one another. These represent some early attempts to assemble large and deep beams from smaller timbers. They can be seen as precursors of con-

7.12 One of the benefits of designing and building traditionally connected timber structures is the opportunity to celebrate the exposed structure. While almost any member can be fashioned from timber, another material is sometimes more appropriate. The chains from which this scallop-shaped loft is hung will be incorporated into its handrails. (Benson Woodworking Co., Inc.)

7.13 Wrought iron tension chord splice hardware for a covered heavy-timber railroad bridge, seen from above. (Benson Woodworking Co., Inc.)

temporary glulam beams. Simply stacking small beams on top of one another does not yield the structural capacity of a single large beam of the same overall depth. The layers must be shear connected to one another to generate composite action. The trick is to interconnect the beams without unduly weakening them. Many forms of shear connectors have been used to prevent sliding at the abutting planes. Because it is relatively simple, lovely, and effective, the most popular shear connector is the key. The shear forces on the keys are a function of the shear forces in the beam, the key's position along that beam, and the size and number of beams being keyed together. These issues are addressed with standard structural principles. The wooden shear key must be sized to resist crushing and shearing, while leaving enough material at the faces of the beams to resist the same forces. Furthermore, the beams must be clamped to one another to resist the prying action generated through the twisting of the eccentrically loaded shear keys. This clamping is usually provided by bolts or lags, which further increase the capacity for composite action, while further reducing the net section. (See Fig. 7.14.)

Another issue to consider in detailing keyed beams is the species and grain orientation of the shear keys themselves. The temptation is to fabricate the keys from material similar to that in the beams and to orient them with the grain running longitudinally, with the beams. Small pieces of wood, like typical keys cut with this grain direction, tend to be brittle. A much more common method to fabricate keys is to use a hardwood with high-compression strength perpendicular to grain and to run

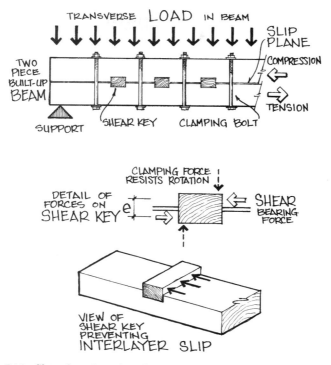

7.14   Shear keys have long been used to build up deeper beams from smaller timbers. They generate composite action between the separate layers of the beam by preventing interlayer slip through their shear resistance. (Brian J. Smeltz, Benson Woodworking Co., Inc.)

the grain in the keys horizontally and transverse to the beam's axis. This means that the key is loaded in rolling shear, not wood's strongest property. As in the case of knots (Chap. 2), keys are likely to shrink more than the pockets or notches made for them. Therefore, the keys need to be longer along the beam axis, which also helps to reduce their prying action. As in the spacing of bolts along the grain (Chapter 6), the cost is in making certain that there is adequate capacity left in the shear faces in the beams, between the keys. The actual degree of composite action generated by these keys is also an uncertain issue, governed by the relative stiffness and bearing of the shear keys.

Connections between Members at Angles with One Another

Mortise and Tenons

By far the most common and best known of the traditional joinery connections is the pegged mortise and tenon. The tenon is cut to fit in the mortise and is held there by peg(s). The tenoned member may also be housed into the face of the mortised member. Usually, these connections join members that are perpendicular to one another; but they are also used in other orientations, knee braces being one common example. Another common variation on the mortise and tenon is to carry the tenon completely through the mortised member. Once the tenon is exposed on the far side of the mortised member, other connectors can be used to hold the tenon in, through keys being the most common.

The pegged mortise and tenon connection has shear, compression, and tension capacity. A mortise and tenon can even provide some resistance to bending, but only in a tight-fitting connection. Because most connections are cut in freshly sawn timber, which will dry and shrink in service, *the reasonable designer does not count on moment resistance in mortise and tenon connections.* A well-executed housing can greatly increase the shear capacity of a mortise and tenon connection, because the entire cross-section of the tenoned member can then resist shear at the typically highly sheared end. If the tenoned member is not housed but simply butt fit against the mortised member, only the tenon cross-section can resist shear forces at the connection.

Through Tenons with Keys

Sometimes, very large beams need to be connected to large posts, and in such a way as to provide a great deal of resistance to tension. A classic example of this situation occurs at the connection of a beam to a post, where the two are being driven apart by the strong action of a brace helping to support the girt loads, as in the central aisle of the Dutch barn. The often immense anchor beams resisted the large forces generated in producing and storing hay. They also served as key elements of the barn's lateral load resistance, being knee-braced from below to the posts of the larger central aisle. Knee braces are far more effective in compression than in tension, because of their undersized tenons. Compression knee braces act to shear and to withdraw the braced girt from the post, at the connection that completes the triangle that has the knee brace as its hypotenuse. The tension induced in the girt at the post can be significant, as can the bending in the post caused by the bracing action. The keyed through tenon—with its long, through and keyed tenon—generates tremendous withdrawal resistance in the girt but without further weakening the post with peg holes. (See Fig. 7.15.)

Knee Brace Joints

I draw two distinctions between so-called *post-and-beam* structures and *timber-framed* ones. The first difference is in their connection methods. Post-and-beam structures can be characterized as "lapped

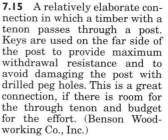

7.15 A relatively elaborate connection in which a timber with a tenon passes through a post. Keys are used on the far side of the post to provide maximum withdrawal resistance and to avoid damaging the post with drilled peg holes. This is a great connection, if there is room for the through tenon and budget for the effort. (Benson Woodworking Co., Inc.)

and stacked"—all the connections being cut with a radial arm saw and the pieces held to each other with lags, steel angles, and straps. Timber-framed structures use the more traditional interlocking connections described in this chapter. The other big difference between the two heavy-timber structure types is in their lateral load-resisting elements. Post-and-beam structures rely on the building's sheathing—either panel products or diagonal boards—to resist the lateral loads. The sheathing is attached to the timbers, which resist gravity loads only. Timber frames, on the other hand, are built with knee braces to keep the frame upright in the face of wind and earthquake. The knee braces keep the framed corners square by acting as axial members themselves and inducing bending and axial forces in the members to which they are connected.

Most knee braces are oriented at 45° to the frame's principal members. This is the most efficient position for resisting reversible loads and the most effective for a given brace length. It also makes the joinery identical at both ends and the brace itself reversible. This angle means that the most easily cut mortise, perpendicular to the timber's surface, makes for a pretty short tenon in most knee brace depths. These short tenons really only have room for one small peg. The bearing capacity of the typical knee brace, however, is quite high. My testing and analysis has led me to conclude that knee braces in tension are only one seventh as effective as the opposing ones, in compression. While repairing, restoring, and remodeling old timber-framed structures, I have often found knee braces in which the "relish" has sheared out from beyond the peg hole. Timber-frame designers ought to make certain that the frame is up to handling the imposed loads, considering only those knee braces that are in compression. Among other things, this means that there ought to be opposing pairs of braces, to resist lateral loads from opposite directions.

One question that comes up regularly is whether knee braces help the frame's beams carry vertical (gravity) loads. The temptation is strong to assume that they do. It is hard to imagine that the knee braces do *not* help the beams to resist their gravity loadings. On the other hand, knee braces are framed into posts and timbers that are generally cut green. This means that they will shrink away from each other, and the knee brace bearing surfaces will separate from them, at least somewhat, as they dry. While knee braces certainly provide support, perhaps the best approach is to ignore them and appreciate their contribution to damping vibrations and reducing deflections. If the frame is being cut from dry, stable material, such as timbers cut from recycled old timbers, I will occasionally design the beams as though their spans were being reduced by the knee braces. At the most theoretical edge of timber framing, we might discuss whether to establish the lengths of the braces based on the undeflected or deflected profile of the beam.

Timber-framed knee braces that are against exterior walls are at the core of a very interesting topic—the relative load sharing between frames and their cladding. Modern timber-frame buildings are often clad in foam core panels. These panels have OSB (oriented-strand board) as the outside skin, rigid foam as the core, and gypsum as the interior skin. Many builders are even using OSB as the interior skin, as well, with these so-called structural panels being attached to the timbers through an intervening layer of gypsum board. The skins are usually bonded to one another with splines at the panel butt joints, making the panel skins a uniform layer, broken only by windows and doors. This is a very efficient cladding system, both in the labor required for its installation and in its eventual thermal performance. It is also potentially very strong and rigid in resisting in-plane shear loads. Among the problems involved in engaging the OSB in resisting lateral loads is the distance from the frame to the *outer skin,* and the presence of the intervening layer of gypsum board finish between the timber frame and the *inner OSB skin.*

The most damning description of timber framing that I have heard was made by a midwestern architect who dismissed the newly reborn craft as "expensive wallpaper." If this assessment has any validity, it is in describing those knee braces that are up against the paneled exterior walls of two-story buildings. Although I suspect that exterior wall knee braces never feel much load, the testing and analysis required to characterize and model the load sharing between a three-dimensional timber frame and the attached and perforated panel walls promises to remain at the edge of research for a few more decades. For now, I either analyze the frame as though it had tarps stapled to it in the wind, or detail the panel attachment to the frame to be positive and reliable—as might be required by most building codes.

Dovetails

Pegged mortise and tenon connections are the most common joints for those members that are perpendicular to one another and that share vertical planes. These are generally the major members of the timber frame. The secondary members in the floor and roof planes are supported from other members in their planes. The gravity loads, which are the major concern for designers of these members, generate forces that are transverse to the plane of the connected members. The simplest way for one member to support another against this transverse loading is simply to lay the supported member across the supporting member. This connection is tough to beat, so long as some connection is made to prevent the supported member from sliding or bouncing off its support. This method also reduces headroom in the structure or requires longer posts to maintain it. It also leaves "shelves," or the exposed tops of the supporting members, near the ceiling of the room below. The way to avoid these shortcomings is to frame the supporting and supported members with their tops flush with one another.

The two most common ways to frame timbers that lie in a horizontal plane, and have flush tops, are with dovetails and tusk tenons. The dovetail is arguably easier to cut and unquestionably easier to install than the tusk tenon. There are some significant structural penalties paid for this simplicity, however. A dovetail without a housing hangs solely on the dovetail itself. The shear stresses are magnified by the reduced cross-section and further magnified by cutting off the bottom of the dovetail. This increases the amount of material left below the supported member, in the support. It also increases the shear stress in the dovetailed member, by acting as a bottom face notch with its attendant shear stress magnification. (See Fig. 6.19.) The supporting member is also unduly compromised by cutting in the dovetail connection. A lot of material is removed to allow the dovetail to settle in from above, particularly if, as is often the case, dovetailed members are supported on opposite sides at the same cross-section. The thin rib of wood between the two dovetails at the top of the supporting member is prone to crushing or buckling. Some dramatic dovetails, illustrated in eighteenth-century books, actually cut all the way across the top of the supporting member, apparently to maximize the length of the dovetail. Some building codes explicitly preclude any notching at all in the top surface of the middle third of any bending members. This prohibition alone is enough to make designers consider the alternative, the tusk tenon.

A tusk tenon does not have to be located exactly at the bottom of the supported member, but it often is. If the supporting member is sufficiently deep, this means that the mortise to receive the tusk tenon is removed near the neutral axis of the supporting member. Like holes in the webs of steel beams, these mortises exact little penalty from the notched member. The commonly cut housing that goes with tusk tenon joints does not remove much material and can even be tapered to zero at the top of the members. The pegs to hold the tusk tenons are optional and easily replaced with concealed straps crossing the top of the joint. Given similarly sized members, a pegged tusk tenon supports the member better and does less damage to the supporting member than the unpegged version, even though it removes more material. If the tusk tenon is not located at the bottom of the supported member, the issue arises as to how the member is supported—on the tenon with its shear stress concentration considerations, or on the housing. The safe thing to do is to overcut the underside of the tenon to assure that the entire member bears on the bottom of the housing. The tenon can be seen as a safety, if the housing crushes or breaks away. Installing a tusk-tenoned member involves spreading the supporting members and, then, squeezing them together over the tenons, while the tenoned member is temporarily supported in place. Assembling buildings that are framed with tusk tenons, therefore, requires detailed preplanning and an adequate number of temporary supports.

These assembly issues can be tough enough to drive the designer to dovetails, even with their structural compromises. There are better ways to cut dovetails than the traditional European methods, previously described, imported by our colonists. The Japanese still cut them, with their highly mechanized methods, but cut them very large and shallow, doing minimal damage to the supporting members and leaving the supported members' cross-sections nearly intact at their supports. An even simpler method is to butt cut the supported member and support it in a deep enough housing to provide the required support, even after any expected shrinkage. These joints should be supplemented with steel straps or toe-nailed spikes, to prevent the supported members from sliding out or lifting up in earthquakes.

Individual Pieces

Pegs

Pegs, or trunnels, are the most common element of traditional timber joinery. Most systems of all-wood joinery have relied on pegs, although the European systems more so than the Oriental ones. Modern pegs are usually turned into round cross-sections, although they can be ripped into polygonal cross-sections. Some early pegs were reeved, or split from branch wood. Pegs have also been "extruded" by driving roughly split pieces through sharpened holes in metal plates. The most common diameter has always been about 1 in (25 mm), although structural applications have been found for pegs from ¾ (18 mm) to 3 in (76 mm) in diameter. Pegs are usually hardwood, and often oak. Early pegs were cut from dense branch wood, but most modern pegs are made from carefully dried trunk wood. Pegs are typically very high-quality material, as close to the "small, clear specimens" used to establish basic material properties for a given wood species as one can find in structural applications. Pegs are simply too important, too small, and too cheap for a builder to consider messing around with low-quality material.

Many early framers left the tapered pegs full length in their holes, with enough left to hang coats and carcasses on. It is still a good idea to leave the pegs sticking out, or proud, in finished buildings. If they are cut flush, even carefully, a lot of visual damage is easily done to the surrounding finished timber faces with the saw. As the timbers dry and shrink across their grain, the pegs do not dry, nor shrink along their length. Pegs sawn off flush, therefore, tend to end up proud of the surface anyway. But they are not very proud, and the minor variations in shrinkage, and peg exposures, can detract from the overall quality of the framing. Some framers precut their pegs to leave about a half inch showing at either end, and even prechamfer their pegs to improve their appearance in the finished structure.

Round pegs are usually driven into holes of only slightly smaller diameter. A few thousandths of an inch will make a peg "tap in" satisfyingly and keep it in place during erection. Timber framers have also long argued about drawboring their pegs—deliberately misaligning the peg holes in the mortise and in the tenon to prebend the peg as it is driven into place. The hope is to induce a tendency for the bent peg to pull the tenoned member into the mortised member as the timbers dry. This is intended to mitigate the gaps that open in joinery as timbers shrink after assembly. It is a delicate art; too much misalignment can induce premature failure of the peg or splitting out of the relish beyond the peg in the tenon. Inadvertently misaligning the holes in the wrong relative direction will tend to force the connection apart. I am not convinced that the practice is worth the effort spent on it. The only certain thing about drawboring pegs is that people will still be getting worked up over it, pro and con, for years; much as the topic has already been debated for centuries.

How strong is a peg? Not as strong as the same-sized bolt, but "stronger than you think." The absence of pegs from the prevalent North American building codes is probably the clearest single challenge to the designer of traditional joinery. Pegs were being replaced by other connection methods just as standardized building codes were starting and industry testing and scientific research methods were first applied to building methods. With the exception of a very few articles and pilot research programs, little has been done in North America since that time to establish peg capacities and behaviors. Recent European and eastern publications address some specific and different connections, but are written in a different language and measurement system (i.e., metric).

The effort involved in getting pegs "accepted" by our building codes seemed pretty daunting, at least until the release of the NFPA's 1991 *National Design Specification for Wood Construction*. That was the first edition of our "timber code" to embrace and use the European Lateral Yield Theory model for establishing the capacity of mechanical, "dowel" type connectors. Basically, its equations model individual failure mechanisms for lag screws, bolts, nails, and screws. The designer plugs in the wood species, connector diameter, and thickness specifics into the equations. The minimum capacity of the germane failure mechanisms controls the allowable design loading. Accepting this model put North American timber engineering in tune, not only with much of the rest of the engineering community, but with twentieth-century scientific methods. It also opened the door to fairly straightforward determination of peg capacities and eventual acceptance of them in the building codes. The smooth bearing surfaces of pegs function much like the unthreaded portions of bolts. (See Chap. 6.)

Previous NDS editions had simply published tables for allowable loads on bolts, lag screws, nails, and screws. These inflexible tables were loosely based in equations, backed up by fairly wildly extrapolated results of tests made in the 1930s. The tables were popular with many designers, however, if only for their simplicity. The new method allows designers to model infinite combinations of species, thicknesses, and connector diameters. Designers can even plug in different bolt stiffnesses. Why not plug in connector stiffnesses, which reflect those one might expect in a wooden dowel or peg? It remains only to establish what those parameters are, not a trivial matter. The equations include the yield stress of the connectors, not a value as easy to describe and establish in wood as in most steels. There are also indications that the inherent differences in the interaction between a wooden dowel and a hole in wood and the same hole filled with steel may mean that some additional failure modes will need to be investigated.

This intrinsically different behavior for pegs, compared to bolts, includes their initial stiffness. Pegs are not as stiff as bolts of the same diameter; they are made from material that is innately more flexible. However, with their slight flexibility and with their thermal and moisture responses similar to the wood members into which they will be driven, pegs can be driven into snugger holes than bolts. This means that the pegs take load earlier, with less "slop" take-up than with bolts. The reduced stiffness of each connector, the tighter holes, and the practice of drilling all the holes for timber connections *after* the members are aligned, means that load sharing is likely to be much more even for a group of pegs than for a group of bolts of the same diameter.

The testing and research required to expand the European Lateral Yield Theory to include pegs is already underway at a few universities in North America. The results and knowledge are expanding as I write. My best advice is to seek out the current state of this program, either through the Timber Framer's Guild of North America or the AFPA. At the certain risk of being dated, I offer an example of the current state of the art of using the European Lateral Yield Theory to determine the capacity of a peg. (See Fig. 7.16.)

7

Tenons

A pegged mortise and tenon consists of these three elements: pegs, a mortise, and a tenon. We have already discussed the peg and calculated its capacity. Mortises can be the weak link of the three, but that is very much the exception. Tenons, on the other hand, are prone to structural problems and deserve some attention. Specifically, the ends of tenons are prone to shearing out on the far side of the peg holes. The basic issue is the required end distance in the tenons, past the peg holes. The loaded end NDS requirement for bolts is 5 diameters for tension in hardwoods and 7 diameters in softwoods. Using the common 1-in (25-mm)-diameter peg and the 4-in (100-mm) tenons that fit in the standard 8 × 8 in (20 × 20 cm) post, there is no way to meet the NDS specified end distance, even in hardwoods. (In hardwoods, a 1-in (2.5-cm) peg would require a 4-in (10-cm)-edge distance in the mortise and a 5-in (12.5-cm)-end distance in the tenon; in other words, a 9-in (22.5-cm) tenon would be required in a member only 8 in (20 cm) wide.) The question, of course, is whether the minimum dimension with steel bolts is sacred with wooden pegs. The NDS allows for modest reductions in the end and edge distances, with commensurate reductions in the allowable loads on the bolts. It seems reasonable to allow these reduced space constraints when using pegs, since we only

DOUBLE SHEAR   LOAD DURATION FACTOR = $\boxed{1.00}$

MAIN MEMBER (TENONED MEMBER)

SPECIES $\boxed{\text{E.W.PINE}}$

$F_{e\parallel}$	4,050	psi
$F_{e\perp}$	1,400	psi
t_m	2.00	inches
load angle	0	degrees
$F_{em} =$	4,050 psi	

SIDE MEMBERS (MORTISED MEMBER)

SPECIES $\boxed{\text{E.W.PINE}}$

$F_{e\parallel}$	4,050	psi
$F_{e\perp}$	1,400	psi
t_s	3.00	inches
load angle	90	degrees
$F_{es} =$	1,400 psi	

PEG(S)

SPECIES $\boxed{\text{R. OAK}}$

F_{yb}	9,000	psi
$F_{c\perp}$	820	psi
diameter	1	inches

DERIVED PARAMETERS

R_e =	2.89	
K_o =	1.25	
k_3 =	0.87	

NOMINAL PEG VALUES

8.3-1: $Z_1 = Dt_m F_{em}/4K_o$ 1,620
(tenon crushes)

8.3-2: $Z_2 = Dt_s F_{es}/2K_o$ 1,680
(mortise crushes)

8.3-3: $Z_3 = k_3 Dt_s F_{em}/1.6(2+R_e)K_o$ 1,080
(peg snaps in tenon)

8.3-4: $Z_4 = D^2[2F_{em}F_{yb}/3(1+R_e)]^{1/2}/1.6K_o$ 1,249
(peg bends)

8.3-"last": $Z_5 = MIN(t_m, 2t_s)DF_{c\perp}$ 1,640
(peg crushes)

SPACING REQUIREMENTS

L/D =	2.00	
k_3 (steel) =	1.59	
8.3-3 (steel) =	1,977	pounds
8.3-4 (steel) =	2,793	pounds
BOLT CAPACITY =	1,620	pounds
CAP_{peg}:CAP_{bolt} =	67%	

LOAD ∥ TO GRAIN

EDGE	rec.'d	1.5D	1.50 inches
END:			
comp.	min.	2D	2.00 inches
"	rec.'d	4D	4.00 inches
"	dia. req'd.	3.33	3.33 inches
tension:			
hardwood	min.	2.5D	2.50 inches
"	rec.'d	5D	5.00 inches
"	dia. req'd.	4.17	4.17 inches
softwood	min.	3.5D	3.50 inches
"	rec.'d	7D	7.00 inches
"	dia. req'd.	5.83	5.83 inches
PEGS IN A ROW			
	min.	3D	3.00 inches
	rec.'d	4D	4.00 inches
	dia.req.'d	3.67	3.67 inches
BETWEEN ROWS		1.5D	1.50 inches

LOAD ⊥ TO GRAIN

EDGE			
loaded	min.	4D	4.00 inches
unloaded	min.	1.5D	1.50 inches
END	min.	2D	2.00 inches
	rec.'d	4D	4.00 inches
	dia. req'd	3.33	3.33 inches
PEGS IN A ROW			
	min.	1.5D	1.50 inches
BETWEEN ROWS			
	min. dia.	2.50	2.50 inches

CAPACITY PER PEG =   1,080   pounds

7.16   A spreadsheet illustrating the use of NDS Lateral Yield Theory equations to calculate the shear capacity of a wooden dowel. This method is not believed to have been approved yet in any building codes. (Benson Woodworking Co., Inc.)

rarely encounter a peg that is loaded as heavily as the same-sized bolt. There are several reasoned ways to arrive at a set of geometrical constraints that will generate the peg capacity, but none is yet code approved and all will certainly be revised before they are.

Rods and Straps

From medieval cathedrals to industrial revolution buildings, heavy-timber structures have long used rods and straps, particularly in long trusses. (See Fig. 7.17.) They are much preferred over simple gusset plates and through bolts. That particular connection method is a carryover from steel structures. But wood is not isotropic; it is very susceptible to tension perpendicular to the grain. Those forces are easily generated in plated connections because of tendencies for the gusset plates to rotate under eccentric loading. Other problems involve timbers that shrink in position and induce splits. Finally, even when protected from weather, temperature fluctuations, in the presence of natural humidity, can condense enough moisture against the timber surface to precipitate decay.

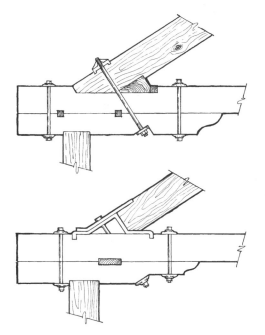

7.17 A few of the hundreds of relatively standardized connection details used for heavy-timber trusses during the early Industrial Revolution. These connections were often patented and marketed by the mass producers of the cast iron hardware involved. (Brian J. Smeltz, Benson Woodworking Co., Inc.)

Splines

Splines, or "free tenons," replace two opposing tenons pegged into a post with a single spline that passes through the post and is pegged to the opposing timbers. They are often used when four timbers need to connect to a post at the same point, with a high spline between one pair of timbers passing through the post above the low spline between the other pair. Splines were very rare in Europe and, therefore, in early New England and recent North American timber framing. Splines were more common in the East, with its more elaborate joinery. Splines have gained in popularity with North Americans, because they make so much structural sense. There are quite a few good reasons for increased use of splines in North America. Splined joints can be stronger than the simple mortise and tenon joints they replace. Splined timber frames can be easier to fabricate and to erect. Finally, many see aesthetic opportunities with splined joinery, which can not be exploited with mortise and tenons. (See Fig. 7.18.)

Splined joints are stronger than the mortise and tenon joints they replace because the members are better connected to one another and the post is not weakened as much as with mortises and peg holes. The members are connected to one another with a spline, which can be of

7.18 A connection illustrating the use of a through hardwood spline. These splines are pegged to horizontal members on all sides of the post, and to the post itself. If horizontal members are opposite one another, the spline can be continuous, and not pegged to the post. This connection method offers many structural improvements and can look better than a pair of mortise and tenon joints. (Benson Woodworking Co., Inc.)

hardwood, even in a softwood frame. When the members are loaded to withdraw from the post, the pegs are loaded parallel to the grain in both spline and member, which is a much stronger direction than perpendicular to the grain, as they are when pegged into the post. Since the spline fits into a slot in the members, end distances in both the spline and the member are readily available in lengths that meet NDS specified minima. Because the post is only housed for the timber, and through-mortised for the spline, peg holes in the post can be avoided. In connections where four members encounter a post at the same elevation, the damage done to the post to house, mortise, and peg the four members can be quite severe, and the pegs are often crammed into the post and near the tenon ends so tightly that installation is not worth the effort. Although splines are mostly relied on for their tension capacity, they can also be detailed and installed to help resist shear forces at the member ends. This potential shear capacity of the spline can be fairly important, because splines located at the bottoms of a member do reduce the bearing area available at the member end, for a given housing depth.

One reason European framers did not use splines was the tooling and labor required to build them and to cut the slots into which they fit. If you are working with hand saws, augers, and chisels, it is a lot easier to cut a tenon and excavate a mortise than it is to mill out a spline and cut a long trench. Modern heavy woodworking equipment makes spline production quite straightforward. Timber framers have adapted or built tools to cut the large slots in the ends of heavy timbers to install the splines.

Frames that rely on splines are often much safer to assemble and erect than those with tenons, for a couple of reasons. Members with long tenons on both ends need to be installed while the receiving members are spread apart. For subassemblies prebuilt and erected with cranes, this is not generally a problem. The challenges arise when the crew is trying to install timbers off the ground, between tall floppy assemblies that are spread far enough to admit the tenoned member. Splined members, on the other hand, slide into place if the available length is just a bit longer than the housed length. Furthermore, if the spline is a lower spline and is already in place, it forms a holding bracket for the next timber being installed, which makes for very safe and convenient raisings. Of course, some timbers have splines in their tops as they intersect at the posts. These timbers are held in by the housing alone, not a safe situation without supplementary support, before the structure is pegged. Timber framers try to design their joinery so that the high splined timbers are preassembled and pegged before being raised with cranes. As with any complex assembly, careful planning is needed to avoid framing yourself into a corner during erection.

Finally, splines can add a great deal of expressed craftsmanship to a structure. From the viewpoint of the occupant, a pegged mortise and tenon joint, up against an exterior wall, could just as easily be a simple housing and square-cut timber toenailed into the post and perhaps reinforced with a steel strap on the outside. There is no way to tell the difference, unless the pegs poke through. Splines, like pegs, almost always show. Since flush is such an unforgiving tolerance, many framers leave their splines, like their pegs, proud of the beam surface, chamfering and even carving the parts of the spline that are left revealed in the finished structure.

Splined timber-frame joints can make a building stronger, cheaper, safer, and more lovely when they are completed.

Summary

Basic Principles of Sound Joinery Design

This chapter has already covered some preferred practices in traditional joinery. For examples: Housed connections are generally better than butt fit joints, and splined joints can be much superior to the mortise and tenon type. There are other issues less clear cut (pun intended) but at least as critical, to weigh in designing good traditional joinery. Designing traditional joinery is all about the art of compromise. A stronger joint has thicker tenons and more and bigger pegs, but this weakens the mortised member, perhaps to the structure's detriment. The real art in designing traditional joinery lies in striking the right balance among the mutually exclusive options. (See Fig. 7.19.) The strongest joint is not worth much if it breaks the weakened members.

When Using Steel, Avoid Concealed Bearing Faces or Concealed Threaded Connectors

The simplistic, all-inclusive steel gusset–plated connection, riddled with bolt holes, has not proven generally pleasing to my eyes. A common reaction has been for some architects and owners to allow that, while they understand how steel might be necessary, they don't want to see any of it. One approach is to invest some time and money on exposed steel when it is legitimately required. Another is to try to conceal steel fasteners or wooden bearing faces. Early authors warned against hidden bearing faces and against joints that require simultaneous and uniform bearing on more than one matched pair of bearing faces. The main issue for all is the uncertainty. If the bearing faces are hidden, how can the craftsperson be sure that they are bearing properly, let alone any person interested in inspecting the structure, either weeks or centuries after the craftsperson has decided that it is adequate? Concealed steel fasteners can be done safely and with taste. It is also possible to get too clever in trying to conceal steel and inadvertently create a nightmare. I

7.19 Knee braces are one defining element of traditional timber-framed structures. The braces provide resistance to lateral loads and contribute to gravity load resistance. They can also add interest to a building, as do these two interwoven planes of knee braces. (Benson Woodworking Co., Inc.)

detailed a seemingly not-too-complex assembly that had a custom nut concealed in beams hanging above a remarkable great room. The special nuts were threaded for a different diameter from each end. I was *right there* when the suspended beam suddenly dropped out of that nut, even after I had *felt* the bolt fully engage it. I am uncertain, to this day, what happened. And that is the problem with concealed connections; nobody can see what is happening.

Spread Out Joints (Compromise with Eccentricity)

One of the toughest aspects of designing traditional joinery is trying to fit several timbers into the same place. Sometimes, particularly in the compound joinery of intersecting roof planes, designers can feel as though they are trying to jam an awful lot of angels onto the head of a pin. It is possible, for instance, that the post top at the bottom of a valley rafter can be the locus of six intersecting members, all trying to land in one spot and to be joined to one another. These are the toughest connections in any structural medium, but can seem nearly impossible in traditionally joined wooden structures. (See Fig. 7.20.)

7.20 Traditionally connected members are more easily joined if they do not all terminate at the same point. Carrying at least one member through the "connection zone" means that there will be some solid timber into which the terminating members can frame. These peeled log posts support the second floor girders and continue on up to support roof plates. Those roof plates continue past the post tops. The joinery is cut into solid timber for the full height of each post. (Benson Woodworking Co., Inc.)

The designer's first efforts ought to be in eliminating, or at least spreading out, some of the members that are trying to connect at one point (i.e., so that their lines of action do not intersect at that point). The price paid by spreading the members can include increased eccentricity at the connections. This can be too big a price to pay when working with heavy timbers and traditional joinery, but often is not. I try to move jack rafters out of the valley rafter foot region, for example.

Another strategy is not to end all the members at the crowded joint, but instead to carry at least one of the members solidly through the connection zone. (For a similar strategy, see Chap. 8 at "Continuity.") At least that will put some solid wood, not just fragile end grain, in the connection zone. Usually, at least one member can pass on by the critical point without exiting the structure or getting in the way of some other member. At the valley rafter foot, for example, you can often extend either an eave plate or a principal rafter right on past the post top. This gives a solid place to cut the joint to accept the valley rafter. (See Fig. 7.21.)

7.21 Valley rafter shown supported on a larger rafter that continues across the supporting post below. The continued rafter provides sufficient wood in which to cut a rather unavoidably complicated mortise. (Benson Woodworking Co., Inc.)

7

Traditional Joinery

Reinforce Tough, End-to-End Joints in Large Timbers with Steel

I have been asked, in a public forum, about the "standard joint for a post top to principal rafter bottom." The response to that question, then and now, is that "if you are counting on the mortise and tenon interconnection at the ends of two long 8 × 8 in (20 × 20 cm) timbers to play much role in actually holding the building together, you have bigger troubles than we can address here." There is simply a limit to the amount of connection strength that can be generated within the small overlap zone at the ends of members. Timbers get fragile at their ends, too, with all those end grain connector distances, for example. The available joint strength may not be very large at these joints. It is the designer's responsibility to ensure that either: (1) The joint does not *need* to be so very strong there, because the rest of the structure is carrying the loads just fine, or (2) The joint can handle the intended design loading, because sufficient reinforcement has been provided. (See Fig. 7.22.)

7.22 When a principal rafter lands on a post top, two large members are being joined at their very ends. While no metal-free joints are as strong as the members themselves, many traditional timber joints have more than enough capacity for the forces that arise in well engineered structures. One version of such a connection is shown. (Benson Woodworking Co., Inc.)

Skepticism Is Not Always a Bad Thing

One thing that the purchaser and patron of the traditional joinery art must appreciate is that, just because we North Americans do not have a long and steady legacy of proven joinery methods, that does not mean that our craftspeople don't have faith in the joints they know. How many times have I heard "I'm no engineer, but . . . ?" How many times have I had pointed out just how many old buildings survive with these joints, proving that they all work, all the time. Pointing out that we can no longer see many of the joints that failed does as much good as launching into spiels about building codes and responsible practice. There is no accepted code of standard joinery practice on this continent. There is no long-standing body of work that proves that all traditional joinery is foolproof. There are, however, tools that can be used to analyze, model, test, and predict performance, and engineers who can apply what they know to responsibly design structures that involve unknowns. Traditional joinery has been proven effective and enduring in a lot of places and by a lot of people. The art of traditional joinery is not just in the past; it has a promising and significant future. (See Fig. 7.23.)

(a)

(b)

7.23 Modern North American timber-framed buildings have been mostly residences. The growing commercial use of the technology includes this corporate headquarters in Vermont. (*a*) Dodecagonal stair tower during erection. (*b*) Detail of hanging coopered King Post. (Benson Woodworking Co., Inc.)

The Future for Traditional Joinery Methods

My introduction linked traditional joinery methods to the development of tools. The evolution of woodworking tools has not stopped with the worm drive saw, the power hand planer, or the cordless drill. We are now using computers to design timber structures that rely on traditional methods for their connections. Some of those computer-generated designs are going directly to numerically controlled timber-cutting machines, where the individual timbers are fabricated with only minimal hand work. Reducing the hand labor makes these methods more economical, but does not reduce the need for sound design of the connections. These new tools make feasible joints that are nearly impossible to cut with hand-held power tools. (See Fig. 7.24.) Tenons with circular cross-sections, which can fit into simple, drilled mortises, are one good example of a joint readily cut with a machine but difficult to do with hand tools. (See Fig. 7.25.)

(a)

7.24   Large and sophisticated numerically controlled timber cutting machines, such as those by Hundegger of Germany, greatly reduce the shop labor required to craft traditional joints, while making cuts with remarkable speed and precision. Although there are fewer than a dozen of these tools in North America, there are nearly a thousand operating in Europe, and almost as many similar Japanese-built machines in Japan. (*a*) Joinery machine with attached planer. (*b*) Examples of the joints the machine is capable of producing. (Hundegger USA, L.C.)

Some milling possibilities
Average less than 10 seconds per milling operation

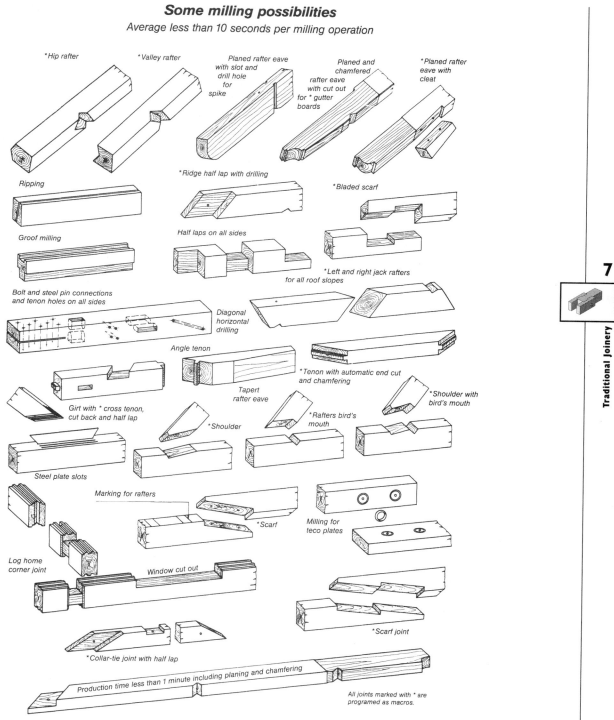

*Hip rafter

*Valley rafter

Planed rafter eave with slot and drill hole for spike

Planed and chamfered rafter eave with cut out for *gutter boards

*Planed rafter eave with cleat

Ripping

*Ridge half lap with drilling

*Bladed scarf

Groof milling

Half laps on all sides

*Left and right jack rafters for all roof slopes

Bolt and steel pin connections and tenon holes on all sides

Diagonal horizontal drilling

Angle tenon

*Tenon with automatic end cut and chamfering

*Shoulder with bird's mouth

Tapert rafter eave

Girt with *cross tenon, cut back and half lap

*Shoulder

*Rafters bird's mouth

Steel plate slots

Marking for rafters

*Scarf

Milling for teco plates

Log home corner joint

Window cut out

*Scarf joint

*Collar-tie joint with half lap

Production time less than 1 minute including planing and chamfering

All joints marked with * are programed as macros.

7

Traditional Joinery

(b)

7.25   Even though they are easy to mill, straight prismatic timbers are not always the most appropriate choice. Nature grows trees in shapes far more intriguing than that. This wandering cherry log is built into and helps support a great room added to the end of a former schoolhouse located on a Vermont hillside. (Benson Woodworking Co., Inc.)

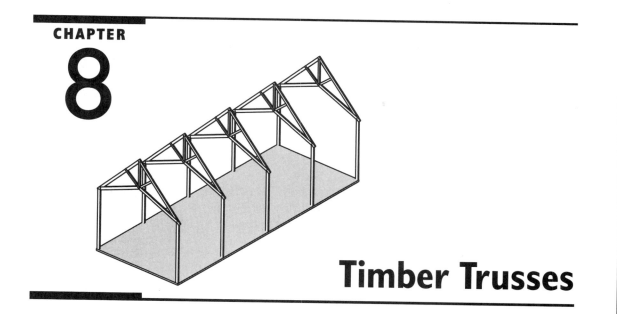

Timber Trusses

Synergy: *The interaction of elements that when combined produce a total effect that is greater than the sum of the individual elements.*[1]

Introduction

Many advances in timber construction have involved new methods of overcoming the inherent length limitations of sawn timbers. Over many centuries, ingenious designers and builders have developed numerous ways of spanning distances greater than the heights of the tallest available trees. One method, previously discussed, involves producing a laminated beam by laying up a large member from a number of smaller members whose joints are staggered. While glulams theoretically can be manufactured to virtually any length or size, their rectangular cross-sections make them relatively inefficient. Although the use of weaker stock near the neutral axes reduces their cost, it has no such beneficial effect on weight. As discussed in Chap. 5, the longer (and consequently the deeper) the beam, the higher the percentage of its strength is expended accommodating its own weight. Furthermore, a glulam that is extremely deep or long can't be disassembled for shipping, because its structural behavior derives primarily from its continuity, from the fact that its multiple layers will only act compositely if they are held together with adhesives applied and cured in the controlled conditions of the fabricating plant. The constraints associated with transporting glulams the significant distances between fabricating plants and building sites also puts practical limits on their size—and therefore their maximum span.

To the extent that they put less material in the webs than at the flanges, wood I-beams are more efficient than glulams. If the knockouts in their webs are removed, their efficiency—their ratio of span to weight—becomes even greater. Unfortunately, the very slenderness that makes such members efficient increases their tendency to buckle with increased depth. Their structural capacity increases at a slower rate than their depth—i.e., they become slightly less efficient as they become deeper.

The ideal long-span timber-framing device would, therefore, be one whose assembly could take place in the field, minimizing the impact of shipping constraints, and whose material would be distributed only where needed within its cross-section, maximizing structural efficiency. The assembly that meets these requirements is the truss. The behavior of a truss is the epitome of synergy. We shall see why as we define trusses and examine how forces are distributed within them.

What Is a Truss?

Different authors have different definitions for the word *truss*. To one, it is a "framework of linear elements that achieves stability through triangular formations of the elements."[2] To another, it is a "structural spanning device in which the loads are translated into axial forces in a system of slender members."[3] Actually, these definitions are complementary: The triangles formed by the slender linear members of a truss translate the loads into axial forces in those members.

Trusses rely on triangulation to maintain their geometries. From experience, you probably already know that a unique triangle is formed when three sides of known length are arranged to make a closed figure. A four-sided figure, on the other hand, can be transformed from a rectangle to an infinite number of parallelograms without changing the lengths of any of its sides. The more sides in the original figure, the more ways it can be transformed.

As discussed in previous chapters, timber connections are generally unable to resist rotation. Because timber trusses achieve geometric stability through the arrangement of their members, rather than through the rigidity of their joints, they take advantage of timber's strengths. Steel and concrete, by contrast, can be engineered with joints having significant moment resistance. Consequently, their long-span structures can be composed of shapes other than triangles.

To help achieve its full structural potential, a truss must be loaded at its joints. (See Fig. 8.1.) Otherwise, the truss members will be subjected to combined loading, either bending and axial tension, or bending and axial compression. Gravity, acting along the entire span of the truss, causes bending in each nonvertical member under its own weight. Fortunately, the dead load of a timber truss—the load induced by gravity—is usually a small enough fraction of the total load that it does not control the overall design, as long as the other more major loads are

8.1   A timber truss engineered solely for panel point loading. The rigidity of the truss is shared with the columns integrated into it.

applied at the joints. Although bending of trusses under their own load can usually be ignored, bending of top chords under uniform floor or roof loads cannot.

While a timber structure may include timber trusses, functional and budgetary constraints make it unusual for the entire structure to be trussed. Lateral stability of the untrussed portion(s) of the structure can be achieved with shear walls, knee braces, or other devices (see Chap. 9). Knee braces can create stable triangular areas within a timber frame, but are often connected to columns and beams at midlength, causing bending in them. The closer knee braces are connected to the ends of the columns and beams, the less bending will be induced, and the more the *overall* structure will behave like a truss.

Understanding Truss Forces

Let us look at the simplest truss, one consisting of a single triangle. For purposes of this discussion, assume that both ends of the base rest on supports and that the top is vertically loaded. Sound familiar? If you've ever seen a gable roof with collar ties at the attic floor, it will be. To understand the relationship of the forces to the pitch of the roof, we are going to build a little structural model. (See Fig. 8.2.)

Materials and tools:

- Two balsa wood sticks, ⅛ in thick × ¼ in wide × 6 in long
- One large paper clip

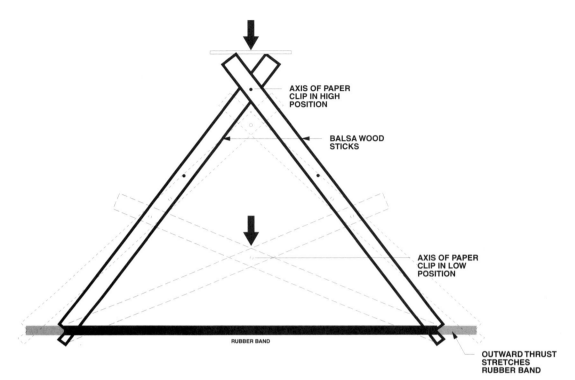

8.2   A working truss model. Loading the apex induces compression in the top chords and tension in the bottom chord.

- Two easy-to-stretch rubber bands
- Scissors
- A smooth table top

Construction:

1. Partially untwist the paper clip.
2. Using the end of the paper clip, punch two holes in the wide face of each stick, one at midlength and the other about ½ in from an end.
3. Notch the wide face of the other end of each stick with the scissors.
4. Make a triangle with one of the rubber bands as the base—looping it around the cut ends of each of the sticks—and with the sticks as the other two sides.
5. Insert the straightened end of the paper clip, as a hinge, through the upper holes of the sticks, and *retwist it so that it does not work loose.*
6. Stand the truss upright on the table top.

WARNING: Because of the significant forces that will build up in the rubber band, it is a good idea for you and any observers to wear safety glasses whenever you operate this model. Do not operate it until and unless the paper clip is securely fastened in place.

This model is intended to illustrate the forces in a trussed gabled roof, and how they vary with changes in the slope of the rafters. The sticks represent the rafters in compression, and the rubber band, the collar ties in tension. Stretch the other rubber band with your hands. Notice that the farther it is stretched, the harder it is to stretch farther. Push down on the top of your truss model. Notice the effort is takes to stretch the rubber band at its base. Then, move the hinge to the holes at midlength of each stick. Again push down on the top of your (now smaller) truss, and keep going until the rafters are resting on the table top. Deform the truss from its high shape to its low shape a couple of times.

Notice that the vertical load required to stretch the rubber band actually *decreases* as it lengthens (i.e., as the slope of the rafters decreases). This is the opposite of what we found when trying to stretch the rubber band directly. The elongation of the rubber band is a manifestation of the amount of tension in it. Saying it takes less effort to stretch it as the roof flattens is equivalent to saying that a given vertical load will produce more tension in the collar ties in a shallow roof than in a steep one.

Recalling high school mathematics, forces can be broken down into components; and, conversely, the components can be combined to yield overall forces, or resultants. In a roof of a given span, the total vertical component of the load on the rafters—including dead, live, and snow loads—will be relatively constant over a reasonably broad range of common roof slopes. By contrast, the horizontal component of the force in the rafters increases as the slope decreases. Thus, the resultant in a steep roof is steeper than in a shallow roof, and the ratio of the length of the resultant to the magnitude of the vertical load is smaller in a steep roof than in a shallow roof. *In fact, the resultant is always parallel to the member itself.*

Consistent with experience, our experiment shows that the rafters need resist less compression and the collar ties less tension in a steep roof than in a shallow one, all else being equal. That's not all: Although not evident in our small lightweight model, each nonvertical truss member is also subject to bending under its own weight. Therefore, the lower the roof slope, the greater the bending in the rafters.

In our previous discussion of combined members (Chap. 5)—those subject to bending and either axial tension or compression—we saw how that portion of their structural capacity assumed to resist bending could not also be used to resist axial loading, and vice versa. Along the length of a combined member, bending moments vary—reaching a maximum at midspan under a uniform load such as gravity—but axial loads are constant. Therefore, the longer the given span, the larger the per-

8

Timber Trusses

centage of a combined member's total structural capacity required to resisting dead load bending. Because bending moments increase as the square of the span, the structural capacity left to resist axial forces decreases rapidly as the span grows. To minimize sagging and maximize axial load capacity, therefore, the bottom chords of medium-span trusses are often hung from the roof peak at midspan, reducing the bending moments by the square of the subdivision, or to one fourth of their magnitude without the hanger. At truly long spans, additional intermediate truss members may cut the span of the bottom chords in thirds or quarters, reducing bending by even larger factors.

External and Internal Forces

Although timber trusses may look different from solid timber beams and girders, the structural purpose of all of these members is the same: to transfer loads from the points of application to the points of member support, without collapsing or displacement. Although timber structures must normally be in static equilibrium with respect to external forces, it is instructive to digress for a moment, and consider the case of the seesaw.

A seesaw in use is far from static. Nevertheless, the first thing you want to do when you get on it, before you ride it, is to put it into equilibrium. The heavier person moves closer to the fulcrum, while the lighter moves further away. Balance is achieved when the product of the weight of one rider and his or her distance from the fulcrum is equal to the product for the other. The vertical load at the fulcrum is the sum of the weights of the riders. Think of the seesaw as a beam with two loads balanced over one support, having an overall span composed of two parts, whose ratio is set to achieve equilibrium. Draw a sketch of it. Now turn it upside down. (See Fig. 8.3.)

Your sketch now looks like a simply supported beam supporting one concentrated load. The load that had been the fixed fulcrum can now move, and the riders who had moved to balance themselves are now fixed supports. Yet, the ratio of the reactions at the ends is still inversely proportional to the ratio of their distances from the intermediate load.

In truss design one of the objectives is to concentrate loads at its joints or *panel points*. In truss analysis, the geometric relationship of the loads to the supports is crucial since, as we have seen, it affects the *external reactions* and, as we shall see, it affects the *internal forces* within the truss itself. Unlike the model used in the experiment earlier in this chapter and the seesaw previously discussed, most trusses have more than one interior joint. Furthermore, the distribution of loads among the panel points may vary. In a roof, these variations can result from the effects of snow or wind; in a bridge, from the movement of pedestrians or vehicles.

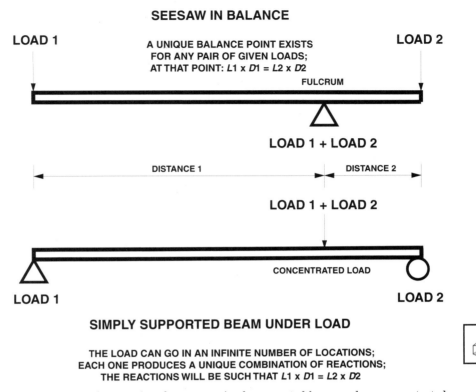

SEESAW IN BALANCE

LOAD 1

A UNIQUE BALANCE POINT EXISTS
FOR ANY PAIR OF GIVEN LOADS;
AT THAT POINT: $L1 \times D1 = L2 \times D2$

LOAD 2

FULCRUM

LOAD 1 + LOAD 2

DISTANCE 1   DISTANCE 2

LOAD 1 + LOAD 2

CONCENTRATED LOAD

LOAD 1

LOAD 2

SIMPLY SUPPORTED BEAM UNDER LOAD

THE LOAD CAN GO IN AN INFINITE NUMBER OF LOCATIONS;
EACH ONE PRODUCES A UNIQUE COMBINATION OF REACTIONS;
THE REACTIONS WILL BE SUCH THAT $L1 \times D1 = L2 \times D2$

8.3 A structural comparison between a simply supported beam under a concentrated load and a seesaw in equilibrium.

Graphical Analysis

In the middle of the nineteenth century, James Clerk Maxwell, a Scottish physicist, developed a graphical technique for determining the internal forces in a truss.[4] Although other analytical techniques are capable of greater precision, Maxwell diagrams are generally far easier and quicker to generate and are, therefore, more helpful in developing a general understanding of truss behavior, than the mathematical techniques often favored by structural engineers. Because trusses must often be analyzed for a number of different load combinations, the speed of analysis can become very important. Just as a bar graph can reveal numeric relationships otherwise hidden within the contents of a spreadsheet, a Maxwell diagram can reveal force distribution relationships otherwise hidden within a table of truss member forces. Examining several of these diagrams side by side, one starts to understand how the geometry of a truss affects its response to load.

Our first diagram will be based on our previous gable roof model. Draw the truss with a span of 24 units (feet, meters, or whatever) and a height of 16 units, and label the external forces, assuming 10 units (pounds, kilograms, or whatever) at the peak. From the seesaw discussion, it is obvious that a load at midspan produces two equal reactions of 5 units at each of the supports. Using uppercase letters, start at the left end of the truss and travel clockwise around its perimeter, labeling each space between the external forces (loads and reactions) in alphabetical order. Then, start at the leftmost triangle of the truss and travel to the right, continuing the labeling with each internal triangle. The completed truss drawing is known as a *space diagram* (see Fig. 8.4). The lettering sequence is conventional, but arbitrary. As you'll see, the goal is simply to make sure that each force has a *unique* two-letter name.[5]

Consisting, as it does, of 3 external forces and 1 triangle, our truss will require $3 + 1 = 4$ letters, A through D. Because all of the external forces are vertical, a plot of them, the *load line,* will be vertical. To generate it, again travel around the perimeter of the truss. On the way from A to B, a downward force of 10 is encountered. In a new diagram using lowercase letters (another convention), draw a line of length 10

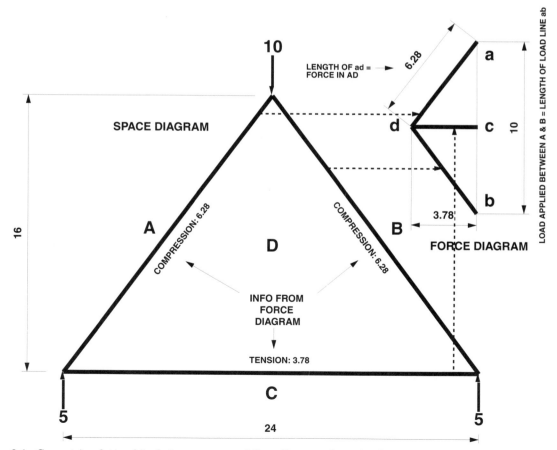

8.4   Geometric relationships between space and force diagrams for a simple truss.

at any scale from *a* down to a point *b*. For the upward force of 5 units encountered between B and C, draw from the existing point *b* up to a new point *c*. When graphed, the upward force encountered between C and A will return us to point *a,* closing the *external force polygon.* If the polygon does not close, it indicates that the truss is not in equilibrium with respect to external forces. In that case, recheck that the sum of the reactions is equal and opposite to the sum of the loads.

By definition, a truss consists of one or several triangles. Recalling that the resultant for each internal force is parallel to the member carrying it, the *internal force polygon* for a truss will never be a straight line. The directions and magnitudes of the lines in the external force polygon are those of the applied loads and reactions; the loads and their directions are known, and the reactions are calculated from them. In the internal force polygon, the directions are those of the truss members and the magnitudes are the forces within them; the directions are known—they are those of the truss members themselves—and the magnitudes are revealed in the Maxwell diagram. The lowercase letters along the load line of the external force polygon correspond to the uppercase letters around the perimeter of the space diagram. The lengths of the lowercase lettered line segments in the completed internal force polygon of the Maxwell diagram represent the loads in the two-lettered members of the space diagram.

To draw a line of the internal force polygon, go to the space diagram. Pick a member, such as the left-hand top chord of the gable, and identify its two-letter name: AD. In the Maxwell diagram, draw a line *ad* parallel to AD, starting at point *a* of the external force polygon. The left end of this line will terminate at a new point *d,* once we have determined where it is. Repeat this procedure for the other members in the space diagram. Unknown point *d* will be located at the intersection of the internal force lines of any two members with *D* in their names. In our example, and indeed in any truss with just one panel, *D* is part of the name of each of the three members. More generally, unknowns can be usually located by working your way around the members converging at each joint, plotting them on the force diagram as you go.

When testing the gable model, our impression was that the horizontal component of the load increased as the gable flattened. Our Maxwell diagram might shed some light on this phenomenon. Draw and label a new space diagram with the same span as in the previous example, but with a reduced height. Adding to the same Maxwell diagram as in the previous example, draw the internal force polygon using dashed lines. On completion, notice how making the truss shallower dramatically increases the forces in the bottom and top chords, confirming our model experiment.

What happens if we compare two trusses of the same shape but different sizes—if we compare a truss made shallower by holding the height and increasing the span, to one made shallower by holding the span and decreasing the height? Surprise! The *configuration* of the

Maxwell diagram is unaffected by changes of scale; it has the same shape for a larger truss as for a smaller one with identical loads, layout, and height-to-span ratio.

The Maxwell diagram has yielded the internal forces in each of the members, but we must still determine their *senses*—i.e., which are compressive and which are tensile. The procedure to do this is simple: Put your index finger on either end of any member in the space diagram. For example, start with the lower left joint. Name any member coming into that joint by traveling clockwise around the joint. Thus, the sloping top chord is AD, and the horizontal bottom chord is DC. Start with AD. In the Maxwell diagram, put your index finger on point *d,* corresponding to the lower left corner of the same line in the space diagram. For member *AD,* moving from *a* to *d* in the Maxwell diagram is toward your finger, indicating compression. Now, go through the same procedure with line DC. Moving from *d* to *c* is *away* from your finger, indicating tension. Had you started at the right end of the space diagram, the bottom chord would have been CD, rather than DC, but the sense of the member would still have been tension. From our experiment with the gable model (and from our everyday experience), we suspected that the top chord was in compression and the bottom in tension, but it's good to have a technique for confirming it, especially as we begin to get into more complex truss shapes.

Having established the magnitudes and senses of all of the internal forces, it is time to go back to the space diagram and label each member with its corresponding force and sense. (Remember: the internal forces in the Maxwell diagram are measured using the same scale as the load line.) Although the Maxwell diagrams were the same for the two trusses of similar appearance but different scale, the implications of their sizes should now be evident: The top chords of the larger truss must be engineered as much longer columns than those of the smaller one. Size, in other words, does matter for members in compression.

The problem with the labeled space diagram is that the three classes of information it displays cannot be correlated at a glance. Forces are shown numerically, senses alphabetically, and truss configuration geometrically. To help visualize the relationships between these classes, and understand their design implications, I recommend that, in addition to (or instead of) the alphanumeric information, the thickness of each member in the final space diagram be shown in proportion to the force within it, that compression be shown with a different type of line than tension, and that zero-force members be shown with still a third line type. These enhancements will not only permit the viewer to perceive instantly what is happening within the truss, but will also ease comparisons among different trusses and load combinations. Notice how dramatically the legibility of the space diagram improves. (See Fig. 8.5.)

The desirability of midspan support for a long bottom chord was discussed earlier. Let's use graphical analysis to gain a better under-

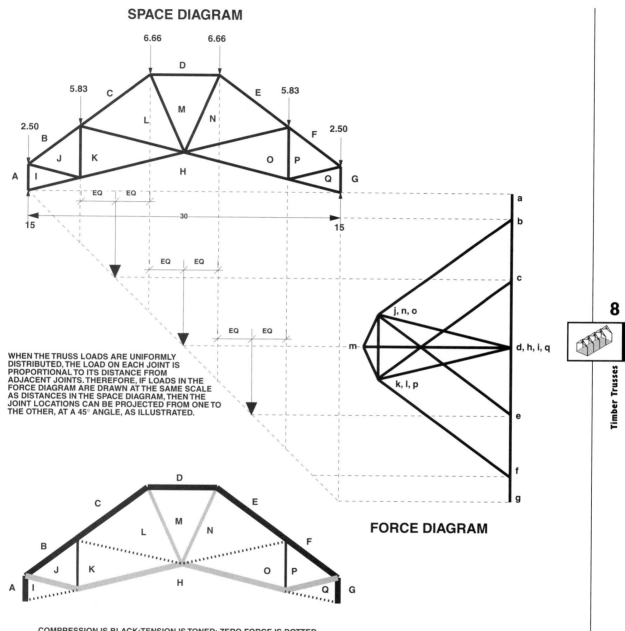

SPACE DIAGRAM

WHEN THE TRUSS LOADS ARE UNIFORMLY DISTRIBUTED, THE LOAD ON EACH JOINT IS PROPORTIONAL TO ITS DISTANCE FROM ADJACENT JOINTS. THEREFORE, IF LOADS IN THE FORCE DIAGRAM ARE DRAWN AT THE SAME SCALE AS DISTANCES IN THE SPACE DIAGRAM, THEN THE JOINT LOCATIONS CAN BE PROJECTED FROM ONE TO THE OTHER, AT A 45° ANGLE, AS ILLUSTRATED.

FORCE DIAGRAM

COMPRESSION IS BLACK; TENSION IS TONED; ZERO FORCE IS DOTTED
LINE THICKNESS, PROPORTIONAL TO FORCE

8.5 Development of a magnitude and sense diagram for a timber truss. This is the truss described and illustrated in great detail in Chap. 14.

8

Timber Trusses

standing of this aspect. Draw a simple truss with a vertical web member at midspan (see Fig. 8.6, position 1). Assume a load of 1000 units at the peak. Label the space diagram and then generate the Maxwell diagram. Notice that points *d* and *e* coincide. The length of line *de* and, therefore, the force in member DE, is zero. Zero-force members can occur when the member in question has been used to divide a large triangle into smaller triangles. A quick look at the space diagram confirms that that is exactly what has happened here. As long as the external forces are held constant, removing a zero-force member does not effect the distribution of internal forces elsewhere in the truss.

The dead load of the bottom chord is a uniform load, inducing bending throughout its length. Ignoring bending for the moment, it is obvious that one half of that load is carried by the hanger, and one quarter goes directly into the supports at each end. Since the hanger is the only member at the bottom midspan joint having a vertical component, it ends up carrying the entire midspan load of the bottom chord, a fact that will now be confirmed with a Maxwell diagram.

Shifting the midspan load from the top of the truss to the bottom has a fascinating effect on the respective Maxwell diagrams (see Fig. 8.6, position 5). The redistribution results in a proportional increase in the internal force in the hanger *de,* but does not affect the lengths and, therefore, the forces in the top or bottom chords. To make the relationships even more apparent, generate a Maxwell diagram for the three intermediate conditions between these extremes, and make a flipbook—an animation—of the entire set.

Our next set of illustrations shows the effect, for trusses of identical height, of moving a given load *w* horizontally relative to a given span *L*. (See Fig. 8.7.) The Maxwell diagrams reveal an entirely different picture: As long as the load stays within the supports, the length of the load line remains constant; as soon as the load gets outside the supports—i.e., is cantilevered—the load line lengthens. Simultaneously, the senses of members AD and DC reverse. When there is only one load and it is located over a support, there is no truss action, as can be seen from its Maxwell diagram. As the load is moved in equal steps to the right, the width of the Maxwell diagram goes from "positive" to "negative" in ever larger increments.

Timber Truss Connections

Once the magnitudes and directions of the forces in all of the truss members have been identified, preliminary member sizing can begin. But, as explained in Chaps. 6 and 7, final sizing is often controlled by the connections. To guide the development of the connections, you must first decide on their intended character.

I think of connections in timber trusses as being either *indirect* or *direct.* The most common means of indirect load transmission are bolts,

SPACE DIAGRAM   **FORCE DIAGRAM**

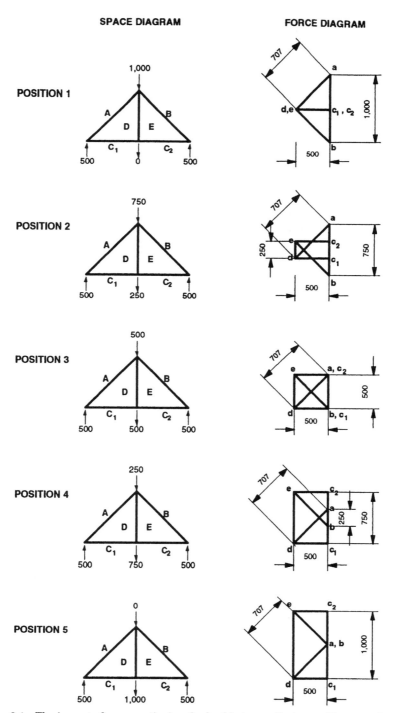

8.6 The impact of reproportioning the load between the top and bottom of a truss. Photocopy the force diagrams and assemble them into a flip-book. When you run through it, the force diagram appears gradually to turn inside out.

SPACE DIAGRAM

FORCE DIAGRAM

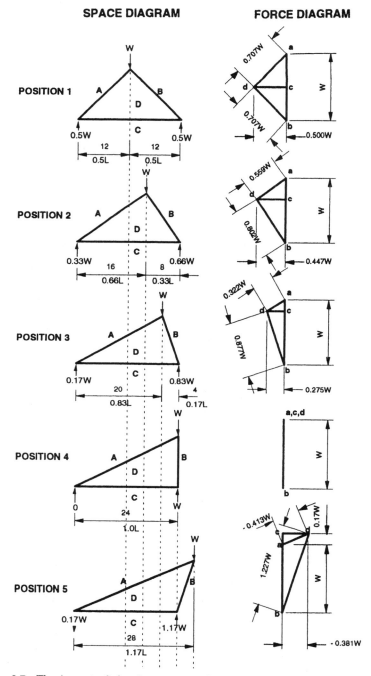

8.7   The impact of skewing a truss of constant base and height. To understand what is happening to the members, pick a specific one and observe how its length varies from one force diagram to the next. Like Fig. 8.6, this diagram is instructive when assembled into a flip-book.

timber connectors, steel side plates, and combinations thereof. In direct connections, wood bears directly on wood; in indirect connections, loads are transmitted between timbers indirectly, through the aforementioned intermediaries. Direct connections are either *unidirectional* or *bidirectional*. Unidirectional connections are those that can transmit solely tension or compression, while bidirectional connections can transmit both.

To help understand the relationship between the geometries and capabilities of some typical direct connections, consider the joint where the lower end of the kingpost connects to the horizontal bottom chord of a simple King Post truss. (See Fig. 8.8.) The King Post is always in tension in service under the dead load of the bottom chord. To protect the trusses from erection damage, assume that the fabricator decided to wrap each one with waterproof coverings, and instructed the erector not to remove the coverings until just before starting the roof deck. Prevented from lifting the trusses from under their top chords, the erector had no choice but to install his or her slings under the bottom chords. Graphic analysis would show the King Post to have been a compressive member during erection. This is an example of a load reversal. Load reversals can be caused by differences between temporary and permanent loads, or by differences between various combinations of permanent loads. When all of the external forces reverse direction—as in this case—the senses of all of the members change. More typically, however, senses change selectively, resulting either from live loads that are

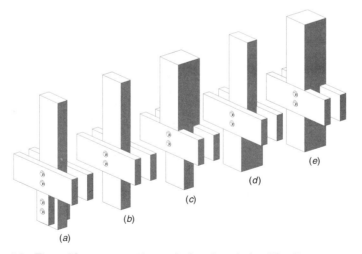

8.8 Five midspan connection variations in a timber King Post truss: (*a*) bearing blocks to support downward (gravity) load from spaced bottom chord. (*b*) bolts for perpendicular-to-grain load (from gravity or erection) from bottom chord. (*c*) shoulder to resist upward load during erection. (*d*) shoulder to carry gravity load in service. (*e*) shoulders for erection and gravity loads in opposite directions.

nonuniform (on a roof, for instance, under snow drifts or wind gusts) or that are moving (on a bridge, for instance, under the load of a person or vehicle).

In the preceding example, unaware of the contractor's methods and means, the designer planned only for tension and, therefore, detailed the King Post to be supported on bearing blocks (see Fig. 8.8a) or to be ripped down for most of its length, leaving a shoulder at its base just wide enough to carry the bottom chord (see Fig. 8.8d). The resulting connection would be unidirectional for tension. If the connection were designed exclusively to resist the construction load reversal, the bottom chord would bear directly on the *bottom* of the King Post's shoulders (see Fig. 8.8c). The resulting connection would be unidirectional for compression. However, to handle the load reversal that will occur between construction and completion, it is essential that the connection be bidirectional. Creating a pocket by overlaying the unidirectional end conditions developed by the designer and builder yields a workable bidirectional connection (see Fig. 8.8e). Recognize, however, that this joint may loosen as the members dry, since shrinkage along the grain *of* the pocket will be less than that across the grain of the member *in* the pocket. This will not be a problem if load transmission is handled solely by bolts (see Fig. 8.8b).

You can't actually put timber directly into tension. In Fig. 8.8e, tension in the King Post is produced only after the bearing surface of its shoulder is compressed, and the shear plane between the shoulder and the rest of the member is stressed. However, since these stresses stay within the member, I consider this to be an example of a direct connection.

By definition, direct connections require no additional hardware (except to keep the pieces aligned). For that reason, they can be quite economical. Where joints consist of more than two or three members, however, direct connections can become extremely complex, especially if there are multiple tension members. Prior to the advent of timber connectors, it was common to avoid this problem by using iron or steel rods for tensile members, with timber used only for compression. It is still common today (see Fig. 8.9). The cross-sectional area lost in accommodating the diameter of the rod is far less than that lost to create a wood shoulder. Furthermore, connections in a truss with compression members bearing directly on one another can be much simpler than those in trusses whose members must pass and be housed into each other to resist tension (see Fig. 8.10). As discussed in Chap. 7, connections in traditional timber joinery are primarily of the direct type.

Indirect timber connections can be configured in nearly endless ways. To accommodate joints involving many members, however, most are either *layered* or *plated*. Seen in elevation, the members in a layered connection are in multiple planes, pass in front of or behind one another, and are bolted or fastened together (see Fig. 8.16). In a plated

8.9 Composite trusses, with compressive top chords of timber, tensile bottom chords of steel cable, and compressive struts of structural steel. This truss is shown in context in Fig. 5.11. Wolf Trap Farm, near Washington, D.C.

connection, the members are all in a single plane and are sandwiched between steel side plates (see Fig. 8.1). These plates overlap the end portions of the members and are bolted through them. Their overlaps are a function of the required end distances (Chap. 6); their thicknesses are a function of the loads. At a minimum, steel side plates are generally ¼ in (6 mm) thick.

As has been demonstrated, a truss is most efficient when loaded and supported only at its joints. Just as bending is undesirable in its members, eccentricities are undesirable in its joints.

- *In elevation,* the neutral axes of all of the members at each joint should meet at a single point. This also applies to members projecting beyond the plane of the truss, including purlins and braces. Otherwise, the connection will be subject to rotation.

- *In plan,* the forces should be symmetrical about the longitudinal axis of the truss. Avoid bolted connections involving just two members in single shear. Such connections are subject to twisting. Although each

8.10 Composite trusses, with compressive top chords and vertical struts of timber, and tensile bottom chords and bracing of steel rods. This truss is shown in context in Fig. 1.5.

of the outside timbers in connections with more than two layers will be in single shear, they will balance each other if together they comprise part or all of the same truss member and, therefore, carry the same portion of the load in the same direction.

Plated trusses can easily achieve concentricity and, because they consist of single plane, are naturally symmetric about their longitudinal axes. Layered trusses can also be concentric and symmetric, but achieving the latter requires careful planning.

Consider again the joint where the King Post meets the bottom chord. In a layered truss, one of these would be the main member and the other, the side members. Under the permanent load described, the post would be in tension. Because overall tensile capacity is solely a function of net section, the slenderness ratio of the post is of no concern. Under construction load, however, the King Post would be in compression. If slenderness ratio is an issue, I recommend making the King Post the main member and giving it a square cross-section. For the side members, make the bottom chord of a pair of narrow and relatively deep timbers. To minimize weight and maximize bolt efficiency in double shear, these should be narrower than the main member; to minimize sagging under dead load, they should also be deeper.

As the number of connected members increases, the connection grows—in thickness if layered, in radius if plated. In a given joint of a layered truss, the minimum number of layers l is approximately twice the number of members m (where each member may consist of multiple submembers). More precisely, $l = 2m - 1$. With a minimum nominal thickness of 3 in (75 mm) for a layer to qualify as a heavy timber, a timber connection with many layers will be very thick indeed. What this formula doesn't tell us is the total number of layers needed for the entire truss.

Layering strategies should be developed from the top down. Start by examining the entire truss for patterns in the truss connections and in the member forces. Ask yourself these questions:

- Do all of the top or bottom chord joints have the same number of submembers?
- How are forces distributed among the web members?
- Which members are subject to load reversals?
- Which members are only in tension or compression?
- Are there any combined members?
- Which joints will be loaded by out-of-plane members?
- Which members can continue through the joints?

Although triangulation is what gives trusses their geometric stability, it can make the development of a layering plan very difficult. Take our simple triangular truss, where each joint has two members. Our formula reveals that the minimum number of layers in each joint is 3.

In a plated truss, the radius or reach of the side plates is a function not only of the number of members, but of their layout. Members converging toward the center of a connection can get closer to it before running into each other if they are uniformly spaced. The radius required for a steel side plate large enough to reach the worst case—the members separated by the smallest angle—can be calculated by adding the required end distances to the following formula, where r is the radius, c is the width of the members, and a is the controlling angle:

$$r = \frac{c}{2\left(\sin\frac{a}{2}\right)}$$

As you would expect, increases in member width c or decreases in the controlling angle a cause increases in the required radius. (See Fig. 8.11.)

The angular relationships among a joint's members also affect their structural capacity. Consider the ends of two simple traditionally joined gable trusses with different slopes, each with the heel of its compressive top chord notched into the top of its tensile bottom chord. We

(a)   (b)

8.11   A row of three-dimensional trusses. Each truss is x-shaped in plan. Midspan connector consists of a short section of steel pipe with projecting plates. These are sandwiched between the two layers of dimension lumber of which each member is composed. Cylindrical openings are sized to accommodate spherical lighting fixtures. Projections from tops of members prevent birds from alighting there. (a) Canopy in context. (b) Canopy in detail. New Jersey Transit Railroad Station, Elberon, NJ. The Goldstein Partnership, architects.

already know that the internal forces in the shallower truss are greater than in the deeper one. However, since the compression design value across the grain is less than that along the grain, the direct bearing capacity of the end joint in the shallower truss is greater than that of the deeper one. This raises an interesting question: If changing the slope of the roof simultaneously increases or decreases both the internal forces and the capacity of the end joints to resist them, what is the optimum slope? Unfortunately, generating an answer is more complicated than it seems: As the slope becomes shallower, the point where the top and bottom chord *centerlines* converge moves further away from the point—the notch—where the top and bottom chord *members* meet, unless the notch is so deep that little of the bottom chord remains. To avoid eccentricity without creating too deep a notch, only a portion of the end of a shallow truss' top chord will be in bearing, making it difficult to make the comparison that would answer the aforementioned question.

Fasteners and Connectors

Perhaps the most important differences between layered and plated connections relate to the arrangement of their fasteners. In the former, there is typically a single group of fasteners arrayed around the center of the joint, with each fastener passing through every layer. In the latter, there is typically a single steel connector radiating from the center of the joint, with a separate group of fasteners serving the member at the end of each arm; each fastener passes through a pair of steel side plates and the member sandwiched between them. The design implications of these differences are a function of the number of members at the joint and the number of fasteners serving each one.

As discussed in Chap. 6, the design of layered connections is analogous to a zoning problem: The truss represents the neighborhood; each joint, a building lot, and the portion of the joint where fastening is permitted, the buildable (boltable) area.

When dealing with trusses, it doesn't take long to realize that a geometrically acceptable and structurally adequate bolt arrangement is hard to develop when the members at a joint are oriented at more than two different angles. Whereas the grid of bolt rows along and across the grain can simply be skewed like a parallelogram when joining two members at an angle other than 90°, there may be no grid capable of accommodating three or more different member orientations. Furthermore, the boltable area of the resulting connection may be so small or of such an odd shape that the number of fasteners it can accommodate is inadequate to resist the internal forces. For such joints, therefore, I recommend the use of the largest possible split rings or shear connectors, all on a single bolt.

Continuity

Timber trusses often are the most reasonable structural option for long spans in heavy-timber buildings, either because one alternative—sawn timber—is not available in the required length, or another—glulam—is too heavy or too costly. Nevertheless, designers should not feel compelled to limit the lengths of the timbers in the truss to the lengths required between joints. In fact, there are several good reasons for assembling the truss out of the longest practical timbers:

- *Control of deflections:* Although truss connections are assumed to act like hinges, significant bending resistance can be developed by extending timbers beyond the truss joints. As in the case of cantilevers, structural continuity decreases deflection. Keep in mind, however, that this benefit comes with a cost, namely, that such members will have to be engineered for the combined loading induced by deflection, however small it turns out to be. In other words, the members made continuous to reduce deflection must, as a consequence, be engineered for combined loadings. (See Fig. 8.12.)

- *Simplification of connections:* The complexity of timber truss connections often seems to increase as the square of the number of con-

8.12 Short-span timber truss loaded at its third points. The continuity of the bottom chord limits overall deflection. Under equal vertical loads, the diagonals missing from the middle third are zero-force members. "The brothers felt strongly that wood, because of its nature, required the building up of separate parts . . . Joints were treated as design elements and contributed to the enrichment of the whole." (Randell L. Makinson, in McCoy, p. 111.) The Gamble House, Pasadena, CA. Greene & Greene, architects.

nected members. Because they reduce the number of individual members to be connected, timbers that run *through* joints are much easier to accommodate than those that terminate *at* them.

■ *Reduction in assembly labor:* Timber truss assembly is labor intensive. Fewer pieces means less measuring, cutting, fitting, drilling, and fastening. Fewer pieces also means the potential for less variation and, therefore, a more consistent appearance.

The degree to which structural continuity can be achieved and the manner in which it is are a function of three factors: the configuration of the truss (specifically the number of members aligned from one truss panel to the next), the dimensions of its members, and the connection strategy. Continuity is a straightforward matter in a layered truss, because each layer has its own right-of-way. In plated trusses, by contrast, continuity of one member prevents continuity of others. This is more a theoretical problem than a practical one, because most truss joints have no more than one pair of members aligned on opposite sides; These pairs usually occur at intermediate joints in top or bottom chords. Where two pairs are aligned—as at the center of an x-brace— one pair consists of zero-force members.

Splices

In solid-timber framing, it is desirable for the beams and joists to be continuous over as many *spans* as possible. In timber trusses, as previously discussed, it is desirable for members to be continuous through as many *panels* as possible. Recognize, however, that practical spans for members subject to bending tend to be significantly shorter than for identical ones loaded in tension. (While each portion of the cross section of an axially loaded member can be heavily stressed, only the extreme fibers of a rectangular bending member can be.) Thus, a solid timber continuous over two or three spans when used as a beam or joist may only be long enough to span one panel when used as the bottom chord of a truss. Planning truss connections, therefore, requires advance knowledge of the availability and costs of the required member sizes, in various lengths.

When a timber extends beyond a truss joint, but is not long enough to reach another one, a splice is required. For members that will be in compression under any reasonable loading scenario, splices should be avoided, due to their relatively low buckling resistance. For members that will always be in tension, however, splices are not only structurally sound, but, by reducing the number of members converging there, may actually simplify the design and assembly of certain complex truss joints.

As an example, consider a multipanel gabled truss with a horizontal bottom chord having joints at its quarter points. For reasons which will be discussed later in this chapter, the tension in the bottom chord is highest in the first and fourth quadrants and lowest in the second and third. To accommodate these different forces and develop some continuity, the bottom chord could be composed of a pair of timbers for the first and last thirds of the span, spliced to a single member occupying the middle third. Not unusually, in a truss without splices, more members converge at the midspan connection than at any other. Therefore, the fact that the bottom chord under this approach is continuous through that joint *and* is composed of a single layer there will simplify the design and engineering of the midspan connection. Supporting the middle portion on the cantilevered ends of the outer portions is helpful, too, in minimizing the magnitude of the moments induced in the bottom chord by gravity.

Splices should be considered an integral part of the vocabulary of timber trusses. Although they are most commonly used in bottom chords longer than the available timbers, they may also be used at one or both ends of short tensile web members, as part of an overall connection strategy.

Deflection

Timber beams and timber trusses behave differently under load. To illustrate, assume that both are supported at their ends, and that the bottom chord of the truss is the same size as the beam. While tension will be distributed uniformly over the entire cross-section of the bottom chord, it will be present only in the lower half of the beam, varying from zero at its neutral axis to a maximum in the wood fibers of its bottom surface.

Assume that both structures have ceilings attached directly to them, and that their deflection criteria will therefore be identical at $l/360$. Assume also that the truss will have enough joints in its bottom chord that its faceted deflected shape will approximate the arc described by the deflected beam. The formula:

$$c = 2r \left(\sin \frac{a}{2} \right)$$

gives the chord length c for an arc of angle a and a radius r.[6] (This formula is the same as that shown earlier for truss member convergence, only solved for a different variable). By inspection, it is evident, for a given angle a, that the chord length increases in direct proportion to the radius—that the chord length scales up with the radius. So, too, does the deflection, the distance d between the midpoints of the chord and arc.

The allowable deflection of the truss will, therefore, be in the same ratio to the allowable deflection of the beam as the span of the truss is to the span of the beam. Timber trusses, of course, are customarily used for longer spans than timber beams. Enlarging the end conditions of these structures reveals that the above ratio will also hold for the elongation of the bottom of the truss under load compared to that of the bottom of the beam. A truss with 4 times the span of a beam of equivalent depth will need supports capable of accommodating 4 times the elongation. Restraint of such movement could lead to crushing of the end grain of the restrained members.

Unlike a beam, a truss can be supported near its top, under its top chords. In that case, its supports must be able to accommodate *shortening* of its top chord under load. Under uplift conditions, the same will be true of the supports for the bottom, though generally to a lesser degree due to the inexorable downward force of gravity.

Under a given deflection criteria, then, the lateral movement to be accommodated at the supports varies linearly with the span of the member, all else being equal. Unfortunately, all else is not necessarily equal. If the bottom chord of the truss is discontinuous, there is the potential at each joint for slippage under load, specifically where slightly oversize holes have been drilled to accommodate bolts. To minimize this effect, use tight-fitting connectors such as split rings.[7] While slippage of $\frac{1}{16}$ in (1.5 mm) per joint may seem negligible, total slippage will be considerable if there are many joints. Furthermore, total slippage can occur purely under the dead load of the truss. Depending on the location and shape of the truss, the resulting deflection could be unsightly.

Cambering is, therefore, worthwhile in long-span timber trusses for the same reason as in long-span glulam beams, but it is achieved in a different way. Because beams are subject to bending from live loads, their deflections can be significant. By contrast, since the only load inducing bending in individual truss members is their own weight, their deflections are usually negligible. In a truss, because deflection *of the assembly* is the problem, cambering *of the assembly* is the solution. In a flat floor truss, for example, the line connecting the centers of the joints in the bottom chord should be a faceted approximation of the smooth large-radius arc of a cambered glulam beam.

For a truss to deflect, the members at each connection must rotate slightly relative to one another. If the members at each connection are arranged with their centroids converging at a single point and are joined by a single bolt, split ring, or shear connector centered on that point, then the connection will act like a true pin or hinge, enabling unrestricted rotation. If, however, the connections are plated and the plates are welded to one another, then there will be a degree of fixity. As the truss deflects, the rotation of the members at each joint will be inhibited by the fixed geometry of its plates, inducing cross-grain tension, and potentially causing splitting. Because the deflection of a truss, like that of a beam, is roughly proportional to the ratio of its span to its depth, one way to minimize this problem is to avoid the use of unreasonably shallow trusses. Another is to secure the end of each member to a plate using a single fastener. The best is to avoid fixity altogether by letting the arms of the plate rotate with the members to which they are bolted.[8]

Timber design values in compression parallel to grain are considerably higher than those in tension for a given species and grade. By contrast, the longitudinal deformation of a given member will be identical regardless of the sense of the axial load. Just as it is customary for glulams to employ higher grade material in their bottom laminations than in their top laminations, it is sometimes necessary to employ different grade material in the bottom chords of flat timber trusses than in their top chords. Recognize, however, that unbalancing the construction of the truss in this way may shift the neutral axis up or down relative to the geometric centerline.

The change in length ΔL of an axially loaded member of length L, cross-sectional area A, and elastic modulus E, under force F, is given as:

$$\Delta L = \frac{(F)(L)}{(A)(E)} \quad [9]$$

In a long uniformly loaded parallel-chord timber truss with a significant number of regular panels, the lengths and forces in the segments of the top chord will be similar to those in the corresponding segments of the bottom chord. In this scenario, L will be constant in *all* chord segments, and F will be similar in *corresponding* chord segments. The deformation of corresponding chord segments will, therefore, vary only as the inverse of the product $(A)(E)$. The simplest way of avoiding differential deformation is, therefore, to make corresponding top and bottom chord segments of similar species, grade, and cross-section.

The allowable compression value for the top (compression) chord may be somewhat lower than the tabular value, depending on the member's slenderness ratio. This, in turn, depends on the manner in which the top chord is braced. With continuous lateral bracing, such as that provided by a timber floor or roof deck nailed directly into the top chord, the allowable value may be the same as the tabular, but the axial capacity could be lessened due to combined loading. If the deck bears only on purlins framing into the panel points of the truss, the top chord segments will be free of bending, but laterally unbraced.

Truss Geometry

Although timber trusses can be fabricated in a nearly infinite number of sizes and shapes, several of the more common configurations bear the names of their developers, such as Howe and Pratt. Variations intended for floor construction are known as flat Howe and flat Pratt trusses.[10] These truss configurations are distinguished by the orientation of their diagonal web members but, more important, by the senses of the forces in all of their web members. Under uniform top chord loading, whether it has a gable or flat form, a Pratt truss has tensile diagonals and compressive verticals, whereas a Howe truss has compressive diagonals and tensile verticals. (See Fig. 8.13.)

These sorts of differences can have significant implications with respect to the design of timber trusses. For a given span, depth, and panel spacing, the length of each and the total length of all *vertical* web members in a gabled Howe will be the same as in a gabled Pratt, but the length of each and the total length of all *diagonals* will be significantly less in the Howe than in the Pratt. On the other hand, the web members in compression in the Howe will be longer and will, therefore, have to be more substantial than those in the Pratt, to prevent buckling.

As previously discussed, Maxwell diagrams can be drawn at any scale. At a given scale, increasing the *loads* on a truss of constant *size* causes a proportional *enlargement* of the diagram—and, consequently, a proportional *increase* in the force in each member—but the overall truss shape and, therefore, the actual length of each member, stays constant. Increasing the *size* of a truss under constant *load* has no effect whatsoever on the diagram—the components of which are, after all, a function of loads and truss geometry, not member lengths—but will increase the *length* of each member and, therefore, the *buckling tendency* of those in compression.

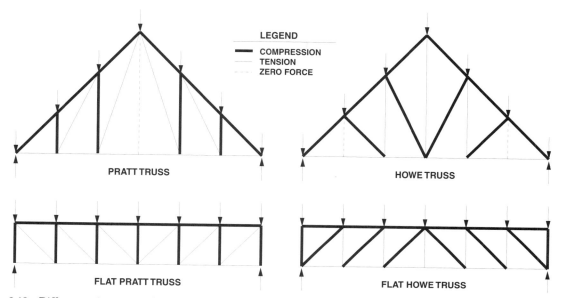

8.13 Differences in senses of internal forces in trusses of similar shape. The compressive web members in the Pratt trusses experience higher forces than those in the Howe trusses, but are shorter. Because buckling is a function of both load and length, each of these truss forms has one factor working for it and one working against it.

8

Timber Trusses

When comparing different trusses, especially those with similar panel spacing and span, it is often useful to take advantage of the fact that member forces are proportional to panel loads within each truss form. Substituting a variable for each of the panel loads yields force coefficients in the completed Maxwell diagram. The force in each member is then simply the product of its coefficient and the panel load.[11]

Let us apply this principle to our simple one-panel roof truss. Assume a single midspan load w on each of 3 trusses, each with the same span but a different slope. (See Fig. 8.14.) The load lines and the locations of points a, b, and c along them will be the same in all cases. As we have seen, the shallower the roof slope, the higher the forces in the top and bottom chords. In the Maxwell diagram for each truss, consider the right triangle of which the force line for each top chord is the hypotenuse. The "height" of this triangle is $w/2$. Since the slope of the hypotenuse in the diagram is the same as the slope of the top chord of the corresponding truss, the length of the "base" in the diagram must bear the same relationship to the "height" $w/2$ as the length of the bottom chord bears to the height of that truss. Therefore, the shallower the truss gets, the further *in* its space diagram is inscribed, and the further *out* its Maxwell diagram is circumscribed, compared to the respective diagrams of a deeper truss. Overlaying the corresponding space or force diagrams makes this relationship particularly apparent. But why, you may ask, it this important?

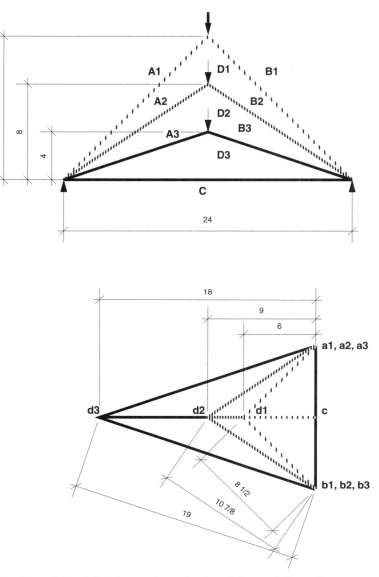

8.14 The relationship between truss depth and internal forces. Reoperate the truss model described at the beginning of this chapter. The increase in leverage evident as it flattens out is quantified in this diagram.

It is important because the designer can control the final appearance of a given truss only after gaining a basic understanding of the impact of changes in loading, slope, depth, and span on the forces in—and ultimately the sizes of—its members. But how do you come up with an overall truss form in the first place?

Think of a timber truss as a long timber beam. As in a beam, loads on a truss are resisted by reactions at its supports. As in a beam, shear is highest at the ends, and moment is highest at midspan, assuming a simple span and uniform loading. At any point along the length of a beam, its resistance to bending is a function of the square of its depth. The same is true of a truss. By convention and for economy, the typical timber beam is mass produced with a rectangular cross-section of constant size for its entire length. By contrast, a timber truss, like a glulam beam, is usually a custom product unique to its project. Production runs are limited in number. While timber trusses are ordinarily composed of conventional sawn or glulam members, their overall forms and dimensions are limited only by the imaginations of their designers. While the economics of solid beams mandate that the maximum moment experienced at any one point set the depth for the entire span, the economics of timber trusses allow the depth at each point of the span to correspond to the moment at that point.

To illustrate how the relationship between the moment diagram and the truss form affects member forces, consider a beam with a concentrated load w at each of its three quarter points, spaced a distance $l/4$ apart (see Fig. 8.15). Each reaction will be $1.5w$, the shear diagram will be stepped, and the moment at the first and third quarter points will be three quarters of the moment at midspan. Now, draw a simple gabled truss with joints at the midpoint of each top chord. Superimpose the beam's moment diagram, using a vertical scale such that the point of maximum moment coincides with the peak of the truss. Generate the Maxwell diagram.

Start a new drawing with the beam's moment diagram. Add a gambrel truss with panels at the same spacing and with the peak at the same elevation as the gable truss, but with the elevations of the first and third quarter points corresponding to the moments at that point. Generate the Maxwell diagram and compare it with that of the gabled truss.

Notice that the maximum forces in the top chord are lower in the gambrel than in the gable. Also, the forces in all top chord segments of the gambrel are relatively constant, while those of the gable are higher at the ends (where the truss is shallower) and lower in the middle (where the truss is deeper). In fact, the chord force is highest where the ratio of moment to truss depth is highest.

From this example, a general principle can be stated: The closer the shape of the truss matches the moment diagram of the controlling load combination, the more consistent will be the forces within the chords.

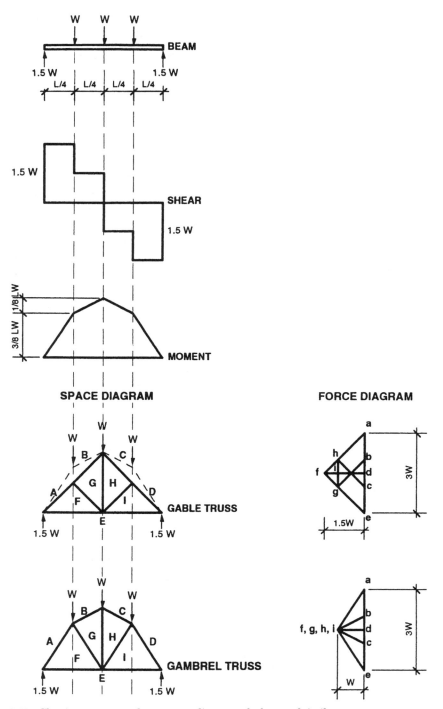

8.15   Shaping a truss to the moment diagram of a beam of similar span.

(For a more comprehensive discussion of this subject, see Ed Allen's "Finding Efficient Forms for Trusses," Chap. 16 of *Design of Building Trusses,* by Ambrose.)

Design Issues

Whether or not part of a truss, timber framing has three roles to play, each with its own requirements. The smallest size a member can be is the largest of the sizes satisfying each of these three roles:

- First, as part of a *construction type*—heavy timber—whose members must be at least large enough to assure adequate fire resistance

- Second, as part of a *structural system,* whose members must be at least large enough to accommodate the applied loads and the intended connections

- Third, as part of an *architectural conception,* whose members may be whatever size is necessary to achieve the designer's specific esthetic objectives

Although manipulating timber sizes and shapes for purely architectural purposes may reduce overall structural efficiency, it is a legitimate exercise. After all, the character of the completed timber structure derives directly from the layout and proportions of its members. The justification is really no different from that used for *entasis,* the slight convex curving of columns by the Greeks and Romans "to overcome the optical illusion of concavity that characterizes straight-sided columns."[12]

Structural honesty, in this context, might be seen as letting the sizes of the members follow from the forces within them. (See Fig. 8.16.) If applied to our gabled truss, the top chord would be heavier at the connections and lighter at midspan. In the gambrel, the cross-sections of the top chord segments would be more consistent than those of the gable (but perhaps not constant, due to the buckling implications of their varying lengths). Other possible architectural hierarchies include making all members the same size (akin to scantling[13]), making all compressive members one size and all tensile members another smaller size or, as so successfully done by Charles Rennie Mackintosh, using size for visual emphasis. Distinctions could also be made between the widths and thicknesses of the members. In general, I prefer to use the smallest possible sizes whenever possible. In the case of the building on the cover of this book, the members were sized to satisfy the minimum requirements associated with the construction type; purely from a structural perspective, some were underutilized.

Although they comprise a relatively small portion of a timber truss, the impact of connections on the overall appearance cannot be overemphasized. It is not surprising, therefore, that truss designers go to such

8.16   Typical timber and iron truss, Delaware Aqueduct, Lackawaxen, Pa. John A. Roebling, engineer. Each 28-ft (8.4-m) truss was suspended from iron cables by U-bolts, 5 ft (1.5 m) from its ends.
"The weight of the towpaths and bracing, and the pressure of the water against the trunk sides
acting through the inside diagonal struts, all bore downward on the cantilevered outer ends of the
floor beams, materially counteracting the stress imposed by the water load at the center and further lowering the total stress in the beams. These transverse beams were finally reduced to pairs
of 6 × 16s, spaced every four feet." (Vogel, p. 5.)

great lengths (pun intended) to embellish their connections. The quantities and spacing of bolts, the shapes, finishes, and colors of side plates,
the proportions and arrangements of the layers, and the locations and
orientations of members in the third dimension all contribute to the
rhythm and character of the truss system and to the architectural
space in which it is located. (See Figs. 8.17*a* and *b*.)

Summary

Timber trusses can be more efficient for intermediate spans than sawn
or glulam members, and can span farther than their practical limits.
Triangulation makes trusses geometrically stable and translates the
external loads applied to them into internal axial forces. Bending of
individual members can be avoided, and truss efficiency maximized, if
loads are applied only at panel points.

Graphical analysis enables member forces to be determined without
calculation, by relying on the fact that the resultant of the force within
each member is always parallel to the member itself. This technique

(a)

8.17 While economy of means has been one of the concerns of this chapter, the use of trusses can also lead to economy of erection. Some trusses can be thought of as assemblies of smaller trusses. In the truss shown, each half consisted of a trussed rafter composed of spaced wooden compression members and steel rod tension members. After erection, the pairs of trussed rafters were joined with a horizontal steel rod, creating a larger truss. Loaded only at its peak and supported only at its ends, it experiences only axial forces. To emphasize that its sole purpose was to support the ridge beam, a gap was left between the top of its top chords and the underside of the roof structure. (*a*) Trusses over main hall. (*b*) Structural details: Siegel Residence (Reprinted courtesy of *Fine Homebuilding*, The Taunton Press, Newton, CT.). Eliot & Risa Goldstein, architects. Nandor Szayer, structural engineer.

starts with a space diagram—whose dimensions correspond to those of the proposed truss—and ends with a force polygon—whose dimensions correspond to the forces in its members. The results yielded by graphical analysis are consistent with the quality of the construction of the Maxwell diagram. Although precision can be increased by enlarging the diagram, it is inappropriate for the precision of the forces to be greater than the relative certainty of the applied loads.

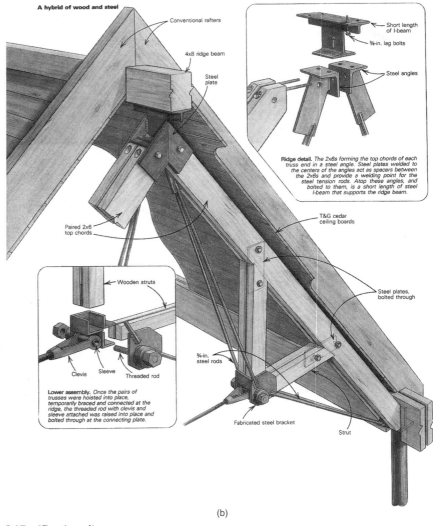

A hybrid of wood and steel

Conventional rafters

4x8 ridge beam

Steel plate

Short length of I-beam

⅜-in. lag bolts

Steel angles

Ridge detail. *The 2x6s forming the top chords of each truss end in a steel angle. Steel plates welded to the centers of the angles act as spacers between the 2x6s and provide a welding point for the steel tension rods. Atop these angles, and bolted to them, is a short length of steel I-beam that supports the ridge beam.*

Paired 2x6 top chords

T&G cedar ceiling boards

Wooden struts

Clevis   Sleeve   Threaded rod

Lower assembly. *Once the pairs of trusses were hoisted into place, temporarily braced and connected at the ridge, the threaded rod with clevis and sleeve attached was raised into place and bolted through at the connecting plate.*

¾-in. steel rods

Steel plates, bolted through

Fabricated steel bracket

Strut

(b)

8.17   *(Continued)*

   The speed with which truss forces can be graphically determined encourages the comparison of various truss silhouettes, panel layouts, and overall dimensions. Facility with graphical analysis enables the architect to lead the truss design process, rather than leaving it to the structural engineer.

   Understand the capabilities of timber trusses by exploring each of the variables related to their design, including span, shape, panel spacing, bracing, hierarchy, and connections. Understand the implications of various load combinations, giving special attention to load reversals. Superimpose the moment diagram on the space diagram to help identify controlling members; recognize the structural penalties associated with significant differences between their shapes.

When designing trusses, remember that they must satisfy simultaneously three sets of requirements: code mandated, architectural, and structural. While the code requirement can be satisfied in only one way—by sizing the members and detailing the connections so they qualify as slow burning—the structural and architectural requirements can be satisfied in as many ways as can be imagined by the designers.

Notes

1. Flexner, p. 1929.
2. Ambrose, p. 423.
3. Allen, *How Buildings Work,* p. 241.
4. There is some disagreement on this point. Ferguson (p. 150) reports that the graphical technique of analyzing trusses was developed at approximately that time by Karl Culmann of the Federal Technical Institute in Zurich, Switzerland.
5. Ambrose, p. 51.
6. Packard, p. 760.
7. Ambrose, p. 253.
8. TCM, p. 7–695.
9. Ambrose, p. 256.
10. Packard, p. 317.
11. Ambrose, p. 399.
12. Harris, Cyril, p. 184.
13. Mark, p. 191.

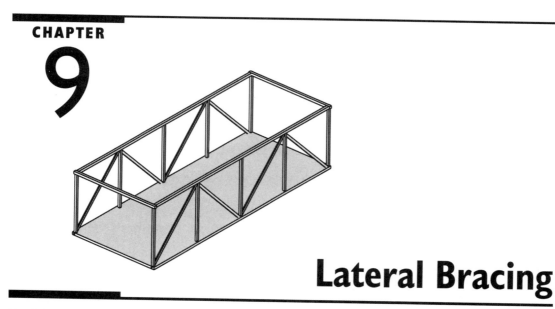

Lateral Bracing

Fred Severud, PE

And the rain poured down and the floods came and the winds blew
and struck against that house and it caved in, and its collapse
was great.

MATTHEW 7:27

Introduction

The design of a structure must consider loadings from several directions: vertical, horizontal, and inclined loads. Vertical (gravity) loads come from the weight of the building components (dead load) and the weight of snow, ponded rain, and building contents, including occupants (live load). Generally speaking, gravity loads are applied to the roof and floor systems and transferred to vertical elements (columns and walls) that deliver the load to the foundation. Horizontal and inclined loads are the result of wind and seismic (earthquake) forces.

All of these loads must be resisted by the framework of the building as a whole. (See Fig. 9.1.) The framework consists of all the structural components: columns, frames, beams, bracing, floor and roof diaphragms, sheathing, and any other components that participate in resisting or transmitting forces. In most structures, part of the framework carries lateral loads; part carries gravity loads; and sometimes, a part is involved in carrying both. This chapter considers the interactions between the members in these three categories.

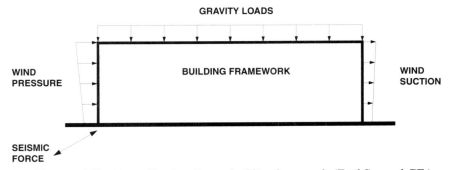

9.1 Types and directions of loads acting on building framework. (Fred Severud, PE.)

Load Paths

As loads are applied to surfaces of the building, they create forces in the supporting members, which are then transferred to the framework. For example, wind on a wall puts the wall studs in bending, and the reactions of the studs put loads on the edges of the floor assemblies at the top and bottom of the stud. A floor assembly acts as a diaphragm to transfer the load to the building frames that must be in the appropriate locations and orientation to resist them. Similarly, vertical loads, such as the weights of occupants, are applied to the floor surface, which is carried by the floor beams. The beam reactions are carried by bearing walls that deliver the load to the foundation; or to girders, which are carried on columns sitting on the foundation. If the forces in the members exceed their capacity, failure results. Failure can be evidenced by too much deflection, or by collapse. Design of a structure, then, involves determining the failure loads for the various components of the structure, and selecting members with sufficient strength to resist all probable loading conditions without reaching failure.

Of course, the design must also consider economy. The best design is worthless if it is so expensive that no one can afford to build it. Due to uncertainties in loading conditions and variations in materials and workmanship, the stresses around which the structure is designed (the *allowable* stresses) are set well below the stresses that would cause failure (the *ultimate* stresses). The ratio between the ultimate and allowable stresses is the *factor of safety*.

Code Requirements

The local building code at the site of a project will determine the vertical and lateral forces for which we must design. The purpose of a building code is to set minimum standards for the protection of the occupants. Most local codes are based on one of the standard (or model) building codes, such as the International Building Code (IBC), the BOCA (Building Officials and Code Administrators) Code, the Uniform Building Code

(UBC), and the Canadian Building Code (CBC). Many large cities, such as New York and Chicago, have their own stand-alone codes. The examples in this chapter use IBC or BOCA. Factors in other codes may differ from those shown, but the general relationships and concepts are similar.

Keep in mind that code requirements are minimums; not all situations are, can be, or should be covered by codes. There is no substitute for careful calculation, good planning, and informed judgment.

Wind Loading

Several factors must be considered when evaluating wind loads, such as building location, building shape and height, exposure (amount of shielding), importance, and wind gusts.

Building location determines the basic wind load at ground level. For most areas, wind contour maps are included in the local or model code. Lines of equal velocity—"wind contours"—indicate the basic wind speed. Wind pressure is determined from the equation $P_v = 0.00256\,v^2$, where P_v is the resulting pressure in pounds per square foot (psf) from wind at a speed of v miles per hour (mph). For points between contour lines, the designer may interpolate to find the design wind. Several factors are noticeable:

1. The highest winds occur near the coasts, diminishing rapidly as you go inland. As an example, on the East coast of the United States, design velocities decrease from 110 mph (180 kph) on the Carolina coast to 70 mph (110 kph) just 150 miles (240 km) inland, while in Alaska, the same decrease occurs in less than 50 miles (80 km).

2. There are "hot spots" in mountainous regions and along shores of large inland lakes, where local topography causes extremely sharp increases in wind velocities. Pressure resulting from wind resistance is released as the wind spills over the mountain or passes over the lake. The areas where this occurs are indicated by shaded regions on the map.

3. Large, open, relatively flat areas, such as the plains states in the United States, experience high wind velocities, due to the lack of obstructions.

Building shape is reflected in constants that modify the basic wind pressures. Both windward (pressure) and leeward (suction) forces must be considered, as well as internal pressures. Wind load design addresses not only the structure, but also the cladding, the roofing, and the other components exposed to wind forces. The silhouette of a building affects *local* wind conditions just as a mountain affects *regional* wind conditions. Therefore, buildings with sloped roofs have special design coefficients that vary based on the number and angles of the roof planes, as well as on the sizes of the overhangs and the number of building sides (if any) that are open. (Obviously, the covered bridges discussed in Chapter 15, with their open ends, experience very different wind loads than the library discussed in Chapter 14, with its glazed ends.)

Building height is reflected in factors that increase with elevation. In most codes, the varying pressures are shown in a table. Fig. 9.2 shows how the pressures increase for a typical case, a building 60 ft (18 m) high, with a basic wind speed of 100 mph (160 kph), and a moderate amount of shielding. Height is shown in feet and pressure is shown in pounds per square foot (psf). In addition to its effect on design pressures, building height may also determine the type of design that is acceptable. Low-rise buildings (mean roof elevation less than 60 ft [18 m] or less than the least horizontal dimension) generally may be designed for lower wind pressures than taller buildings. Most timber buildings fall into the low-rise category.

Exposure has four categories in many codes: Exposures A, B, C, and D. The most severe is Exposure D, which is at the edge of a body of water greater than 1 mi (1.6 k) wide, with no shielding. Exposure A is in an area shielded by tall buildings in a large city. Exposures B and C are between the extremes of A and D.

Importance of a building is reflected in the factor *I*. Essential facilities—those involved in emergency response or used for shelter or treatment in emergencies, such as hospitals and those that normally

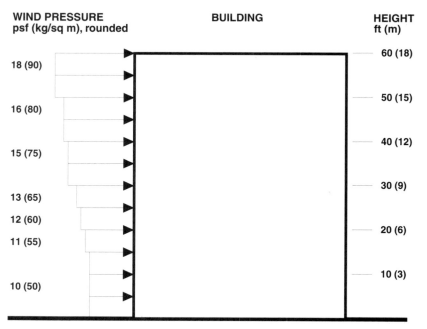

9.2   Code-mandated wind pressures as a function of building height. (Fred Severud, PE.)

house a large number of people—have the highest importance factor, while lesser uses have lower factors.

Gusts are considered using gust response factors. These are a function of exposure and building height.

Total wind loading on the building is the basic wind load, multiplied by all the above factors. For example, for a building 60 ft (18 m) or less, $P = P_v I K_h [G\,C_p - G\,C_{pi}]$

where P = total wind load at a particular elevation

P_v = basic velocity pressure (from map, based on topography)

I = wind load importance factor (from code, based on occupancy)

K_h = exposure factor (from code, based on local shielding)

$G\,C_p$ = product of external pressure coefficient and gust response factor

$G\,C_{pi}$ = product of internal pressure coefficient and gust response factor

Because most timber structures are 60 ft (18 m) or less in height, this or a similar formula will usually apply. The wind loading for a building frame will be applied at the appropriate locations and at the intensities calculated from the above data. In calculating the effect of wind on a structure, wind loads are usually applied as concentrated loads at each floor of a lateral force-resisting frame. When more than one frame resists lateral forces in a structure (the usual case), the loading will be applied to each frame according to its relative stiffness. This is based on the fact that the more flexible frames will deflect under smaller loads so that the stiffer frames will end up resisting the rest. Due to the floors' stiffness, the frames can be considered "linked" together, so that they all deflect the same amount. In this way, a three-dimensional structure can be represented by several two-dimensional frames, tied together by inextensible links.

9

Seismic Loading

Philosophy of Design

Because it is impossible to predict accurately the magnitude of earthquake loadings, seismic design must be somewhat conservative. However, if we designed structures to be completely safe against damage during the greatest quake that could possibly occur at a site, the cost would be prohibitive. Therefore, when designing for a major seismic event, the normal approach is to protect occupants against death and injury, but to accept some structural damage to the building. The degree of damage that is considered acceptable depends on the usage of the building (see "Importance" in the following).

Factors

As with wind loading, the seismic lateral loads depend on several factors, including building location, importance, building shape, framing system, and foundation soil profile. The major difference is that seismic loading is a function of the magnitude and distribution of the *mass* of the building, acted upon by accelerations applied through the foundations, whereas wind loading is a function of the *silhouette* of the building, acted on by external forces.

Location

The theory of plate tectonics states that the surface of the earth is made up of several moving "plates." These plates float on the underlying layers, which act as though they are plastic or semifluid. At their boundaries, the plates tend to collide and grind against each other, building up potential energy. Movement or failure of part of a plate edge releases that energy in the form of an earthquake. The resulting seismic forces act like an underground explosion, shaking the earth and any buildings supported by it horizontally and vertically, in a complicated way. Shock waves can be felt for hundreds of miles, especially in a severe quake. The closer the structure is to the Fault Line (the plate boundary), the higher the seismic forces. For a given geographic area, past earthquake history is reflected in seismic zone maps, which give applicable acceleration factors. Seismic maps are similar in concept to wind contour maps mentioned previously. Areas that have experienced the most seismic activity have the highest factors; areas with little or no seismic history have the lowest factors; and intermediate areas have correspondingly intermediate factors.

These accelerations may be used to develop static load factors or to create dynamic response models. Just as concentrated loads can be converted into equivalent uniform loads, dynamic seismic loads can be converted into equivalent static loads. Static load factors approximate the dynamic forces generated in an earthquake and may be used in lieu of more complicated dynamic analysis of the structure. However, in severe quake areas, and with buildings deemed to be highly important, a more rigorous analysis may be necessary. This is usually done using three-dimensional computer models and time-history force input, taken from a previous earthquake in the area.

Along with the time-history input, the natural frequency of the building must be determined. The most severe damage and danger to occupants is when resonance occurs, where quake excitations are in phase with the frequency of building oscillations. The result is a build-up of force with each successive pulse. Most quakes have a long natural period (in the range of fractions of a second to several seconds). Low-rise buildings (including most timber structures), generally have shorter natural periods; therefore, resonance is less likely to occur with them than with taller, more slender, and more flexible buildings.

Importance

Buildings are generally divided into three Seismic Hazard Exposure Groups, depending on use and occupancy of the buildings. Importance, as defined for seismic hazard, is similar to, but not the same as, that for wind hazard. Buildings that are essential for postearthquake recovery are listed in Group III, buildings that have a substantial risk to the public—due to their occupant load or, in the case of power stations, due to the consequences of their failure—in Group II, and all others in Group I. Examples of Group III buildings are fire, rescue and police stations, hospitals, buildings containing hazardous materials, primary communication facilities, and power-generating facilities needed for postquake recovery.

Based on the effective peak velocity-related coefficient and Seismic Hazard Exposure Group, the building is assigned a Seismic Performance Category, from a low of A to a high of E. This category will determine the criteria for designing the seismic force-resisting system for the building. As you would expect, structures in the higher importance categories must be designed for higher loads. If a building has several occupancies, the importance factor will correspond to the most restrictive.

Building Framing System and Shape

The construction types providing lateral stability in conjunction with the Seismic Performance Category determine the design methods that may be used. The types of construction are: *moment resisting* (those using space frames to resist lateral loads), *shear walls/braced frames;* and *dual* systems (those using a combination of moment resisting frames and shear walls/braced frames). The basic difference between moment resisting and braced frames is that braced frames have nonrigid connections (idealized as "pins") and diagonal braces, whereas moment resisting frames' connections are considered rigid, so that no bracing is required. Shear walls are elements that must be stiff in the direction of loading; since we don't know this direction in advance, they must be provided in the major directions of the structure to allow for loading in any direction. In all cases, lateral stability must be provided for the full height, breadth, and depth of the building. (See Fig. 9.3.)

The interaction between horizontal and vertical lateral force resisting elements is critical. Consider a long narrow warehouse, for example. In one approach, braced vertical frames could be provided at regular intervals along its length. If these were too costly or interfered with the use of the building, the lateral bracing could be moved to the perimeter. In this alternative approach, the roof could be designed as a horizontal diaphragm running the length of the building, tied into shear walls at its ends. Just as the mass at the flanges of a beam is more effective in resisting bending under vertical loads than the mass closer to the beam's centroid, shear walls at the perimeter of a building

9.3   Commercial building with diagonal timber bracing on the outside, in Aspen, CO.

are more effective in resisting twisting under lateral loads than those closer to the building's center of mass.

Lateral restraint systems of different types may be combined within a given floor or may carry between floors. Because they are solid and can interfere with space planning, internal shear walls are usually located next to fixed elements, such as stairs or elevators. If shear walls are required on the perimeter, their locations must be coordinated with window and door openings.

A building under lateral loads acts like a cantilever fixed at the ground. The lateral load capacity of such a structure needs to be greater near the bottom than near the top. (The situation is somewhat analogous to gravity loads, which accumulate as you descend.) In a seismic event, the response of the building will be a function not only of the *horizontal* distribution of its mass, but also of its *vertical* distribution. It is better for the *mass* to be concentrated near the bottom than near the top of the building.

While structural continuity is easier to achieve if the vertical bracing system is the same type and in the same location on every floor, this is not always possible. Architectural requirements may mandate the use of a more open lateral bracing system on the ground floor than on upper floors. Alternatively, shear walls may have to be offset to accommodate rooms of different sizes.

Yet another issue is whether the bracing is inherent to the structural frame or provided by another system. Timber construction can go either way. In some buildings, the lateral resistance is provided by timber frames braced with knees and other diagonals (see Fig. 9.4); in others, the resistance comes from rigid concrete or masonry shear walls or from stair or elevator shafts. Sometimes, the bracing consists of a combination of these two approaches. Regardless of the specific systems employed, the connections between the elements resisting lateral loads and the rest of the structural frame are critical. Design must allow for shrinkage of the timber frame and for differential thermal movement between the timber frame and other structural elements. When properly oriented, connectors with slotted holes can enable adjacent systems to move independently in certain directions but together in the direction in which bracing is required.

Building regularity or irregularity in shape must also be considered. The ideal structure for seismic design is symmetrical. Offsets in shape, as in "L"-shaped buildings, tend to be more adversely affected by seismic loading. The complex motions caused by earthquakes make the separate wings move independently. This causes them to tear apart at reentrant (inside) corners, as where wings meet. To prevent this, separate the wings; make two buildings instead of one, or design the reentrant corner to resist the resultant forces. An alternate solution, which lessens the effect, is to "soften" the corner, perhaps by introducing a diagonal portion. In Fig. 9.5, the plane skewed in plan represents the preferable boundary of the building.

9.4 Unusual lateral bracing in a barn which was recently renovated into a private club. The structure looks something like, and probably behaves something like, a rigid timber frame. Park Avenue Club, Florham Park, NJ.

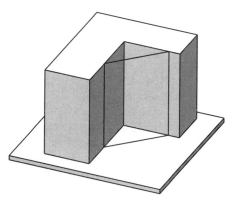

9.5 Problematic reentrant building corner and preferred alternative configuration. In a building subject to seismic forces, as in a timber beam subject to bending forces, stress concentrations can be minimized if reentrant corners are avoided. (Fred Severud, PE.)

Foundation Soil Profile

The rigidity of the foundation soil has a major effect on the seismic design forces. The site coefficient S may range from $S_1 = 1.0$ to $S_4 = 2.0$ S_1 represents rock or stiff soil conditions; S_4 represents soft clays or silts; S_2 and S_3 are intermediate conditions. Basically, structures on rock or very stiff soils are much less likely to be damaged than the same structures on softer soil. Testing of buildings in various soil profiles has confirmed that soils that transmit vibrations at high speeds (high-frequency soils) cause less damage than those that do so more slowly. The denser or harder the soil, the higher the frequency. Bedrock is ideal; soft silt and clay soils are the worst; medium-to-dense sand is intermediate. Stiffer soils, like stiffer buildings, are less prone to resonance.

Building Components

Unlike wind design, seismic design must consider the safety and attachment to the structure of nonstructural components of the building, such as elevators, ceilings, window walls, mechanical and electrical system components, and other materials and equipment. These elements must be attached to the structure in a manner that minimizes damage to essential services during a seismic event. Because some are substantial in size and weight, they could also be dangerous to occupants if they were to break loose during such an event.

Total Seismic Loading

Total seismic loading consists of the dead load of the building and/or building components, multiplied by the applicable factors for structure, location, and site. Since they tend to move independently of and out of phase with the structure, live loads are not considered contributors to seismic loading conditions.

Design may be based on the *Equivalent Lateral Force Procedure* (ELFP) or the *Modal Analysis Procedure* (MAP). In both methods, the base of the building is considered fixed. In the ELFP method, as in graphical analysis of trusses (Chap. 8), the equivalent static forces are applied to the appropriate joints of the structure, and the resulting member forces are calculated. In the MAP method, the building is "modeled as a system of masses lumped at the floor levels with each mass having one degree of freedom; lateral displacement in the direction under consideration." While the ELFP method may be applied using relatively simple hand calculations, the MAP procedure requires the use of computers. Which type of analysis is appropriate depends on many factors, including:

- Importance of the building (high importance factor requires modal/dynamic analysis)
- Complexity of the shape (complex shapes require modal/dynamic analysis)
- Degree of seismic risk (buildings near fault lines are more likely to experience seismic events, so require modal/dynamic analysis)
- Building mass (heavier buildings experience higher forces, so require modal/dynamic analysis; timber buildings are usually light enough for static analysis)
- Building height (taller buildings experience higher forces, so require modal/dynamic analysis; timber buildings are usually low enough for static analysis)

Combined Loads

Based on statistical analysis, it is not probable that the structure will experience all types of loads simultaneously. Therefore, all codes allow increases in allowable loading or stress for unlikely load combinations. For example, when wind and earthquake loadings are added to gravity loads, a reduction factor is applied to the lateral loads. The reduction factor differs with different codes and with the type of design methods applied.

For both of the following approaches, the following notations apply:

D is dead load, defined as the weight of all materials of construction included in the building, i.e., all permanent gravity loads.

L is live load, except roof live load, including any permitted live load reductions. Live load is defined as ". . . loads produced by use and occupancy of the building. . . ." Live loads "do not include construction or environmental loads such as wind load, snow load, rain load, earthquake load, flood load or dead load." Based on probability theory, live loads may be reduced on the basis of area; as area supported increases, the percent of live load applied decreases.

L_r is roof live load, including permitted live load reductions, if any.

S is snow load.

W is wind load, load due to wind pressure.

E is combined effect of horizontal and vertical earthquake-induced loads.

With IBC, using strength or *Load and Resistance Factor Design* (LRFD), structures and all portions thereof must be designed for the most critical effects from the following combinations of factored loads:

$$1.4D$$

or

$$1.2D + 1.6L + 0.5(L_r \text{ or } S)$$

or

$$1.2D + 1.6(L_r \text{ or } S) + (f_1 L \text{ or } 0.8W)$$

or

$$1.2D + 1.3W + f_1 L + 0.5(L_r \text{ or } S)$$

or

$$1.2D + 1.0E + (f_1 L \text{ or } f_2 S)$$

or

$$0.9D +/- (1.0E \text{ or } 1.3W)$$

where f_1 = 1.0 for floors in places of public assembly, for live loads in excess of 100 psf (4.9 kN/m²), and for parking garage live loads
    = 0.5 for other live loads
    f_2 = 0.7 for roof configurations (such as sawtooth type) that do not shed snow
    = 0.2 for other roof configurations

If working strength design or *Allowable Stress Design* (ASD) is used, IBC requires that structures and all portions thereof must be designed for the most critical effects from the following combinations of factored loads:

$$D$$

or

$$D + L + (L_r \text{ or } S)$$

or

$$D + (W \text{ or } E/1.4)$$

or

$$0.9D - E/1.4$$

or

$$D + 0.75[L + (L_r \text{ or } S) + (W \text{ or } E/1.4)]$$

Increases in allowable stresses specified in the appropriate materials section of the code may not be used with these load combinations except that a duration of load increase shall be permitted, as indicated in the chapter of the code referencing wood.

Advantages of Timber Framing

The driving force in seismic loading is the mass of the building. As the earth shakes, inertia makes the mass want to continue to travel in the direction it started to move. In the vicinity of the building, however, the earth changes direction cyclically during the quake. A lighter structure, then, can follow the earth motion more closely than a heavier one. Since timber structures tend to be lighter than structures made of other structural materials, they tend to be less sensitive to seismic loading. This provides a distinct cost advantage, since the added cost for lateral loading is minimized.

As previously noted, there are "duration of load" increases for timber framing, which do not apply for any other structural materials. This means that the allowable stresses may be increased for loads that are only applied for a short duration. When load duration is other than "normal," NDS recommends that the allowable stresses be multiplied by the following factors:

0.90 for permanent load (more than 10 years)

1.15 for 2-month duration (as for snow)

1.25 for 7-day duration

1.60 for wind or earthquake

2.00 for impact

(These factors apply for Allowable Stress Design. If Load Resistance Factor Design is employed, the factors are different, but the concept is the same.)

Because wind and earthquake loadings are of relatively short duration, the effect of these provisions is to recognize that a timber structure is relatively more resistant to lateral loadings than a structure of another material with similar design factors. In other words, there is less increase in the cost of a timber structure caused by lateral loads than there is in a structure of other materials.

Types of Lateral Support Systems

Moment resisting frames are made of rigidly connected members, i.e., the original angles between the members are not allowed to change due to loading. In timber structures, this can be accomplished by using laminated arches or special connections between members which prevent relative rotation.

Braced frames use the principle of triangulation to provide rigidity. A triangle is the most stable geometric form; any other shape made up of "stick" members can change *shape* without requiring that the members change *length*. Braced frames are made up of triangles, similarly to bridge trusses. If each panel of the frame consists of a triangle, the frame is stable without rigid connections. (See Fig. 9.6.) This greatly simplifies the design of the connections, since no moment (bending) need be transmitted. This also simplifies the design of the frame; since the members take only axial forces (tension or compression). For examples of different types of bracing, see Fig. 9.7.

9.6 Diagonal bracing of timber-framed hayloft. New glulam columns, girders, and braces support timber beams from a dismantled railroad trestle. Bassett Barn, Cambridge, VT, Butternut Construction, Inc.

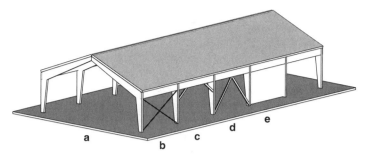

9.7 Examples of common lateral bracing systems for timber build-ings. (*a*) Rigid frames. (*b*) Tension braces. (*c*) Knee braces. (*d*) Diagonal braces (chevron configuration shown). (*e*) Shear walls. Through appro-priate engineering, it is usually possible to achieve adequate overall lateral rigidity without bracing every bay. (Fred Severud, PE.)

An instructive example of a combination of roof trusses and bracing is the Montville, NJ, library, designed by the author of this book and engineered by myself (see Chap. 14). The main structure is composed of 30-ft (9-m)-wide gabled trusses, on a 10-ft (3-m) spacing. The main trusses are "scissor" type, which produces a vaulted ceiling effect. The side bays are 15 ft (4.5 m) wide and have shed roofs. The outside bays have a flatter slope (3/12) than the top chords of the central bay (9/12) but match the slope of their bottom chords. The plan of the library is essentially cruciform, with four wings around a central atrium. Between each pair of trusses is a system of diagonal braces providing lateral restraint in the other direction.

In heavy-timber construction, the main roof trusses must be braced at the bottom chords, truss to truss. (Top chords are normally braced by the diaphragm action of the roof decking.) In this library, the longitudinal braces form chevrons, or inverted "vees" between the main trusses. This allows lighter framing, since the purlins are thereby supported at mid-span. In addition, the chevrons repeat the vaulted effect longitudinally and eliminate the need for horizontal members at the bottom chords of the roof trusses. Near the tops of the columns are knee braces in two directions. Those oriented laterally brace each roof truss; those oriented longitudinally brace the entire group of trusses. The knee braces also cut the lengths of the columns, making them stiffer for lateral as well as for axial loads. For a photograph of the framing system, see Fig. 9.8.

One benefit of heavy-timber bracing is that it can be left exposed, because it is inherently fire resistant. Heavy timber will char in an intense fire, but will take a considerable time to ignite. Steel, on the other hand, will buckle and lose strength as it is heated. Steel must therefore have a fireproof cover of masonry or other insulating mate-rial. This fireproofing adds cost to the structure, adds another trade, and adds weight.

9.8 Three levels of chevron bracing. To avoid combined loading of the bottom chords of the trusses, the braces spring exclusively from panel points. New Public Library, Montville, NJ. The Goldstein Partnership, architects. Severud Associates, structural engineers.

Shear walls are vertical, and can be of concrete or masonry or of timber or stud construction with diaphragms of timber or structural panels. An advantage of concrete shear walls (block or cast-in-place) is that they can also serve as fire walls (see Chap. 4). If plywood diaphragms are used in heavy-timber construction, the thickness of the plywood must conform to heavy-timber requirements, i.e., not less than 1⅛ in (30 mm) in thickness. Alternatively, tongue-and-groove decking may be used. Decking is usually placed diagonally to the support members.

Sheathing, the outside wall structural facing, is used to transmit lateral loads to the force-resisting system. Diagonal tongue-and-groove decking, plywood, or particle board, may be used for sheathing. To qualify as part of a system of lateral restraint, sheathing must be sufficiently thick and must be attached to framing in accordance with code requirements.

Diaphragms are floor or roof systems designed to transmit lateral loads to the frames or shear walls, acting in compression. They are generally horizontal or nearly so. To produce the strongest system, all edges of an individual panel are attached to framing members or blocking attached to framing members, thus producing a *blocked diaphragm*. For a sketch of a blocked diaphragm, see Fig. 9.9. Since buckling is a major factor in all members partially or entirely in compression, blocking is very effective in stiffening the relatively thin plate of plywood or similar material that forms the diaphragm, especially at its edges. As with columns (Chap. 5), the greater the slenderness ratio of the diaphragm, the lower the load at which it will buckle. Therefore, floor diaphragms with widely spaced framing need heavier decking not only to resist bending under gravity loads but also to resist buckling under lateral loads. Due to their low buckling resistance, unblocked diaphragms may not be used in the most severe building classification, Category E.

Buckling can occur when axial compression is applied to a slender member; at the critical load, the member will become unstable and deflect laterally. The key factor in buckling is slenderness, or thickness in relation to length. The more slender the member, the lower its critical buckling load. This effect was quantified first by Euler in 1759. He proved mathematically that the critical load for elastic buckling is $\pi^2 EI/L^2$. Since π is a pure mathematical constant (approximately 3.14159) and E is an elastic constant for a given material, the remaining quantities are I and L. L is the length, and I is a function of the geometry of the member—I/L is higher for members whose thickness is larger in relation to length. For example, if E for a particular grade of timber is 1,200,000 psi (8,274,000 kPa), L is 10 ft (3 m), and the member is 3½ in (90 mm) square, the critical load P is 10.3 k (4.68 tonnes). If the cross section is reduced to 3 in (75 mm) square, the critical load becomes 5.55 k (2.52 tonnes). As we can see, a small decrease in cross section results in a large decrease in the buckling load. Also,

9

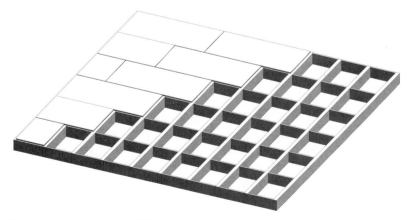

9.9 Isometric drawing of blocked diaphragm. Ends of decking panels bear on and are fastened to support members; edges bear on and are fastened to solid blocking. (Fred Severud, PE.)

since length is squared in the buckling formula, reducing the length by one-half increases the critical buckling load by a factor of 4. In the case of blocked diaphragms, much of the applied load is transferred into the blocking. Since the blocking is much less slender than the diaphragm, this transfer greatly increases the critical buckling load of the assembly, compared to that of an unblocked diaphragm.

Plywood is the most commonly used material for diaphragms in modern timber structures. Before plywood became generally available, diagonal tongue-and-groove boards were generally used. Plywood has several advantages over board sheathing: It comes in larger sizes so requires less labor and nailing; it is cross-laminated so is stronger and more rigid for in-plane forces; and its engineering properties are more consistent, leading to greater structural efficiency and more accurate design. Plywood sheets are typically 4×8 ft (1.2×2.4 m); supports are normally spaced so that plywood end joints occur over them. Whether diaphragms are laid out with the long dimension or the short dimension perpendicular to the support members, typical support beam spacings are therefore 12 in (300 mm), 16 in (400 mm), and 24 in (600 mm). This translates to 8, 6, and 4 spaces per panel, respectively, in the long direction, or 4, 3, and 2 spaces in the short direction. The size and spacing of nails or screws for panels affects the strength of the diaphragm, so the structural tables of the American Plywood Association base the strength of various diaphragms on layout and attachment.

Tension systems can be an economical method of transmitting lateral forces. Tension-only members, such as steel rods or cables, may be used with timber frames (which take the resulting compression forces), but only if they are protected—with intumescent paint, for example—to meet the fire-resistance requirements of heavy-timber construction. Timber may also be used for tension members, but steel's greater tensile strength leads to smaller members. A further problem with timber tension-only members is the reduction in net cross-sectional area caused by drilling for bolt holes (see Chaps. 6 and 7). The advantage of timber for these members, however, is its ability to take other loadings, such as wind, without excessive "flutter" (vibration), and to accommodate stress reversals if they occur.

An example of a steel tension system in a timber building is the utility building the authors designed for the Anheuser-Busch Company in Newark, NJ. The main supports for the building were glulam arches, which carried lateral loads in their planes. In the other direction, the lateral loads were carried by steel tension rods in a "V" pattern. Horizontal compression members of timber span between the arches. Once the building was finished and the necessary forces were developed in the tension rods, the turnbuckles were welded, to prevent any malicious or thoughtless tinkering. The reinforcing steel in the foundation mat took the lateral thrust at the bases of the arches. The result was a very inexpensive and efficient lateral load system. (See Fig. 9.10.)

9.10 Tension braces that alternate direction in alternate bays. Product Distribution Building, Anheuser-Busch, Newark, NJ. The Goldstein Partnership, architects. Severud Associates, structural engineers.

9

Lateral Bracing

Combined (dual) systems use the various elements of the lateral bracing system to resist loading in proportion to their individual stiffnesses. For seismic loads, a moment-resisting frame must resist at least 25 percent of the load, with shear walls or braced frames carrying the balance. There is no such requirement for wind loading.

Summary

Lateral loadings generally are the result of wind or seismic events. Various parts of the structural framework may be used to resist lateral and vertical loads. The applicable building codes give minimum requirements for estimating these loads.

Wind loads vary from place to place; they are functions of location, building shape, building height, exposure, importance, and gusts.

Seismic loads are functions of location, importance, building shape, framing system, building mass, and foundation soil profile. Because quake magnitudes are unpredictable, some structural damage is tolerated, but the design attempts to prevent death or injury to occupants.

Combined loads occur, but the probability is that not all loads will occur *at full intensity* at the same time. Therefore, codes allow for increases in allowable loads under combined loadings.

The **advantages of timber** are that timber-framed structures tend to be lighter than those framed with other structural materials. Because the driving force in seismic loading is mass, seismic loads are smaller in timber buildings than those of steel or concrete. There are also increases in allowable stresses in timber due to the short durations of wind and earthquake loads. These do not apply for other materials.

Types of support systems that can resist lateral loads include moment resisting frames, braced frames, shear walls, sheathing, diaphragms, tension systems, or combinations of these systems.

10

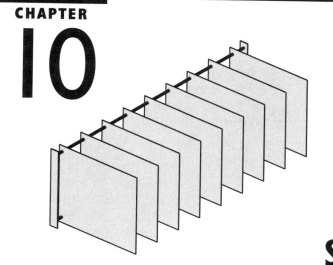

Timber
Specifications

If you don't ask, you don't get.

RICHARD SAUL WURMAN, *Follow the Yellow Brick Road*

Introduction

Drawings and specifications are the medium through which designers communicate their intentions to builders. Contractors rely on the contents of these documents to prepare bids, schedule manpower, and perform construction. Inconsistencies or incomplete aspects in the contract documents may increase bid prices, reduce job site productivity, or lessen overall quality. Worse yet, such deficiencies may lead to change orders and to litigation against the designer.

The drawings and specifications must complement one another. Avoid duplication (what is covered in one should not be covered in the other) but be complete (what is *not* covered in one *should* be covered in the other). The form of the information should be consistent with the form of the document in which it appears, and all information should be conveyed as efficiently as possible. Therefore, details and other graphic displays belong in the drawings, while descriptions belong in the specifications. A picture is worth (putting in the drawings, if it saves) a thousand words (in the specifications).

In some cases, the specifications cover global issues, while the drawings cover local ones. Even if you show a material in every plan, section, and detail, you may not be able to communicate your intention that it be used throughout the building unless you make a general statement to that effect. The proper place for such a statement is in the specifications.

This entire chapter is devoted to specifications, specifically timber construction specifications, because I believe that an understanding of their structure will enhance communications between designers and builders, increasing the quality and cost effectiveness of the resulting buildings. Whether you are an architect, an engineer, or a contractor, I hope the pages that follow give you an increased appreciation of the role of the specifications, so that they no longer sit unopened on the shelf during construction of your next job.

Specifications as Instruction Sets

Every profession has its own unique vocabulary. The vocabulary of building construction is distinguished by the dialects belonging to its individual trades, of which timber construction is one. Timber specifications need not be comprehensible to the layman, but they must communicate clearly and consistently to the timber contractor. Richard Wurman's book, *Follow the Yellow Brick Road,* while not specific to the construction industry, is a valuable reference in working toward this objective.

Wurman identifies five components common to all instruction systems, which I have grouped under three headings:

1. The environment
   a. Context
2. The parties
   a. Instruction givers
   b. Instruction takers
3. The instructions
   a. Form
   b. Content[1]

The environment in which specifications are prepared and interpreted is that of a construction project. More particularly, the project may involve new construction, renovation, or a combination. Our focus is timber construction, but in the context of an entire project, not in isolation. Because the type of construction contract—single general contractor with subcontractors versus multiple prime contractors—determines the roles and responsibilities of the parties, it should also be considered a fundamental aspect of context. But perhaps most critical to a timber contractor's understanding of a particular project are the locations and characteristics of the physical and conceptual boundaries of the timber construction work within the overall project.

The parties who produce and act on construction specifications obviously include designers and contractors, but it would be an oversimplification to stop there. Explicitly or by reference, specifications incorporate instructions prepared by consulting engineers, testing labs, trade associations, code agencies, and others outside the architect's office, and are read and interpreted by manufacturers, suppliers, estimators, attorneys, and others working for or on behalf of the timber contractor. With respect to the production and processing of construction specifications, then, the role of the contractor is a mirror image of the role of the architect: The architect brings together and integrates the input, and the contractor distributes and delegates responsibility for the output.

The instructions are the specifications themselves. With the spread of computerized specifications systems, standardization has come to their form and, to a lesser extent, their content. No matter how conventional the project, however, every set of timber specifications must be tailored to its particular aspects. Instructions, like buildings, are constructed from a set of common components. To Wurman, these are six in number (see Fig. 10.1):

1. Mission: What is the objective?
2. Destination: What will be the final state?
3. Procedure: What are the specific directions?
4. Time: How much time is available to complete the task?
5. Milestones: What landmarks will be encountered along the way?
6. Red flags: How will you know if you veer off course?[2]

In the case of building projects, these components are distributed among the various contract documents.

Mission or objective is given in a brief project description in the general conditions.

Destinations—both general and specific—are depicted in the working drawings.

Procedures, to the extent they affect the quality of the result, are spelled out in the specifications.

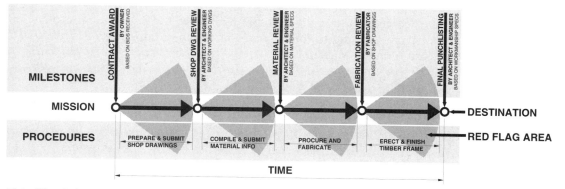

10.1   The six instruction components associated with timber construction.

Time for the entire project is usually given in the bidding documents, but its allocation among the various work activities is left to the contractor.

Milestones are found in the specifications, typically under coordination.

Red flags, if present at all, are in the specifications, under workmanship.

As this chapter is about specifications, our focus is on the components of instructions that are found there: procedures, milestones, and red flags. Procedures define the route, while milestones and red flags are markers that may be encountered along the way. Milestones give the contractor positive reinforcement, providing reassurance that progress is being made; red flags warn the contractor when he or she veers off course.

Because methods and means are the responsibility of the contractor, the wise architect or engineer will be circumspect about the procedures included in the specifications. The designer is primarily concerned with final results; in a large sense (and supported by legal precedent), the contractor's methods and means are none of the designer's business. It would be a mistake, however, to think that the contractor's methods are nobody else's business. Contemporary construction projects take place in a highly regulated environment. Regulations may be promulgated by local and state governments, trade unions, federal agencies, insurance companies, and, of course, the owner. Working hours may be limited by local ordinance or through collective bargaining agreements; access to and availability of the site may be restricted, especially during renovation of occupied buildings; certain procedures may be prohibited due to their risk to persons or property, while others may be required to assure quality.

Scope of Work: General Issues

Specifications are divided into Divisions and subdivided into Sections. Divisions roughly correspond to the major construction trades—Concrete, Steel, Masonry, Carpentry, etc.—while Sections cover particular subclasses of work. A good example of the relationship between Divisions and Sections is afforded by Division 7, *Moisture Protection.* The Sections within this Division include "Waterproofing," "Roofing," and "Caulking." While they all relate to moisture protection, each is located in a different part of the building—at its basement, on its roof, or around its openings—and each involves different materials.

The Sections of the specifications are like the pieces of a jigsaw puzzle—only if they are shaped properly will they fit tightly and properly with their neighbors and produce a coherent picture. The interfaces between adjacent pieces must match one another. A building requiring no coordination among trades would be like a puzzle with square pieces. In the real world, however, construction is highly interdependent, like a puzzle with interlocking pieces. The projection of one piece must correspond with the recess of the next.

The scope portion of a specifications Section is where the shape of a particular piece of the puzzle is defined. The projections of a particular piece are defined in its specification Section, in a paragraph entitled "Work Included." The recesses are covered in a paragraph entitled "Work Excluded." Every aspect of work included in one Section should match the work excluded from another Section. I had to look pretty far afield to come up with a name for these complementary aspects. In poetry, a *couplet* is a unit of verse consisting of two lines that form a complete thought. Since each pair of work included and work excluded statements in the specifications forms a complete thought, I have named them *work couplets*. The circumstances under which they are required are many and varied. One must provide them to:

Coordinate the efforts of two different trades.   Even the simplest building project involves many trades. Although coordination may not be required among all of them—the roofer and excavator, for example—it is usually required among many of them. To identify these conditions, think about the likely construction sequence, and study the wall sections and details. A detail-related work couplet in the Timber Construction Section would include timber roof decking, but exclude dimension lumber nailers and their attachment to the decking. Conversely, the Rough Carpentry Section would include the nailers but exclude the decking. A sequence-related work couplet would, for example, exclude from the Timber Section the column base anchor bolts provided for the timber columns under Concrete.

Clarify the distinctions between the Sections within a given Division. The Roofing Division is a case in point. Roofers are responsible for a wide range of materials and systems, from asphalt shingles to single-ply membranes. Each system involves flashings and other sheet metal, but their details may be very different. If both occurred on the same job, a work couplet would help the bidders understand whether Shingle Roofing and Membrane Roofing would each be responsible for its own sheet metal, or whether one Section would be responsible for all of it.

Clarify the distribution of similar items among two or more sections. Wood is an example of a building material that is used in slightly different forms for significantly different purposes. Large structural members are typically covered in the Timber Construction Section of the specifications. Smaller members, whether used as framing, blocking, or nailers, are normally found under Rough Carpentry. Wood trim and cabinetry belong under Finish Carpentry and Millwork. Similarly, distinctions must be made when different materials are arranged in similar configurations. If there are steel trusses and timber trusses on the same job, for example, work couplets are needed to clarify the scope of truss work both in Structural Steel and in Timber Construction.

Address situations in which materials are furnished by one Section, but installed by another. At its most basic, construction results from the application of labor to building materials. Consequently, the largest and most detailed portions of any specification Section are commonly those describing the quality of the materials and of the workmanship required for their installation. A trade can be responsible for furnishing a material, installing it, or both. By convention, furnish + install = provide. Furthermore, if the specifications do not state that something is to be *furnished* or *installed,* the presumption—the default setting—is that it is to be *provided.* To illustrate, consider a job where concrete block walls are to be reinforced with rebar, but the vast majority of the rebar on the job is for the concrete foundations. Concrete blockwork is specified under Masonry, whereas concrete foundations are specified under Concrete. Each could be responsible for its own rebar, but it would not be unusual for a separate Reinforcing Steel subcontractor to furnish all of the rebar, with installation in concrete the responsibility of Concrete, and installation in masonry the responsibility of Masonry. The resulting work couplets would be distributed among the "Work Included" and "Work Excluded" portions of the affected Sections like this:

Work item	Specification Section		
	Reinforcing Steel	Cast-in-Place Concrete	Masonry
Furnish all rebar	Included	Excluded	Excluded
Install rebar in concrete	Excluded	Included	Excluded
Install rebar in masonry	Excluded	Excluded	Included

Coordinate the interfaces between several trades. (See Fig. 10.2.) Coordination must cover two types of interfaces between trades: physical and conceptual. An example of the former is the interface between curtain walls and the structural members to which they are to be anchored. A work couplet will clarify which trade is responsible for the anchors between the systems and trades. An analogous situation arises when the architect wishes to have a sample constructed of an assembly involving more than one trade. A work couplet could indicate whether the distribution of responsibilities with respect to the sample are the same as those with respect to the actual building. Conceptual coordination can involve any number of aspects, but the most common relate to geometries (dimensions, alignment, spacing, proportion, and arrangement) and properties (composition, finish, compatibility, and color).

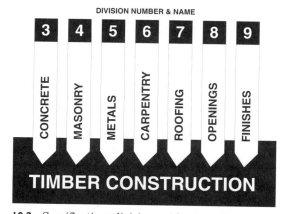

10.2 Specifications divisions with which timber construction must often be coordinated.

The final list of Work Included items for each Section can be thought of as its table of contents. Just as the key aspects of timber construction and its typical design and construction sequence provide a framework and sequence for the chapters of this book, each of the items listed in the Work Included portion of Timber Construction should correspond to a paragraph or provision in its Materials and Workmanship portions. Each item excluded from a given Section should refer to the Section in which it is included. In fact, using these pointers, you ought to be able to navigate through the specifications in a manner akin to browsing through the Internet.

No matter where the architect draws the line between one Section and another the contractor is *not* obligated to follow it. It's not that the architect's boundaries are arbitrary, just that construction methods, means, and sequences are the responsibility of the contractor. While the contractor is obligated to build the entire building, he gets to decide how. In the case of the structural frame, he is free to subcontract fabrication separate from erection, to have the left half of the frame provided by a different sub from the right, the second floor by a different sub from the roof, or the framing by a different sub from the decking. If he so chooses, he could even rent a crane and be responsible for erecting the work of *every* trade.

If the contractor is free to ignore the architect's scope statements, why should the architect prepare them? Based on my experience, the answer is simple: Contractors tend to adhere pretty closely to the jurisdictional lines drawn by the architect, as long as they are consistent with the jurisdictional limits of their applicable collective bargaining agreements. Furthermore, with their references to related work, the architect's scope statements help the contractor organize the work, even if it is divided differently from what the architect imagined. And finally, due to time constraints, the contractor may follow those statements for bidding, even if the decision is made to alter them when buying out the job.

Scope of Work: Timber Construction Issues

Carpenters are responsible for a lot of different things on a job site: framing and blocking, of course, but also the erection and dismantling of concrete formwork, and the installation of doors, frames, and specialties. In fact, on a nonresidential project, the bulk of the work of the carpenters may have nothing to do with wood, even though that is the material with which they are often most closely associated. The common thread running through all carpentry activities, then, is not the material with which the carpenters are working, but the skills and tools they bring to the job.

The scope of the Timber Construction Section should be limited to the timber frame and those materials associated with it. In many cases, deciding where to draw the line between what is included and what is excluded is a matter of judgment and experience. (See Fig. 10.3.)

Typically, for their safety, all other trades must vacate the site while the timber frame is being erected. If the goal is to get the timber contractor off the site so the other trades can return as quickly as possible, nonstructural timber work should be excluded from the timber contractor's work. The overall scope of the timber construction specifications will vary, depending on the following project characteristics:

- Single general contract versus separate timber construction contract
- Shop fabrication versus field fabrication
- Shop finishing versus field finishing
- Standard connectors versus custom connectors

Timber Work Within or Separate from General Contract

The issue here is accountability: Under a single *contract,* the general contractor is the single *contact;* all communications between the subcontractors (including the timber subcontractor) and the design team go through the general contractor. The general contractor coordinates the work of all the subcontractors and arranges for the correction of nonconforming work. Although he has the power to backcharge a subcontractor for the consequences of his defective work, the general contractor suffers the consequences first. If the timber sub rejects the anchor bolts installed by the concrete sub, the contractor has both an obligation and an interest in solving the problem as quickly as possible, particularly if there is a fixed period for construction. Whereas responsibility may eventually be assigned to the offending sub, the contractor's overriding goal will be quick resolution.

Under multiple *contracts,* there are multiple *contacts.* If the timber prime rejects the anchor bolts installed by the concrete prime, respon-

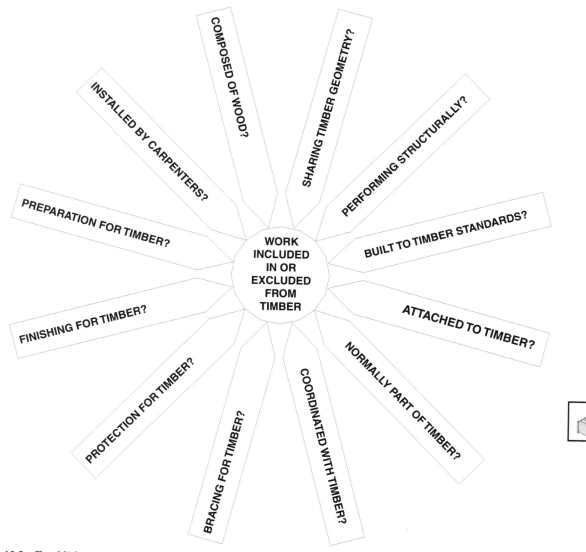

10.3 Establishing the scope of the timber contract.

sibility must immediately be assigned, since each party has a financial interest in placing blame on the other. In the absence of a single general contractor, assigning responsibility falls to the architect or to the construction manager (if there is one), who must base all decisions on the contents of the contract documents.

Thus, while the contract documents must always address matters of coordination between *trades,* those involving a separate timber contractor must also address coordination between independent *contractors.* To cover the work of all trades, the requisite dispute resolution mechanisms are customarily described in the general conditions. In conjunction with such general requirements, I recommend the inclusion of a specific coordination paragraph in the timber specifications. It should require that timber delivery and erection be coordinated with the work of other trades, to minimize disruptions in their respective schedules. And, to avoid disputes like the one just described, the tolerances specified for each interface with another trade, such as between the anchor bolts cast in by Concrete and the column bases set by Timber Construction, should be the same for both trades, but based on those of the less-forgiving material.

Templates

All construction processes are imperfect; tolerances tell you what degree of imperfection is acceptable. If every trade were imperfect in the same way and to the same degree, fit would not be a problem. Tolerances, however, are specific to the trade, the material, and the stage of construction, and are relative to the sizes of the materials—a column might be acceptable if ½ in (12 mm) out of plumb, but the joints in a baseboard will be objectionable if wider than ¹⁄₁₆ in (1.5 mm).

On a construction project, you need a means of communicating accurate information about inaccurate processes. While the *intended* geometry of each piece is spelled out in the shop drawings, its *actual* geometry is all that matters to the workers in the field. In fact, given the idiosyncrasies of different materials, coordination among trades can be more important than the individual performance of any one of them. On jobs with a low level of prefabrication, each trade can accommodate the preceding work, since the work is there to look at and work to.

One of the advantages of large-scale timber buildings is that they are largely prefabricated. This means that on-site preparation for the timber can occur simultaneously with its off-site fabrication. Unfortunately, this also means that fabrication of the timber frame often precedes fabrication of the other systems—curtain walls, metal roofing, etc.—with which it must fit. How do you address this problem? One answer is with templates.

A template is a surrogate for the real item that will ultimately be installed in the field. For the contractor who is unable to visit the site, a template is the next best thing to being there. When a site visit would not yield the information needed to coordinate the work, a template can. Templates also help assure a good fit between the work of different trades. The rules for template origination are simple:

Rule 1: If one party shop-fabricates, templates originate from that party.

Rule 2: If both parties shop-fabricate, templates originate from the party requiring larger tolerances.

Rule 3: Since tolerances usually get tighter as work progresses, templates typically originate from the party whose work comes earlier in the construction sequence. (Existing conditions represent a specific type of such earlier work, typical of renovation projects.)

Because template production can be costly and time consuming, templates should only be called for when the information they convey must be acted on prior to the mating of the actual materials or systems. Anchor bolt templates are common for this reason; the contractor can't afford to wait until the base plates arrive before casting the anchor bolts into the foundations. However, if the work of different trades is joined by connectors capable of accommodating their individual variations, templates are not necessary. Whereas a fixed connector might be composed of a single welded steel fabrication, an adjustable one could have two or more parts bolted together through slotted holes.

Here are some typical templates you might consider specifying in conjunction with timber construction:

Template information	Section furnished by	Section furnished to
Locations of anchor bolts	Timber construction	Concrete
Configurations of arched glulam openings	Timber construction	Windows or curtain walls
Locations of holes in steel connectors	Miscellaneous metals	Timber construction

Specifying for Protection

The loads placed on a building and its components during construction are very different from those it will experience once it is in service, and not just in terms of magnitude. Construction loads tend to be temporary and uncontrolled, often concentrated in areas different from those subject to permanent loads. Service loads tend to last longer and, if the design of the building is consistent with its use, will be distributed in a manner that the frame can resist. The net result is that the *temporary* loads of construction are often responsible for the bulk of the *permanent* damage.

Construction control is the flip side of construction protection; the more control, the less need to protect against the risks associated with the loss or lack of it. It makes perfect sense, therefore, that the contractor should be responsible for providing and balancing both. Even so, there is little a contractor can do to protect against the water stains from a sudden downpour during erection or the dents caused by a rolling scaffold. Sure you could build a tent over the site or wrap every column, but the costs would be prohibitive. The alternative is for the architect to specify for protection.

Specifying for protection means selecting materials and systems capable of making it through construction intact, without damaging adjacent materials. In the case of timber construction, it means visualizing the construction process to identify the stages during which the framing and decking are most vulnerable. Specifically, the contractor should:

- Galvanize all fasteners to prevent against rust stains
- Supply all connectors with corrosion-resistant coatings
- Provide weep holes, for drainage, in all shoe-type connectors
- Apply all preservative treatments *after* fabrication, but *before* shipping the frame
- Ease the edges of all timbers to minimize splintering
- Deliver all glulams to the job site in watertight wrappings
- Lift and brace all members with nonmarring devices
- Cover all roof decking with roofing paper at the end of each work period
- Avoid overstressing the frame and each of its components
- Bring the building up to room temperature gradually, to minimize checking

Specifying for Appearance

To achieve the best possible appearance, the specifier must understand the nature of timber materials and of timber construction. Look at samples of the specified materials and picture them in place. Imagine the sorts of incidents that could happen during erection of the timber frame which would adversely affect its final appearance. Then write provisions to help the contractor avoid them. Take nothing for granted; none of the people actually doing the work will have as comprehensive a vision of the final result as you.

The control designers have over each of the aspects comprising the overall appearance of a timber structure varies greatly. One of the unique aspects of timber structures is that the degree of control is inversely proportional to the sizes of the components. At the extreme of maximum control are mass-produced fasteners and connectors; at the other extreme are the timbers themselves. The fact that the timber structure must answer to two masters—architect and structural engineer—further complicates the picture. Materials satisfying strength requirements may not satisfy appearance requirements, and vice versa. Clearly, the architect and engineer must coordinate their specifications so that both parties will be satisfied with the final result.

Recognition must also be given to the fact that timbers will continue to move for the life of the building, in response to changes in relative humidity. Therefore, standards for fitting tolerances and joint tightness must be given in the context of indoor or outdoor environmental conditions during construction and after conditioning. Pockets in beam hangers and other connectors must be sized to receive timbers during construction, before they have completely dried out. Unless the project is in the desert, therefore, it is inevitable that the timbers will not fit as tightly once equilibrium moisture content is achieved. Where possible, specify that the moisture content of the timbers be slightly less at time of fabrication than anticipated in the completed and occupied building. This will help assure that joints tighten, rather than loosen, during conditioning.

Another common problem relates to differences in appearance between small samples and full-size timbers.[3] The problem of scaling up is not unique to timber, but its specific manifestations are, due to the fact that wood is a natural material. Defects such as knots can vary widely in size and spacing, even within specific grades. Because of differences between earlywood and latewood, color within a single piece and among several pieces can vary widely, even within specific species. These differences also affect the manner in which coatings and finishes are absorbed, further affecting color.

There are so many variables, in fact, that it behooves the specifier to become familiar with and inform the owner of the allowable variations within the specified grades and species, so that the final result does not come as a surprise. In addition, make sure that *multiple* samples—showing the whole range of growth characteristics and grain variations permissible within the specified grades—be submitted to indicate the relative uniformity of the structure's final appearance and that, where applicable, a portion of each have the specified finish applied. Recognize that sawn timber is graded for structural purposes, not for appearance, so that before specifying a particular grade, you should familiarize yourself with the types, sizes, and distribution of defects allowed in it. (Glulam, by contrast, is available in various appearance grades.)

Regardless of the range of acceptable variation, it is imperative that the contractor be responsible for identifying and rejecting materials whose properties fall outside that range. While the architect and engineer also have that power, it is the rare project where either would be on-site as much as the contractor.

One of my goals as an architect is to achieve visual consistency and order, construed as broadly as possible. To achieve this objective, I typically specify that the timber contractor is to orient all connector bolts in a consistent direction; trim off excess bolt lengths, turn all square column bases in the same direction; where appropriate and feasible, align deckboard joints across hips, valleys, and ridges; orient all column laminations in the same direction; and cut all rafter tails to their final lengths in the field.

Specifying for Performance

While you specify for protection to insure that the timber work survives construction, you specify for performance to insure that it pro-

10

Timber Specifications

vides adequate service long afterward. Keep in mind that a timber frame is first and foremost a structural mechanism. (And, as evidenced by the numerous barn frames reused in houses, a sound timber structure can far outlast the building it initially supports.) Timber frames age differently from concrete and steel frames; tailor your specifications accordingly. The question to ask about the frame is not "what will it look like immediately after erection?" but, rather, "what will it look like many years later?"

Recognize that changes in moisture content affect the short- and long-term behavior of timber construction. In the case of custom steel timber connectors, galvanizing should be performed after they have experienced the stresses associated with fabrication; in the case of heavy timbers, surfacing should be performed after they have experienced the stresses associated with loss of moisture. Shrinkage is inevitable, as are the distortions resulting from it. A member surfaced dry—i.e., after shrinkage—is therefore likely to be much less distorted than one surfaced green. Specify with this in mind.

Recognize that the distribution, duration, and magnitude of the applied loads affect the character and magnitude of member deflections. Timber bending members deflect nonelastically under high long-term loads. Deflection is inevitable. If appearance over the life of the building is important, and you are using products, like glulams, that can be tailored to the job, specify that camber be proportional to anticipated deflection. At multiple spans and cantilevers, make cambering provisions appropriate to their more complex deflection patterns. And bear this in mind: Just as the joints of a wooden boat get tighter the longer it sits in the water, the members of an appropriately precambered timber get straighter the longer and more heavily it is loaded.

Quality Control

Because they cover quality control briefly, if at all, the abbreviated specifications typical of residential projects—even those involving timber—are inadequate for most nonresidential heavy-timber buildings. Although descriptions of required materials are necessary, they are not sufficient by themselves to assure the owner that the quality of the finished product will match the quality of the specified projects. One of the key roles of the timber specifications, then, is to bridge this gap (see Fig. 10.4). Therefore, to control quality during construction of a timber building, the specifications should require that:

- All timbers have grade stamps, to enable any discrepancies between the grades specified and those furnished to be identified

- All grade stamps be placed in areas that will be concealed in the completed building, to avoid the time and expense of removing them, and the risk that damage will result

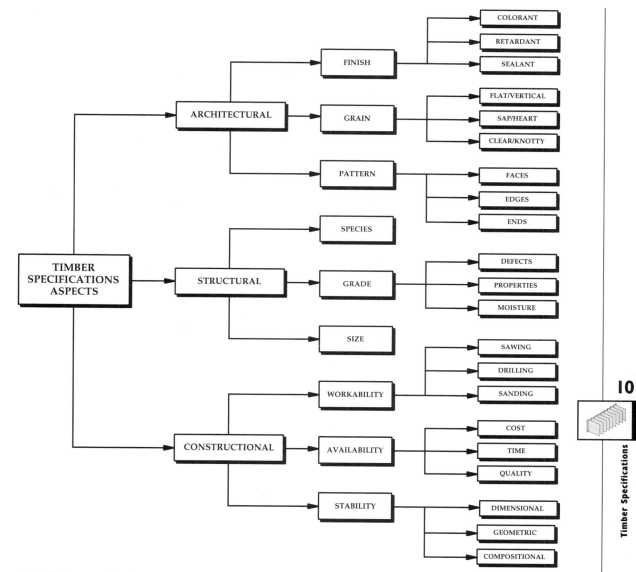

10.4 Timber specifications aspects.

- All timbers be stored under cover and off the ground, to assure that their condition at time of installation is equivalent to their condition at time of fabrication

- All manufactured timbers, such as glulams, be accompanied by documents certifying that they were produced in accordance with applicable codes and standards

Shop drawings play an important part in the quality control process for any prefabricated system. (See Chap. 12.) Their importance is heightened, however, when engineering is part of the contractor's scope. Such is often case with timber construction. Rather than simply showing how the structural engineer's connections will be produced, the contractor must often retain an engineer to translate the schematic indications on the drawings into workable details that can be economically produced.

When specifying timber engineering, define the qualifications of the engineer, the limits of the engineering work, the design criteria, the deliverables, and any follow-up services. The most basic qualification is that the engineer be licensed to practice in the jurisdiction where the building will be located, but it would not be unusual to also require that the engineer have significant experience in the engineering of timber structures. The scope of the timber *engineering* may or may not match the scope of the timber *construction;* explain any differences, and fill any gaps between the responsibilities of the project's engineer-of-record and the contractor's timber engineer. Design criteria can be given by reference to applicable codes, but make sure to identify any special provisions—provisions for future expansion of the building, connections to other systems, or options for disassembly and reuse—that would not otherwise be obvious. Deliverables should consist not only of the shop drawings themselves, but also of as many complete sets of signed and sealed engineering calculations as will be required to secure the necessary building permits. The most common follow-up service would involve an inspection by the contractor's engineer of the substantially complete timber structure, followed by a report certifying that it was constructed in accordance with the contract documents and the authorized timber shop drawings and engineering calculations. State in the specifications that payment is subject to the correction of defects, whether the defects are identified by the contractor's engineer or by any of the members of the design team.

While any experienced detailer is likely to include the whole range of necessary information in the timber shop drawings, the only way to make sure it is included is to insist on it in the specifications. This is particularly so when the information requires coordination with other trades. It is customary for detailers to label components outside the scope of the timber contracts as "by others." If the final detailing of any aspect of the timber frame is a function of the work of other trades, say so in the specs, and tell the reader where to find the requisite information.

Consider the attachment of a curtain wall to an exposed gable-end timber truss. In the absence of guidance from the specifications, the curtain wall engineer might assume that the timber engineer allowed for wind loads in the design of the end truss, and that it can therefore accommodate wind loads from the curtain wall at any point. The timber detailer, by contrast, might assume that the curtain wall manufacturer noticed that the wind-loaded end truss is identical to the unloaded intermediate ones, and that the only reasonable places to attach would be at the laterally braced panel points of the truss. Essentially all that is needed in the specifications is a requirement that these two trades coordinate their work to achieve a particular objective. If attachments are only to be made at certain locations, both parties need to know it up front, not only because it affects their details, but also because it may affect their costs. On this job, the requirement was stated simply as "anchor to trusses only at panel points; anchor to columns only at timber connections."[4] The curtain wall manufacturer could only achieve that objective by transferring loads with heavy and costly structural steel tubes inserted inside the aluminum framing; failure to comply would have induced lateral bending in narrow timbers engineered only for axial loads. (See Fig. 14.12.)

Connections between glazing and framing systems and timber frames must be described fully. If the exact systems are known during detailing, the connections can be shown in the working drawings. If, as in the case of competitive bidding, the systems are not known in advance, the connections can only be shown schematically in the drawings or described in words in the specifications. Include the vertical deflection criteria developed for the timber structure by the structural engineer in the specs for the glazing systems and for any stud framing systems used as infill. Distinguish between instances where the timber frame is supporting other work—such as roofing—in response to loads along all three axes, and those where it may be supporting other work—such as curtain walls—along only two.

Over the past several years I have witnessed several instances where gaps in responsibility between related parties could have been disastrous. While neither of the following examples involved timber construction, either or both of them could have. In the first case, the engineer-of-record for a two-story building delegated to a specialist the engineering of the rigid frames on its second floor, but failed to advise him of the fact that the uses on the first floor required that the entire building be able to resist a higher seismic load than that associated with the uses on the second floor. Had the building been built as initially engineered, the second floor would have collapsed under a seismic load lower than the threshold of the first.

In the second case, a contractor awarded fabrication of structural members to one firm, and of nonstructural members to another, but apparently failed to compare their two contracts to make sure that all aspects of the fabrication were covered. As it turned out, the members the structural fabricator deemed to be structural did not exactly agree with those defined by the nonstructural fabricator, and neither agreed with the nonbinding jurisdictional lines drawn in the Work Included and Excluded paragraphs of the specifications. Because certain items fell through the cracks during estimating, the contractor underbid the job. Not until the architect and engineer reviewed the shop drawings and identified the missing pieces did the contractor realize he had a problem.

Both of these examples relate to quality control. The first illustrates the need for coordination but, more importantly, the fact that no part of a building should be engineered out of context. The second illustrates that, although the contractor is free to define the scope of each trade differently from the specifications, any gaps that result become his problem. To protect the owner in the event that the contractor chose that route, both the structural and nonstructural sections of the specifications included this provision:

> . . . it is of the essence of the Contract that these two Sections coordinate with one another and with the General Contractor to assure that all . . . elements indicated in the Architectural Drawings, the Structural Drawings, or both, are provided to the Owner, under the Contract. In other words, what isn't provided by one Section, shall be provided by the other.[5]

The fact that the specifications are composed of words is the source of much of their strength. Words, after all, are capable of conveying difficult and subtle concepts. When sloppily used, however, they can be ambiguous and confusing. The goal, then, is to make the specifications complete and precise, but construction is so complex that it would be unrealistic to expect that goal ever to be achieved fully.

Recognizing that, in spite of a professional's best efforts, the specifications *will* contain errors and omissions, you should include a provision in the specifications requiring a conference between the designers and builders just prior to the start of timber erection (if not sooner). The primary purposes of this *timber preconstruction conference* are to review all authorized submittals and shop drawings, to review those aspects of the specifications applicable to the field work, and for the contractor to report on the acceptability of the preparatory work performed by others. I have found these conferences to be extremely valuable. Without exception, the process of talking through the job elicits from the erector important questions and comments. Although sometimes they reveal the erector's lack of familiarity with the specifications, more often they point to ambiguities and contradictions in those documents.

To understand the scope of the preconstruction inspection and report required of the contractor, consider this provision from a recent set of timber specifications:

> At least one week prior to moving onto the Site, the Timber Construction Contractor shall inspect the work and verify that . . . the foundations are ready to accept the timber structure . . . Carefully examine all surfaces and check all dimensions associated with preparations made by other trades. In detail, identify in writing all defects, and transmit that report to the General Contractor, with a copy to the Architect. Absence of such notification will act as acceptance by Contractor, of all preparations.[6]

Alternates

One of the most common ways designers try to protect their clients (and themselves) from cost overruns is through bidding alternates. Alternates can be useful, in that they enable the scope of the work to be modified on the basis of the bids rather than solely in anticipation of them. Alternates only accomplish their objective when they are clear in intent and modest in number.

Alternates fall into two basic categories: those that affect other work, and those that do not. This distinction has important consequences, both in the preparation of the alternate's description and in the preparation of the bidder's price. The cost impact of each alternate involving multiple trades must be determined by each affected trade and then compiled by the general contractor. When only one trade is involved, the contractor need only add his or her markup to the trade price.

Alternates can affect the work of other trades without affecting their prices. On a recent project, for example, I wanted a price to switch a portion of the roof deck from heavy timber to thinner tongue-and-groove plywood. The change affected slightly the elevation of the roofing work, but did not affect significantly the quantities of labor or materials. On another job, the actual spans of the nailbase roof sheathing turned out slightly longer than estimated, necessitating the use of longer deck planks. The total roof area stayed constant, but the contractor nevertheless requested an extra. I rejected the extra after determining that the work of installing a smaller number of larger panels was roughly equivalent to the work of installing a larger number of smaller panels.

Alternates that originate with another trade sometimes affect the timber work. A switch from asphalt shingles to slate roofing would not only increase the roofing's *cost,* but also its *weight;* it might even increase the thickness or grade of the roof deck and the sizes of the timbers. Similarly, an alternate involving the elimination of a rated ceiling from a timber building might have to include painting the now-exposed structure and deck with a rated intumescent coating system.

Because timber is often partially or fully exposed in completed timber-framed buildings, it can be thought of as both a structural and a finish material. Thus, any alternates that affect the timber must take into account both its structural aspects—including geometry, grade, and connections—and its appearance. Even so, structural considerations nearly always control, as they must. Fortunately, glulams from the various appearance grades are all capable of meeting the same structural requirements. Therefore, you can take an alternate for a more economical appearance grade, without affecting member size or structural capacity, and thus without affecting the structural work of other trades. A number of possible stand-alone alternates involve finishing. Examples include eliminating field painting of shop-primed steel connectors or eliminating staining of those surfaces of timber framing and decking which will ultimately be concealed.

Once all of the implications of an alternate are understood fully, ask yourself if the potential savings justify the effort required of you to prepare a comprehensive and unambiguous description, and of the contractor to prepare a price. If the potential savings will be small, but the issue important, advise the owner of the approximate cost, and recommend that it be handled through a change order during construction.

Retainage Reduction

In the construction of a timber building, the timber frame is usually one of the first aspects to be completed. Although other trades, such as Excavation and Concrete, start their work before the timber, they usually don't complete their work until after the timber. Especially if the timber contract is separate from the general contract, the timber contractor is likely to request early release of his retainage. I recommend that this issue be addressed in the timber specifications. If the retainage is the only financial incentive for the contractor to finish, consider releasing it only if all timber punchlist work has been completed and your answer to each of the following questions is "no":

- Will the timber contractor have any remaining contractual obligations, whether procedural or substantive, following completion and acceptance of the timber work? (An example of a procedural obligation would be submitting outstanding shop drawings or warranties; an example of a substantive obligation would be removing temporary bracing or protection.)

- Could the environmental conditions during completion of the building reveal defects serious enough to cause you to reverse a previous finding of acceptability? (An example of this would be drying out and checking of the timber frame following building enclosure.)

- Could the nature of subsequent work reveal otherwise invisible defects serious enough to cause a reversal in a previous finding of acceptability? (An example would be if a subsequent coating was inconsistently absorbed due to improper preparation or conditioning of the timber substrate.)

Summary

The timber specifications are the written instructions that describe the scope of the timber work, and the quality of the materials and construction. A comprehensive scope consists of a list of aspects included within the contract, and a complementary list of aspects excluded from it. Each item included within or excluded from the timber contract is excluded from or included in the scope of another trade. Such complementary pairs of spec statements, or work couplets, are a valuable aid both to the specifier—to verify the completeness and organize the contents of the project manual—and the subcontractors—to prepare their bids and coordinate their work with related trades.

Since wooden building materials are used in many assemblies besides those of timber construction, and carpenters are responsible for installing many materials and systems besides those of timber construction—including some that are completely devoid of wood—it is imperative that the limits of the timber specifications be precisely defined. The specifications are only complete, therefore, when responsibility for furnishing and/or installing each aspect of the work is included in the work of a designated trade, excluded through complementary statements from all related trades, and cross-referenced with pointers between the trades. Responsibilities for *furnishing* are outlined under the Materials portion of the specifications, for *installing,* under Workmanship, and for *providing,* under both Materials and Workmanship.

Tailor the general conditions to the selected approach to project delivery; tailor the specifications to the preferred distribution of work between the shop and the field. Emphasize coordination in the contract documents and enforce it during construction, among the various trades and among the parties responsible for the various aspects of the timber construction. Familiarize yourself with the standards and conventions of the industry, to maximize the utility of the timber specifications to the timber contractor. Specify templates whenever geometric information subject to manufacturing tolerances must be shared between trades prior to the mating of their actual materials or systems.

Specify for protection, appearance, and performance. For protection, account for the conditions typical of construction, including temporary loads, precipitation, and damage from erection or by other trades. For appearance, allow for natural variations among timbers, differences between architectural and structural properties, and expansion and contraction of the members from changes in moisture content. For performance, require that the members be dried before surfacing and, at glulams, cambered during fabrication, to minimize the visible impact of shrinkage and deflection.

Establish stringent quality control procedures that enable verification of contract compliance. If engineering is required, specify criteria for the design of the work and certification of the adequacy of its construction. Define the characteristics of the connections between the timber frame and other materials and systems, and indicate which trade is responsible for them. Construe the concept of coordination as broadly as possible when drafting the specifications. Mandate a timber preconstruction conference to confirm that the builder's understanding of the project is consistent with that of its designer.

Use particular care in establishing the alternates and in defining each one; exclude items so small as to have minimal impact on the budget, or so large that the owner is unlikely to authorize them at any cost. Define each alternate clearly, to assure that the prices offered for it are credible and meaningful. Distinguish between alternates involving individual trades and those involving multiple trades. Make sure to specify the default conditions, the conditions that will apply in the event that the alternate is not accepted.

With respect to the retainage held on the timber contractor, consider reducing it on completion of the timber frame, rather than on completion of the entire building. Hold only enough to assure the satisfaction of all close-out requirements and to protect the owner against defects associated with subsequent conditioning and finishing.

Notes

1. Wurman, p. 16.
2. Wurman, p. 170–174.
3. Telecon with Ben Brungraber, January 8, 1998.
4. The Goldstein Partnership, Montville Specs., p. 8A-3.
5. The Goldstein Partnership, Community Center Specs., p. 5A-6.
6. The Goldstein Partnership, Montville Specs, p. 6B-9.

Timber Construction

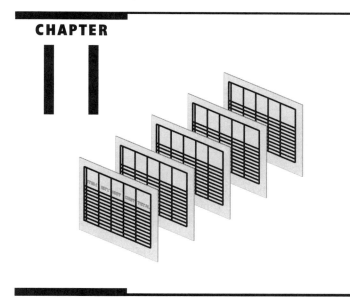

Estimating and Bidding

Experience is a thing that enables you to recognize a mistake whenever you make it again.

FRANKLIN P. JONES[1]

Introduction

Anyone who spends enough time in or around the construction industry hears stories of jobs where the low bid bore no resemblance to the designer's estimate. While specific circumstances vary widely, the root causes of such discrepancies are relatively few and consist of what I call the four Q-factors: *quantities, qualities, qualifications,* and *quotations.*

Let me start by defining terms:

- **Quantities** are work measurements that affect prices. In timber construction, these include lengths and cross-sectional dimensions of members, areas of floor and roof decks, pounds of nails, amounts and sizes of bolts, days of crane rental, and size and number of crews. *Labor and material quantities are derived from the working drawings.*

- **Qualities** are the characteristics of the materials. In timber construction, these include grades of lumber, standards for connectors and fasteners, types of adhesives, finishes, and coatings, and benchmarks for workmanship. *Qualities are given in the specifications.*

- **Qualifications** are the overall conditions governing the work and the associated procedures and protocols. Regardless of the type of construction, these include subcontracting, scheduling, payment,

11

Estimating and Bidding

correction, protection, certifications, insurances, labor, and temporary facilities and services. *Qualifications are covered in the general conditions of the construction contract; some are reserved for the general contractor, others also apply to subcontractors and suppliers.*

- **Quotations** are the prices assigned to the quantities, qualities, and qualifications. The price of each work item consists of either a material component, a labor component, or both. Items purchased for or furnished to the job under the timber contract, for installation by others, have only a material cost component. Items purchased for or furnished to the job by others, for installation under the timber contract, have only a labor cost component. Items *provided*—furnished *and* installed—under the timber contract have both material and labor cost components. *Unlike the previous three Q-factors, quotations are found outside the contract documents, in the marketplace.*

While the form of a construction estimate is basically very simple and unchanging—a spreadsheet where individual work amounts are multiplied by their respective unit material prices and/or labor rates and totaled—its content integrates the ever-expanding experience and judgment of its author. Each bid is an experiment that yields valuable data at the end of the bidding period and throughout construction. Bidding failures and construction problems create the likelihood of lost profits, but also the opportunity for gained knowledge.

Each item in the contract documents has cost implications, either in and of itself, or in combination with others. The quality of the estimate is a direct consequence of the ability of the estimator—whether working for the designer or the builder—to understand and integrate these interrelationships. The purpose of this chapter is twofold: to help *builders* understand what goes into a timber construction estimate, and to help *designers* understand how builders conceptualize and price the work. My ultimate objective is to minimize the dollar difference between what the designer thinks a timber structure *should* cost and what the contractor says it *will* and finds it *does* cost.

Quantities

Estimating any job involves organizing what is shown on the drawings into categories, and determining quantities for each work item (see Fig. 11.4). Here is some advice for the estimator:

- *Avoid the temptation to start the takeoff too soon.* Any time spent upfront planning the takeoff will more than pay for itself, ultimately, in completeness and accuracy. Conversely, don't bother to bid if you aren't going to have enough time to prepare properly.

- *Start by familiarizing yourself with the project and with the manner in which it is described in the working drawings.* Every project is different, and every architect and engineer has a different way of presenting information. Understanding the vocabulary is a critical first step in understanding the design.

■ *Arrange the estimate to parallel the construction sequence.* Work from the bottom up, like a contractor, rather than from the top down, like a structural engineer accumulating loads.

■ *Keep a running list of questions and assumptions.* At the appropriate time, submit to the architect any which remain unanswered after a thorough review of the contract documents, including any addenda.

■ *Use color to distinguish between member types, and between members that have been counted and those which have not.* Markers or pencils work fine when working from prints. If, however, you are fortunate enough to be able to work from a CAD file, such distinctions are best made by exploiting the layering and coloring capabilities of the software. Sharing digital data not only helps convey the architect's design intent to the constructors,[2] but can streamline the contractor's estimation and shop drawing preparation processes.

■ *Differentiate between typical and atypical conditions.* The specifications or general conditions often contain a clause to the effect that a material shown in one location is to be used in "other similar locations throughout the building, unless another material is indicated. . . ."[3] Not everything needed to do the job is shown; the engineer, for example, might label as typical the joist hangers in one particular bay. It is up to the estimator to extrapolate from that information the requirements for the entire building. Failure either to extrapolate or to extrapolate properly is common and can be costly. Become the low bidder by figuring out how to buy out and execute the work efficiently, not by underestimating it.

■ *To assist in checking the estimate, give each work item a sufficiently descriptive name.* An item entitled "joist hangers," for example, might include or exclude any number of related aspects, including purchasing, setting, or fastening. Be explicit! If only purchasing is included, make a note to that effect, and make sure to include a complementary item covering installation. In the event that installation is to be done by others, add an appropriate note in the respective line of the estimate.

■ *Don't forget the fasteners.* Most timber construction is held together with fasteners of various types; timber connectors cannot perform properly without the requisite fasteners. The more costly the fastener, the more critical it is that it be counted accurately. To help in that endeavor, consider including in your estimating spreadsheet specific lines for connectors and fasteners; whenever a work item calls for the former, make sure to indicate the applicable quantity for the latter.

- *Quantify the work using conventional and appropriate units and scales.* While both nails and bolts are installed one at a time, nails are purchased by weight, but bolts by quantity. The scale may be relative to the size of the job. For small-timber buildings or those involving intricate joinery, daily worker output may be measured in number of joints laid out and cut or in number of pieces installed, while for large simple repetitive timber structures, output is likely to be measured in hundreds of square feet of floor or roof covered per day.[4] Field productivity will also be a function of the degree of shop fabrication.

- *Account for the diagrammatic nature of the working drawings.* A responsible estimate allows not just for what they *show,* but also for what they *mean,* especially in relation to the other contract documents. While the specifications don't usually contain quantities, their provisions can affect the quantities taken off the drawings. A provision requiring that glulams outside the building be treated with preservatives, for example, would necessitate the bifurcation of the glulam material takeoff. Similar responses would be called for where exposed timbers must be of a higher grade or have a better finish than concealed ones, or where bearing walls obviate the need for beams.

- *Understand the quantitative implications of the building's geometry.* Architects often wish, for example, to place all exterior soffits at the same elevation, even when the roofs above them slope at different angles. One common means of achieving this objective is to vary the horizontal projection of each overhang in inverse proportion to the slope of its rafters. An example may be shown for one slope, with notes indicating how it is to be adjusted to accommodate the others. Make sure to allow for the resulting variations in rafter length.

The quantities shown in the drawings will match the quantities actually needed if and only if there is no waste. This is typically true of prefabricated structural components, but is rarely the case with timber decking, especially random length material. If, as discussed in Chapter 13, deckboards are installed ragged beyond each end of the floor or roof and cut off later, some material will be wasted in the process of producing the square edges specified. If small openings are decked over initially, to be cut out later, some material will be wasted to accommodate the contractor's preferred construction sequence. And, if there is a chance—and there is always a chance—that some of the boards will be damaged in shipping and handling or will be too warped to use, some material will be wasted to achieve the specified appearance.

In each of the above examples, waste is generated as the contractor either attempts to achieve the objectives stated by the designer or organizes the work to maximize productivity or enhance safety. Both are examples of methods and means. Strictly speaking, these are solely the province of the builder, yet without anticipating and allowing for them, the designer will underestimate the amount of material the contractor will require for the job.

Think of the construction site as a closed system: All building materials delivered to the job site which are not incorporated into the structure must be removed. Timber waste can be hauled away and dumped or recycled, or, occasionally in the case of untreated lumber, sold as feedstock for wood-fired power plants. Thus, the real cost of decking, for example, must include not only the cost of purchasing material sufficient for the specified spans and layout, but also the labor to sort it for quality, place it on the floor or roof, cut off the excess, haul and stack it for collection, and pay for its removal and disposal.

Quantities are also a function of building geometry. Roofs of timber structures can be sloped or, if framed with glulams, arched. Length calculations must account for such nonorthogonal geometries, whether in plan or section. For each piece, an allowance must be also be made for the difference, if any, between its centerline length and its overall length (see Figs. 11.1 and 11.2*a* and *b*). The magnitude of this difference is a function of the end conditions of the member. Plumb cuts maximize the difference, while square cuts minimize it. The same concerns apply to the beveled ends of members oriented at an angle in plan.

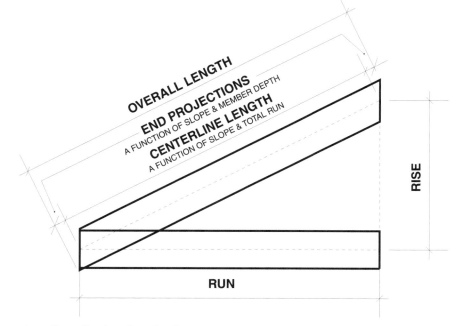

II.I Centerline length and end projections: the two components of rafter length where there are plumb cuts at each end. To order rafter material, round up to next standard size from the sum of these components.

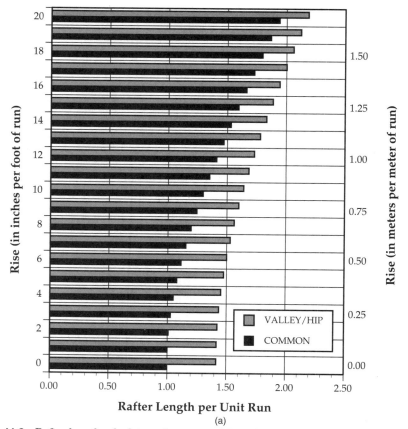

II.2   Rafter length calculators, for common hip and valley rafters of 45° in plan: (*a*) Centerline rafter lengths per unit run of common rafters. Excludes projections of sloping member ends.

Recently, I had the opportunity to review a takeoff by a contractor who, it turned out, had seriously underestimated the lengths of the rafters in a pyramidal roof. He had scaled their lengths from the plans, without accounting for their 45° slopes. Common rafters that in plan looked 10 ft (3 m) long actually needed to be over 14 ft (4.2 m) long. Hip rafters that in plan looked slightly over 14 ft (4.2 m) long actually needed to be over 17 ft (5.1 m) long. This resulted in two types of errors. The first, obviously, was that the actual length of each common rafter was more than 40 percent greater than was assumed. The second was related to the fact that the specified rafter material was only available in 2-ft (0.6-m) increments. Therefore, the common rafters would have to be cut from 16-footers and the hip rafters, from 18-footers. For each common rafter, in other words, the actual length that had to be purchased to realize the length needed for construction was 60 percent greater than was assumed. No wonder the contractor lost money on the job.

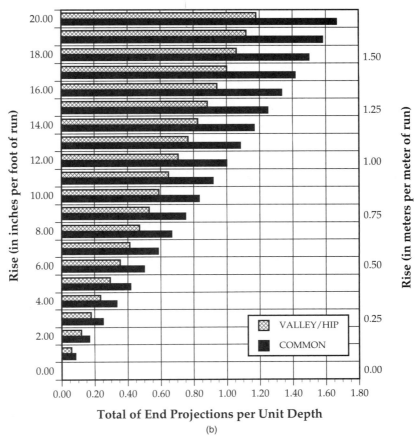

11.2 *(Continued)* *(b)* Total rafter end projection per unit depth.

Consider also the thicknesses and angles of intersecting members and connectors, and any distinctions between the details of intermediate and ends bays. Often, intermediate floor beams span from one column center to the next, while those at the ends extend to the last column's outside face.

There is an important lesson here: When taking off nonorthogonal framing, add two columns to your spreadsheet, one for geometrical multipliers, additions, and subtractions, and one for rounding upward to the next available length.

Productivity factors are to construction labor what waste factors are to construction materials. Each project and each of its aspects has a degree of difficulty. The greater the difficulty of the task, the lower the productivity of the workers performing it. The overall difficulty of a task is the net result of the interaction of a variety of factors. Some, such as the worker's skills or tools, are under the control of the worker or contractor. Others, such as the slope or elevation of the roof—and the attendant safety measures—are a consequence of the building design, and are thus initially under the control of the designer. And others, such as the weather while the work is being performed, are beyond anyone's control.

The time it takes to install a component of a timber frame varies with project size and complexity, but the variation is not linear. Most of the time is spent carrying the component from a stack to the vicinity of its final location, hoisting it, positioning it, and securing it. Within a relatively broad range of component types and sizes, the time required to perform each of these operations will vary only slightly. It should come as no surprise, therefore, that the labor required to install a 4×8 in (10×20 cm) beam hanger is only slightly more than that required to install a 2×8 in (5×20 cm) one, which is half the size. Sometimes, productivity is a function of how output is defined: the cost of drilling a 1-in (2.5-cm)-diameter hole in timber, *per unit volume of material removed,* is one-third that of drilling a ½-in (1.25-cm) hole;[5] and sometimes, productivity is a function of economies of scale: the labor required per thousand board feet or per pound of lumber actually declines slightly as member size increases.[6]

Clearly, the time and effort required to set up each operation can drive the overall cost of erection. In such cases, it may be more economical to install a few large pieces rather than many small ones. On the other hand, the unit costs of purchasing and installation tend to decrease as their quantities increase. Depending on the project's characteristics, the quantities of its materials, and the learning curve associated with their installation, it may be more economical to install many identical small pieces than a smaller number of larger ones.

For the above reasons, it is a good idea to stand back occasionally and reevaluate your proposed approach to construction. Are there alternative methods, means, or sequences that would achieve the same results in less time, or at lower cost? Do the specifics of the job justify the purchase or rental of special construction equipment? Are there other ways of conceptualizing the work, which would generate economies of scale? The sequence assumed at the end of your preplanning should be thought of simply as a point of departure for the start of actual estimating. The estimating process is an opportunity to test and ultimately refine or reject your initial hypotheses.

Even though the nomenclature of the industry is based on nominal sizes, all construction dimensions, structural properties, and weights are based on actual sizes. Because heavy timber is typically installed at a much lower moisture content than when it is cut, it will have shrunk considerably in the interim. Regardless of when it is performed, surfacing reduces net section even further. Consideration of shrinkage leads to two conclusions:

1. For a given change in moisture content, the larger the member, the greater the difference between its nominal and actual dimensions. (This relationship is the basis for every timber sizing table.)

2. For a given member size, the greater the difference between its initial and ultimate moisture content, the greater the difference between its nominal and actual dimensions. This explains (at least in part) why actual *lumber* sizes have decreased as dimension lumber producers have shifted from air-drying to more effective kiln-drying. (However, as previously discussed, there has been no such shift in the handling of sawn timbers.)

When taking off timber materials, keep in mind that quantities are almost always based on the nominal volume of the original pieces, in what is called *board measure*. As its elementary unit, board measure (bm) utilizes the board foot. A board foot is a volume of wood with an *actual* length of 1 ft (0.3 m) and a *nominal* cross-sectional area of 12 in^2 (75 cm^2). Thus, with w_n as the nominal width in inches, t_n the nominal thickness in inches, and l_a the actual length in feet, the board measure of any timber is given, in the English system, as:

$$bm = \frac{(w_n)(t_n)(l_a)}{12}$$ (See Fig. 11.3 for a graph of this relationship.)

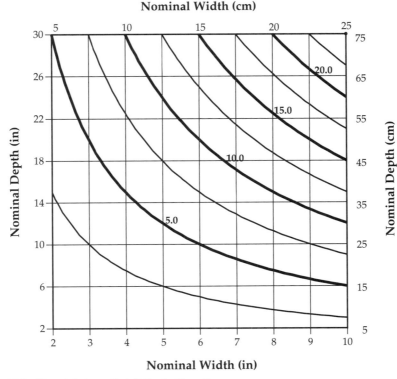

11.3   Board feet per foot (0.3 m) of length.

While initially it may seem odd that timber estimating is based on a combination of nominal and actual dimensions, it is entirely consistent with the way wood shrinks. As discussed in Chapter 2, tangential and radial shrinkage are significant, while longitudinal shrinkage is negligible. Therefore, nominal and actual longitudinal dimensions are essentially equivalent.

Board measure calculations are based on nominal cross-sectional dimensions even when the members are planed, textured, patterned, jointed, or otherwise modified after surfacing. Thus, a given number of board feet of tongue-and-groove decking will cover a smaller area than the same quantity of square-edge decking of the same nominal size. To avoid estimating errors, let your supplier do the takeoff, or give the supplier the takeoff in terms of the areas to be decked, and let him or her convert them into board measure, allowing for waste.

Recognize, too, that the required structural depth of a hip or valley rafter, the top edge of which is beveled, is usually given at its shallowest—and weakest—point. Yet, for estimating purposes, the depth of such a member is the nominal dimension needed to accommodate its deepest point. Finally, because the ratio of actual to nominal dimensions varies with timber size, the weights of two stacks of the same species and board measure will differ, if the sizes of their members differ. To illustrate, consider a thousand board feet (MBF) of 4×4 in and 8×8 in, assuming a weight of 35 pcf for each (see Table 11.1).

Qualities

A good set of timber specifications holds the contractor to a higher standard where the structure will be exposed than where it will be concealed (see Fig. 11.4). Take advantage of such distinctions, but only when it makes economic sense to do so. With few exceptions, timber construction materials are available from multiple manufacturers or suppliers. As with other commodities, larger orders are subject to larger discounts.

TABLE 11.1   Geometric and Volumetric Comparison of Two Timber Sizes

Aspect	4×4 in	8×8 in
Nom. size	4×4 in	8×8 in
Nom. area	16 in^2	64 in^2
Nom. volume per linear ft.	$^{16}/_{144} = 0.11$ cf	$^{64}/_{144} = 0.44$ cf
Actual size	3.5×3.5 in	7.5×7.5 in
Actual area	12.25 in^2	56.25 in^2
Actual volume per linear ft	$^{12.25}/_{144} = 0.085$ cf	$^{56.25}/_{144} = 0.391$ cf
Weight per linear ft	$0.085 \times 35 = 3$ plf	$0.391 \times 35 = 14$ plf
Actual/nominal area	$^{12.25}/_{16} = 0.77$	$^{56.25}/_{64} = 0.88$
Board feet per linear ft	$^{16}/_{12} = 1.33$ bf/lf	$^{64}/_{12} = 5.33$ bf/lf
Linear ft per MBF	$^{1000}/_{1.33} = 750$	$^{1000}/_{5.33} = 188$
Weight per MBF	$750 \times 3 = 2250$ lb	$188 \times 14 = 2632$ lb

---

Quantities and Qualities

Material types

    Framing, bracing, blocking, and decking
    Fasteners and connectors
    Treatments and finishes
    Firestopping, fireblocking, draftstopping

Material adjustment factors

    Slopes and angles
    Standard lengths and quantities
    Shrinkage and waste

Management responsibilities

    Coordination and scheduling
    Management and supervision
    Estimating and purchasing
    Requisitioning and financing
    Administration and close-out

Labor responsibilities

    Fabrication
    Mobilization and demobilization
    Loading and unloading
    Shipping and handling
    Wrapping and unwrapping
    Organizing and shaking out
    Assembly and erection
    Connecting and fastening
    Trimming and finishing
    Repairing and replacing
    Housekeeping
    Debris removal/disposal

II.4   Timber estimating and bidding checklist: Quantities and Qualities. Use in conjunction with contract and bidding documents.

Therefore, be creative while shopping: If the designer, for example, has specified a higher grade of decking except for a small concealed area of the building, compare the cost of a larger order of the higher grade for use throughout the building to that of splitting the order between two different grades. This is one of the few instances where enhancing the quality of a building may actually reduce its cost (and perhaps reduce the risk of a given member being installed in the wrong place).

The designer's failure to standardize is no excuse for the builder's failure to do so. Under competitive bidding, productivity is the key. The quickest way to reduce productivity is to ask the workers to keep track of items that are similar, but not identical. The cost of removing and replacing one improper beam hanger could more than justify the cost of upgrading a few to match all the others, to reduce the chance for error. However, unless the specifications give you the latitude to make such changes, it would be wise to request of the architect during the bidding period, wording permissive enough to enable such changes, *as long as they are consistent with the indicated design intent.*

This last phrase is particularly important, since the architect is usually the sole judge of contract compliance. To ease erection of some complex connections in one of my timber buildings, the timber subcontractor requested the option of fabricating what were shown as single piece connectors, in three separate pieces, to be bolted together in the field. Not until they were detailed fully in the shop drawings was it evident that the joints between the parts would be visible in the final assembly. Referencing the original detail and the specification provision mandating conformance with the indicated design intent, I succeeded in convincing the subcontractor to modify the parts so that the final assembly looked essentially the same as if the connector had been fabricated in one piece.

While the standardization and computerization of specifications has probably improved their average quality, it has also made it appear as if all that is required of the spec writer is to decide which of the standard paragraphs to include. (As emphasized throughout this book, nothing could be further from the truth.) The net result has been an erosion of job specificity and the inclusion of "leftovers" from previous projects, both of which can pose serious problems for the bidders. Custom work, of which timber construction is but one example, demands customized specifications.

This is particularly true when the bidder is responsible for final engineering of the structure and the detailed design of its connections. It is customary in that case for the final sizes, gauges, fasteners, anchorages, and spacings established by the contractor's engineer to supersede those shown schematically in the drawings by the engineer of record. It is therefore possible, although unlikely, that a takeoff based solely on the working drawings will underestimate the job. To help identify such circumstances, the bidder's engineer should be involved, at least preliminarily, during the bidding process. Also remember that wooden members only qualify as heavy timbers if they are at least a certain size; don't plan to submit shop drawings with smaller members, even if they are structurally adequate.

The goal is to understand what will be expected of you, and to price the job accordingly. Therefore, read the specifications thoroughly *before* starting your estimate, and remember to return to the specifications *after* the estimate is complete, to check that nothing has been forgotten. Proportion your labor estimates to the project's degree of difficulty, as measured by its complexity or tolerances. Highlight unique or unusual aspects. In timber construction, these often involve architectural enhancements to what are normally purely structural components. Among the most common are chamfering or routing of member edges, painting or finishing of connectors, and coordination with other building systems. Obviously, each of these has a direct cost associated with it, which must be included in the bid. Less obviously, each may result in indirect costs, all of which must also be included: higher grade material to compensate for the loss of section from chamfering, extra preparation to accommodate special finishes, or increased shop drawing costs to coordinate otherwise independent systems.

Field experience should be a condition of employment for timber estimators. Only then will they know when the benefits *to the contractor* of upgrading the specified materials will outweigh the costs.

Take decking nails, for example. Using galvanized instead of common nails could cost an additional $20 per 100 lbs[7] (45 kg), but imagine the costs avoided at the end of the job, for removing rust stains from the decking.

Qualifications

A document is needed to establish an overall framework for the work, "to spell out clearly and completely the rights and duties of the parties . . . the ground rules under which the project will be constructed"[8] (see Fig. 11.5). The general conditions fill this need. The estimator must understand which aspects of the general conditions will affect the timber contract, and to what extent the risks thereby assumed can be shared with or passed along to subcontractors or suppliers.

Provisions for payment by the owner to the general contractor are usually covered in great detail, as are time limits or conditions on subsequent payments to subcontractors. Unfortunately, the delays inherent in the system virtually assure that subs will get paid slowly. Consider this scenario:

- The fabricator offers to deliver decking just after the first of April, if a check is waiting for him. He refuses to start fabricating without a hefty deposit, but the owner won't pay for material until it is incorporated into the work. To stay on schedule, the timber subcontractor has no choice but to advance full payment for the decking material.

Qualifications

Submittals

    Shop drawings and engineering
    Product literature and samples
    Reproduction and distribution
    Calculations and certifications

General conditions

    Preconstruction and progress meetings
    Coordination meetings and site visits
    Safety meetings and procedures
    Permits and inspections
    Insurances and bonding
    Progress photos and documentation
    Phone, fax, and postage provisions
    Costs for additional drawings/specs.

Temporary facilities

    Temporary bracing and supports
    Staging and scaffolding
    Cranes, lifts, tools, and equipment
    Patterns and templates
    Temporary protection and security
    Temporary power and sanitation

Quality assurance

    Sample construction
    Plumb and level survey
    Anchor bolt survey
    Bolt inspection report
    Record drawings
    Guarantees and warranties

11.5 Timber estimating and bidding checklist: Qualifications. Use in conjunction with contract and bidding documents.

- At the end of April, only half of the material has been installed, so only half can be billed to the owner.

- The owner has 30 days to pay the general contractor, and uses almost all of it, making payment at the end of May.

- In his requisition of June 1, the general contractor must certify that payments to his subs are up-to-date with respect to his May requisition. To do so, he must only pay the timber subcontractor on June 1 for the *installed half* of the decking even though the timber subcontractor advanced payment for *all* of the decking material nearly two months earlier. Payment for the other half is at least a month away.

- Because it is a public job, the owner is only allowed to retain 2 percent of the amounts due the general contractor. By contrast, the timber subcontract allows the general contractor to retain 10 percent from the timber subcontractor. If the timber work represents 20 percent of the overall contract sum, then by the end of the job, the retainage the general is holding, on the timber subcontractor alone, *equals the entire retainage the owner is holding on the general contractor.*

Because of long delays in getting paid and the magnitude of the possible retainage, allow in the estimate for the costs of financing the job. Alternatively, try to negotiate favorable credit terms with the fabricator and erector, especially to protect you in case the general contractor is not prepared to accept delivery when the fabricator is prepared to make it. Make your bid subject to at least two conditions, one limiting the retention percentage held by the general contractor on your work to the retention percentage held by the owner on the general contractor's work,[9] and the other mandating that the general contractor extend the time for completion of the timber subcontract by the same amount granted by the owner to the general contractor for work that affects the timber subcontract. During the bidding period, submit a written request to the architect for a provision requiring timber retainage to be released on final completion of all items on the timber punchlist, rather than on the much later final completion of the entire building. Also ask for reconsideration of any prohibition against payment for materials suitably stored on-site prior to their incorporation into the work.

Lumber is subject to the same sorts of price fluctuations as other commodities. Monitor its price in the newspaper, in trade publications, or on the Internet. If the price is rising, secure a price commitment from your supplier as early in the bidding as possible; if it is falling, wait until just before the bid is due to secure a commitment. In the event that the supplier won't commit to a final price until receipt of a firm order—an order you won't be able to place until and unless you get the job—allow in your bid for whatever price change you or your supplier anticipates in the interval between bidding and ordering.

Safety measures can be costly, so it is worthwhile to clarify with the general contractor, *before bidding,* who will be responsible for those that relate to the timber construction. Those aspects associated exclusively with timber, such as temporary bracing, will normally be the responsibility of the timber subcontractor. Negotiations will therefore be limited to those aspects that will benefit subsequent trades, such as safety railings and toeboards (placed at the edges of floors or roofs to keep materials and workers from falling[10]). Make sure that your estimate accurately reflects the distribution of the attendant costs, including installation, maintenance, *and* removal.

Debris removal is often addressed in both the specifications and the general conditions. Clarification is needed of who will be responsible for debris removal, as between the timber and general contractors, and, if split, the limits of responsibility of each.

You should also negotiate the limits of responsibility with any trades having unusual interfaces with the timber work. For example, consider a project of mine, consisting of a faceted dome framed with glulam arches connected at their low ends to a steel tension ring and at their high ends to a steel compression ring. A custom faceted circumferential duct was hung from predrilled holes in the glulams (see Fig. 5.17*a*). While the timber subcontractor knew that his shop drawings would have to show the holes for the duct hangers, he could not finalize his price until he knew who was going to install the hangers.

The general conditions may mandate that essential subcontractors attend progress meetings even when their work is not underway. In any project where the primary structure is timber, the timber subcontractor's participation at these meetings is essential, at least until the frame is substantially complete. Calculate the number of site visits required beyond those you would plan to make during erection, assign a reasonable value to your time, and add this as a line item to the estimate. Do the same for subcontractor coordination meetings.

Although the requirements for timber shop drawings could be addressed as part of the timber specifications, I have chosen to discuss them here, in the context of the general conditions. Typically, timber fabricators prepare their shop drawings using their own detailers, or subcontract them to engineering firms specializing in such work. Either way, the cost of the shop drawings will be a function, primarily, of the complexity of the job, the types of connectors (standard versus custom), the amount of repetition, and the amount of engineering required. It would not be unusual for the cost of the timber shop drawings to be between 5 and 10 percent of the sum of fabrication and erection.

A set of shop drawings for a large or complex timber structure could easily exceed 100 drawings. Thus, the costs of printing, shipping, and handling could be substantial, especially if—as is likely—several submissions are required before the design team authorizes fabrication. Since the quality and comprehensiveness of the submissions is under the control of the detailer, all of these processing costs should be included in the detailer's fee.

I find it useful to think of the contract documents this way: The working drawings show the timber components and their arrangements; the specifications establish their makeup and the standards for their installation, and the general conditions describe the management services needed to execute the work. Like other types of construction, timber construction involves bringing workers with common skills together in the field to create a single instance of a unique product.

To a great extent, profitability in timber construction derives from recognizing and responding to the need for ample site supervision. In fact, one of the crucial factors in the decision to bid on a particular project should be the availability of a competent project superintendent.[11] Having a deep understanding of carpentry is an essential prerequisite, if the superintendent is to communicate effectively with the workers. At least as important, however, is the ability to implement the estimator's strategy. (That is why it can be helpful to designate as the superintendent for a given job the same person who prepared its estimate.) Because so much of a typical timber frame is prefabricated, efficiently organized erection is the key to profitability. Therefore, an experienced supervisor with an inexperienced crew will probably be more profitable than an inexperienced supervisor with an experienced crew.

Before investing time and energy in preparing a bid, determine how the project fits into your short- and long-term plans. Does it require skills or equipment your firm doesn't already have? If so, can you still be competitive? Is it a job you want or need? Does it play to your specific strengths? Which aspects will you be able to handle in-house, and which will have to be subbed out? Are you prepared for the additional coordination that will result? Will price be the sole determinant for award, or will quality and experience—in years or number of jobs—also be considered? Will bidders be prequalified? Are the contract documents clear enough that you'll be able to price the job with confidence in your understanding of it? Answers to such questions should enable you to decide how to proceed.

Buildings are similar to many other products in that their assembly involves a number of repetitive operations. Buildings are different from other products in several important ways: The workpiece is stationary and the workers mobile; the product is produced once or in very limited quantities, and the makeup of the work crews varies from day to day. To be effective, the superintendent must think of the timber construction very differently from his or her workers. For the carpenter, the key to success lies in mastering each task. For the superintendent, the key to success lies in orchestrating the switchovers between tasks. Designers and builders both focus on intersections, but designers are concerned with those between materials, while builders are concerned with those between activities. The superintendent brings value to the project by managing the tasks in the context of the evolving project. Therefore, allow plenty in your estimate for a good one.

The makeup of the timber contractor's management team is a function of the size and complexity of the project. In most cases, the superintendent will be responsible for field operations, and a project manager or project executive will schedule the job and handle relationships with subcontractors and suppliers. The skills required of managers may sound contradictory. On the one hand, they must be willing and able to create and enforce project schedules. On the other, they must be flexible enough, on a moment's notice, to revise or rewrite those schedules to accommodate changing circumstances, including delays in deliveries or in the completion of preparatory work by other trades. The percentage of the manager's salary charged to the project should equal the percentage of his time dedicated to it.

No matter how good the shop drawings, the fabrication, or the erection, some corrective work is inevitable when dealing with natural materials, such as wood, assembled with human hands. By contract, assign responsibility for each type of repair to the party responsible for the initial defect. Require the fabricator, for example, to replace mismanufactured timbers or connectors. Require the erector to repair members damaged after delivery. Consult your attorney for advice as to whether your subcontractors should be required to protect you as you are required to protect the owner. After passing along as much risk as possible, allow in the bid for corrections of your own, and for administering the corrections of your subcontractors. If you will be assessed liquidated damages in the event of late completion, decide whether you can finish the job by the owner's deadline. If not, allow for the lesser of damages during the extra time you will need or the additional costs you will incur to assure timely completion.

Serious cyclists often speak of the final 10 percent of their ride as requiring as much effort as the first 90. The same can be said of the effort required of the timber contractor (and the GC) after substantial completion: The amount of work is small, but progress is exhausting. Particular care is needed to avoid damaging adjacent finished surfaces. Whereas the amount of work over which to distribute the expense of setting up staging and ladders is high during erection, it is extremely low during punchlist repairs. Consequently, an excessively long punchlist can put any project into the red. I believe that the subcontractors that try the hardest to shirk their closeout responsibilities are those that failed to allow for them in their bids. Don't make the same mistake.

One of the ways of minimizing closeout work is by performing it earlier in the project. In the case of the construction itself, implement and enforce appropriate quality control procedures, to minimize the inci-

dence of errors. Immediately repair those that do occur. On record drawings, register deviations from the working drawings or shop drawings as they occur, or at least while they are still visible. Dedicate one binder to warranties and certifications, and insert copies as they are received, rather than waiting to the end of the job to compile them. Create your estimate in a form that can easily be exported into the schedule of values, to become the basis for your requisitions. This approach will not only reduce errors, but will also increase your overall competitiveness. Allow in the bid for the costs of satisfying all specified or foreseeable closeout requirements.

Quotations

There are several potential sources for costs on which to base your estimate or bid (see Fig. 11.6):

Subcontractors: To the extent that you can rely on quotations from subcontractors, your risks can be reduced. Regardless of the reputation of the firm or your experience with it, get at least one but preferably two other prices for comparison. If none is available, do your own estimate of the work to be subcontracted. Unless you enjoy construction delays and disputes, do not award the job at a price you believe to be unrealistically low, even if the subcontractor believes otherwise.

Suppliers: The above caveats also apply to suppliers. In addition, recognize that timber prices have become quite volatile in recent years, as a result of tightening supply and increasing global demand. Make sure the supplier is willing to guarantee the quoted prices, in writing, for the duration of the timber construction period. If you have confidence that prices will decline between the date your bid is due and the date you would place the order if awarded the timber contract, consider sharing the risk of price fluctuation with the supplier. Also, require that the supplier credit your account for any defective material.

Personal experience: There is no more reliable way to price a job than to start with the actual costs—particularly labor costs—you incurred on similar previous jobs. Adjust them for inflation and for differences in form and content between the projects.

Cost reference books: There are a variety of national construction cost reference books available, all of which are updated at least once a year. Unfortunately, their coverage of heavy-timber construction is limited. Consequently, while they may enable the designer to develop a preliminary budget for the timber frame, they are of limited value to the estimator, except for the most conventional timber structures.

Quotations

Components

 Base bid
 Alternates
 Unit prices
 Allowances

Information sources

 Experience on previous projects
 Trade publications and websites
 Suppliers and vendors
 Fabricators and erectors
 Cost manuals and references

Business decisions

 Depth of relevant experience
 Risk and importance of project
 Relationship to ongoing work
 Bondability and capacity to perform
 Confidence in estimate
 Realistic profit expectations

Contractual matters

 Dispute resolution mechanisms
 Bonus/penalty provisions
 Delay and liquidated damages provisions
 Furnishing/installation responsibilities
 Adequacy of construction time
 Limits of contract
 Form of contract
 Type of contract (general, prime, sub)
 Limits on subcontracting

Project-related concerns

 Site conditions, access, and staging areas
 Project size and difficulty
 Quality of contract documents
 Proprietary aspects of specifications

Market factors

 Likely competitors
 Competitiveness of market
 Material price volatility
 Labor rate stability

Financial considerations

 Payment and change order procedures
 Payment frequency and timeliness
 Retainage percentage/reduction provisions
 Trade and quantity discounts
 Financial viability of owner

Labor-related issues

 Quality/skills of available workers
 Productivity expectations/crew sizes
 MBE/WBE requirements
 Union/prevailing wage requirements
 Relevant provisions of union contract
 Wage increases during contract life

Methods and means

 Construction strategies and sequences
 Standardization and quality control
 Allowances for uncontrollable aspects

Bidding issues

 Bid validity period
 Bid costs (bid bond, prebid conference, etc.)
 Adequacy of bidding period
 Prebid conference attendance

11.6 Timber estimating and bidding checklist: Quotations. Use in conjunction with contract and bidding documents.

Timber fabricators and suppliers: Even if they are also furnishing prices to your competitors, these sources have an interest in maximizing their chances of success. Consequently, they are more than likely to help you learn the ropes.

Regardless of the labor rates, be sure they apply to the job at hand. Public projects in many jurisdictions require that workers in the field be paid so-called prevailing rates, which are essentially equivalent to union wages. Imagine what an impact that regulation could have on the bid of an open-shop contractor. Off-site labor is typically exempt, which is fortunate since most timber fabricators are nonunion.

Putting It All Together

There are several aspects to verify before finalizing your bid:

Scope of work: For a variety of reasons, including union jurisdictional agreements, the general contractor may distribute the work among the subcontractors slightly differently from the way the architect did in the specifications. If any of the work included in the timber specifications is to be handled by another trade, make sure to incorporate it into the conditions of your proposal, and to adjust your bid accordingly.

The big picture: One of the games general contractors like to play involves furnishing to bidding subcontractors only those drawings and spec sections describing their particular work. This may sound reasonable, but it is not. To understand fully both the physical and contractual contexts in which its work is to be performed, each subcontractor must be familiar with *all* of the drawings, *all* of the specifications, *all* of the general conditions, and *all* of the addenda. To be sure that you get the whole picture, request of the architect that all documents be sent directly to your attention, even if it means paying an additional cost. That is a small price for added peace of mind, particularly in the context of overall estimating costs, which can total ¼ percent or more of the value of a bid.[12]

Site conditions and access: Visit the site and familiarize yourself with access to it. Construe "access" as broadly as possible: A nonunion timber fabricator working on one of my structures did not discover until his nonunion driver tried to make his delivery, that site access was effectively limited to union members. As a result, the truck had to go back to the plant several hours away, and return to the site with a union driver. Notwithstanding this example, access typically relates to transporting materials and equipment to the site, and moving them around once they are there. If mobilization for or delivery of the timbers will result in oversize loads or necessitate overtime labor, allow appropriately for same in your bid.

Rounding of quantities: Even though a carpenter may only need an hour or two to perform a particular task, he or she normally gets paid by the day. The same reasoning applies to equipment rentals, though the mobilization and demobilization time for the equipment may be so great that the minimum rental period may be a week or longer. The total for each work item must therefore reflect not just the calculated quantities of labor, materials, and equipment, but also the impact of union rules, purchasing conventions, waste factors, and bidding uncertainties. If using spreadsheet software for estimating, consider codifying these constraints in the respective cell formulas (and don't forget to review them on the next job).

Bidding uncertainties: There are a number of potential uncertainties in any bidding situation; your objective should be to eliminate as many as possible. Perhaps the most familiar sources of uncertainty are ambiguities within the contract documents themselves. If brought to the attention of the design team early enough, most of these will be clarified, through addenda, before the bid date. Unit prices, allowances, and alternates comprise another category of uncertainty, a category I refer to as *optional work*. In each case, the bidder must offer prices for work that may or may not be performed. The difficulty of pricing is compounded if the owner reserves the right to authorize such work at any time during the construction period. After all, the unit price for additional timber decking will be far higher if furnished piecemeal than if purchased simultaneously with the rest of the decking. Although optional work is the last thing bidders want to deal with, it may be the only way for the designers to accommodate budgetary constraints or, on renovation projects to address aspects concealed during design. In conjunction with replacement of a leaking roof, for example, we recently requested that the price for replacing rotted areas of the timber deck be in the form of a unit price, since the extent of the damage could not be quantified until after the existing roofing was removed. On another job, the size of the addition to be built depended on the cost of the job; the base bid was for the smallest acceptable addition, with each additional increment priced as an alternate. Rather than developing a price per square foot by dividing the base building area into the base bid, the smart bidder evaluates each alternate separately, adjusting the prices of materials and the productivity of labor accordingly.

You're finally ready to apply your markup. Balance the desire to maximize profit against the imperatives of competitive bidding. Identify your competitors, and attempt to predict which aspects they will subcontract. Review the results of previous timber biddings. Evaluate the tightness of the market. Assess your strengths and weaknesses and, most important, be realistic.

Summary

A *complete* estimate is one that includes all the work required under the contract; a *responsible* estimate goes much further: It is integrated with a detailed work plan that takes advantage of the specifics of the job to maximize productivity and minimize cost. Getting from a complete estimate to a responsible one requires experience and judgment.

A responsible estimate integrates four types of input: quantities, qualities, qualifications, and quotations or, put in the form of a question: How much, what kinds, in what context, and at what cost?

Quantification begins with a thorough understanding of the working drawings. Familiarize yourself with the project and its vocabulary. Let the construction sequence guide the estimate. Be explicit about assumptions. Color code the members, and check them off as they are counted. Differentiate between typical and atypical conditions. Label work items descriptively. Don't forget to count the fasteners. Quantify with appropriate units and scales. Account for the interrelationships between the drawings and specifications. Understand the implications of the building's geometry. Allow for waste. Adjust productivity to the difficulties of the job. And, most important, keep the big picture in mind.

Quality is covered primarily in the specifications. Standardize wherever possible to maximize field productivity. Explore alternative ways of achieving the specified objectives. When engineering is within the scope, verify that the sizes indicated schematically satisfy the engineering criteria. Recognize that a bidding phase investment in engineering—by the contractor or the engineer—may more than pay for itself during construction. Never use members smaller than the sizes required for the construction type. Allow for both the direct and indirect costs associated with architectural enhancements of the timber structure. Standardize materials and systems to minimize total cost.

Qualify the job in accordance with the general conditions. Establish your tolerable level of risk, and endeavor to shift the rest to parties better able to handle it. Allow appropriately for the costs of financing the job. Define the lines of jurisdiction between the timber contract and other contracts. Include in the bid your share of the cost of safety, temporary works, and debris removal. Assign a value to attendance at progress and coordination meetings. Allocate funds to the preparation, revision, and handling of the timber shop drawings. By contract, delegate to others as much corrective work as is reasonable and appropriate; be realistic in allowing for the aspects under your jurisdiction. Streamline the processing of data and the generation of forms to reduce project overhead.

Quote the job on the basis of information from sources proven to be reliable, be they subcontractors, suppliers, cost reference books, personal records, timber fabricators, or suppliers. Regardless of the source, make sure the information is complete, current, and consistent with the contract. Confirm that the scope of each quotation dovetails with the scope of all related ones. Test each quotation for credibility before accepting it. Verify that the rates for labor assumed in the estimate apply to the rates for labor—union versus nonunion—required in the field. Test your proposed bid prices against your actual costs on previous jobs. Use rules of thumb—such as that job costs will be divided evenly between labor and material—only for general guidance.

After compiling the raw data, verify that allowances have been made for all work items, and that exceptions—your own and those requested by the general contractor—are noted in the proposal. If you haven't already done so, familiarize yourself with the site and its access, and determine the impact on the bid. Adjust quantities to reflect industry conventions. Price optional work separately from required work, recognizing the inefficiencies inherent in it. Finally, account for market conditions in setting your markup.

Notes

1. As quoted in Wurman, p. 300.
2. Sanders, 368.
3. The Goldstein Partnership, Montville Specs., p. G5-C-2.
4. Phone conversation with Ben Brungraber, PE, December 8, 1997.
5. Waier, p. 169.
6. Waier, p. 172.
7. Waier, p. 168.
8. Sweet, p. 499.
9. Bunéa, p. 115.
10. Harris, Cyril, 506.
11. Clough, p. 71.
12. Clough, p. 71.

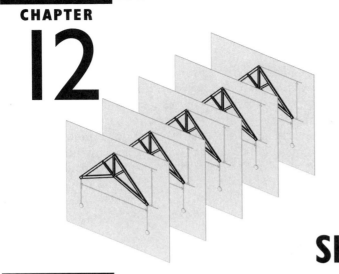

Shop Drawings

. . . the creative process is the process of learning how to accomplish the desired result.

ROBERT W. MANN

Introduction

The great cathedrals of Europe were constructed under the guidance of polymaths. The expertise of these individuals encompassed *mens et manus,* literally "mind and hand." Over the succeeding centuries, roles within the construction industry have grown increasingly specialized. Responsibility for design and construction now reside in separate organizations, and the actual work of construction may be distributed among dozens of others. As buildings have gotten increasingly complex, fabrication of their components has moved increasingly off-site. The level of information in the working drawings has proven inadequate for fabrication of most materials and systems.

The Purpose of Shop Drawings

Shop drawings guide the shop workers; they enable the fabricator to model designated portions of the building on paper, before actually making them in the shop. As the party responsible for methods, means, and sequences, the contractor is the appropriate party to establish how to achieve the results indicated by the designers. Shop drawings for timber, like those for any other building subsystem, have their own particular conventions. Most relate to the fabricator's particular tools and equipment. Standard details have evolved to accommodate the natural properties of wood: its hygroscopic nature, its directionality, its susceptibility to decay, its natural variations, and the various ways it can be sawn.

Shop drawings help facilitate the coordination of related work. While internal consistency is necessary for coordination within the shop, it is not sufficient for coordination between trades; shop drawings must, therefore, address all of the boundary conditions, all of the interfaces with adjacent materials and systems. Except in unusual circumstances, a timber frame is simply an armature, a framework, a chassis, if you will, on which to mount the rest of the building. And, since the frame must be erected quite early in construction the timber subcontractor plays a leading role in the shop drawing process. The locations and details of much of the subsequent work follow from the contents of the timber shop drawings.

On a given job, timber construction may be only one of several aspects under the jurisdiction of carpenters. Carpenters working for one subcontractor may be responsible for attaching nailers to wood decking provided by another. It is not enough, then, for the timber shop drawings to simply show the boundary conditions or interfaces with other work; they must also define exactly which work is in the timber contract, and which is "provided by others." Only if the architect has an understanding of industry conventions will the work included in and excluded from the timber specifications be consistent with the distribution of work designated by the timber contractor in the timber shop drawings. And only if "single-source responsibility" is specified are there grounds for rejecting a division of work different from what is specified.

The primary reason an owner bids a project is to secure the best price for the work. The desire to bid, however, must be tempered by the understanding that there is a fine line between being specific and being overly restrictive or proprietary. Knowledge of industry standards and conventions is necessary, but may not be sufficient, when procuring work that pushes the limits of manufacturing. Competition will be more imaginary than real if each of the bidders ends up subcontracting the same work to the same manufacturer. The lesson here is simple: If manufacture of your timber frame will require capabilities unique to a particular fabricator, be prepared to pay a premium, perhaps a hefty premium, for the work.

There are times when the timber contractor will be asked to furnish materials *to* another trade, or to install materials furnished *by* another trade. To assure a consistent appearance, the contractor may be asked to furnish all the wood decking, even that which will be installed in nonstructural applications by others. Or to minimize opportunities for timber damage by other trades, the contractor may be asked to attach to the timber beams items such as duct hangers furnished by the HVAC subcontractor. In either case, the shop drawings should be clear about the degree of the timber contractor's responsibility for each item.

Shop drawings represent only one category of submittal, the graphic part, corresponding to the working drawings. Others, such as product literature, samples, certifications, warranties, test reports, and, if within the scope of the timber contract, engineering calculations, evidence conformance with the specifications. Still others, such as staging area and crane set-up maps, and fabrication and erection schedules, relate to the general conditions. While this chapter focuses on the role of timber shop drawings, each of these other submittal types should be an integral part of any timber project's comprehensive program of quality assurance.

The Importance of Shop Drawings

The most fundamental difference between residential wood framing and heavy-timber construction is where they are fabricated: on the site versus in the shop. The fabrication setting is both cause and effect. It leads to a very specific project delivery process and follows from the types and sizes of the materials to be used.

Residential wood-frame construction utilizes lightweight, short-span, off-the-shelf lumber products with standard properties, easily worked with conventional tools, erected by hand, and secured with nails, in configurations of great structural redundancy. By contrast, timber construction involves substantially larger and heavier, longer-span, custom products with job-specific properties, which by their very nature must be fabricated under controlled conditions by specialized equipment, erected by crane, and secured with high-strength fasteners, in configurations of minimal structural redundancy. The house framer is usually both fabricator and erector; the timber contractor is rarely both, and may be neither.

The route from mill to job site is different for dimension lumber than for heavy-timber products. To transform a tree into a 2 × 4, it is felled, sawn, dried, and surfaced. Even though it might make stops at the wholesaler and retailer after leaving the mill, each piece arrives at the job site essentially unchanged. With the advent of engineered lumber and glulams, the route from mill to job site has become more circuitous. Lumber for glulams goes from mill to fabricator, where it may be planed, curved, laminated, drilled, mortised, and finished, prior to site delivery.

Finally, the capabilities and responsibilities of the workers in the field, and the consequences of their actions, are different for stick framing than for heavy timber. In stick framing, beams and joists are light enough that they can be lifted by hand, and small enough that they can be worked with portable tools. If a piece doesn't fit properly, it can be modified in the field or replaced by another. If the workers make a mistake or run short, relief is available at the nearest lumberyard. *In stick framing, the time and cost penalties for poor planning and poor execution are small.*

In timber framing, the spans are long and the loads are great, so the pieces are large and heavy. Fabrication requires the use of specialized equipment in a controlled environment. Field work is thus limited to erection. If a piece doesn't fit properly, options for replacement may be very limited; even if you could find a piece of the right size at the lumberyard, it might need to go to the fabricator for working and finishing. In the meantime, the job may have come to a standstill. *In timber framing, the time and cost penalties for poor planning and poor execution can be enormous.*

Shop drawings play a critical role in time, cost, and quality control. The more work done in the shop, the less needs to be done in the field. This reduces overall construction time and the impact of the weather on workmanship, productivity, and progress. Given the proper guidance, workers in a shop environment can produce more and better products than their field counterparts. For materials to be prepared for a potentially distant site before the work to which it will be attached is even present requires a level of dimensional and quality control unheard of in most residential construction.

Shop drawings are necessary whenever products are being customized for a particular project. Since every timber building is unique, timber shop drawings are an integral part of the timber construction process. They describe the fabricator's raw materials and illustrate how he or she intends to work with them to achieve the designers' objectives. (See Fig. 12.1.)

Coordinating Shop Drawings with Other Trades

Although timber shop drawings are primarily a vehicle for communication between the timber detailer and the architect and engineer, they also have several other important roles:

- They should indicate the degree to which the job will be preassembled. (While this is technically a methods and means issue, it does no harm for the architect to tell the detailer that he limited the heights of the trusses so that they would fit on a truck.)

- They must be coordinated with all work that is already in place or that will be in place by the time the timber work is erected.

- They should integrate the fabrication and erection schedules; it does no good to have the columns arrive *after* the beams.

At interfaces between different trades, their respective shop drawings can be dependent on one another. An example would be where curtain walls are attached directly to the structural frame. Since the bulk of both the window framing and the structural framing is usually shop fabricated, their shop drawings should be coordinated prior to starting

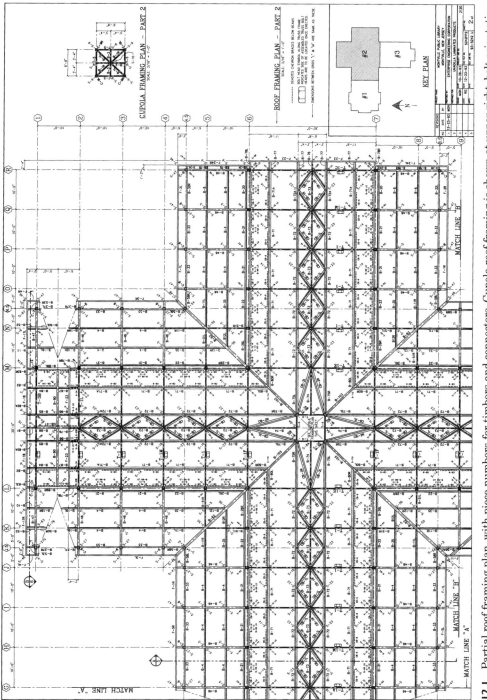

12.1 Partial roof framing plan, with piece numbers for timbers and connectors. Cupola roof framing is shown at upper right, bolt orientation at center right, key plan at lower right. The framing of a typical transept is shown in context, from below, in Fig. 14.6. All the shop drawings shown in this chapter were prepared for the new public library, Montville, NJ. The Goldstein Partnership, architects. Severud Associates, structural engineers. Timber engineer and detailer: Enterprise Engineering Consultants, Ltd., Peshtigo, WI. Timber fabricator, Unadilla Laminated Products, Unadilla, NY. Timber contractor, Dajon Associates, Hackensack, NJ. General contractor, The Conklin Corp., Franklin Lakes, NJ. (Enterprise Engineering Consultants, Ltd., Peshtigo, WI.)

the fabrication of either. That is not to suggest that the trades will be negotiating from positions of equal strength; on the contrary, structural issues must always prevail. One thing is certain, however: No trade will make provisions for another unless required to do so by the contract.

Fabrication Differences between Timber and Steel

Even though both steel and timber members pass first through mills and then through fabricating shops, the value added by each operation is very different for each material. Whether made from scratch or from recycled steel, it takes a tremendous amount of energy and enormously expensive machinery to roll a steel section. To produce a timber 4 × 8 in, by contrast, all you need is a saw.

Another major difference between timber and steel is the stage in the manufacturing process where their strength is imparted. Steel members leave the mill with as much strength as they will ever have. Although steel beams and columns are occasionally welded up from smaller members, most steel fabrication for buildings involves subtraction: cutting, coping, punching, and burning. The end result is a job-specific member with the same or less strength than it had on arrival at the shop. With glulams or engineered wood beams, by contrast, small and relatively weak boards or fibers from the mill are glued up in the shop, creating a product the total strength of which is much greater than the sum of the strengths of the pieces of which it is comprised. Even though subsequent fabrication processes are analogous to those of the steel fabricator, a glulam leaves the shop with far *greater* strength than its components.

The production of curved members also differs greatly between steel and timber. It takes extremely powerful equipment to bend a rolled steel section. Consequently, relatively few steel fabricators do their own bending. Working with thin and flexible boards, the timber fabricator needs only some jigs and clamps to hold the bent laminations together long enough for the glue to dry. Consequently, most glulam fabricators do their own bending.

Differences between Working Drawings and Shop Drawings

All working drawings are abstractions; except for full size details, what is depicted is not intended to be taken literally. Walls are shown by pairs of lines with little or no detail between. Finish materials are shown in detail for small typical portions of the surfaces they are intended to cover. Premanufactured assemblies, such as doors and windows, are shown diagrammatically. And virtually everything is shown much smaller than actual size. Although CAD has changed things somewhat, enlarging a working drawing does not usually reveal any useful information that would otherwise have been hidden. To build from a set of working drawings, then, a contractor must first know how to translate its general small-scale contents into specific large-scale actions. Regardless of the type of project, this translation must occur. But without the requisite knowledge, it cannot.

Residential framing is more forgiving than timber framing, because, in several ways, it is more redundant structurally:

- Joists and rafters are so closely spaced that concentrated loads are shared among adjacent members.

- Nonbearing stud partitions are constructed by the same trade, at the same time, and in the same manner—tight to the underside of the structure—as bearing partitions, so that they often end up carrying loads not intended for them.

- Plywood wall sheathings contribute to structural resistance not only where their contribution is intended—such as in the lateral bracing system—but also where it usually is not—such as at door and window lintels.

Timber framing, by contrast, is relatively unforgiving:

- Joists and rafters are usually too far apart to share loads.

- Nonbearing partitions stop short of the underside of the structure to prevent inadvertent loading and the resulting stresses.

- Wall sheathing, even if of plywood and part of the lateral bracing system, contributes little to the capacity of the lintels at the large openings typical of heavy-timber buildings.

In simple terms, short spans, light loads, and the wealth of structural redundancy in residential wood framing yields a high inherent factor of safety. The long spans, heavy loads, and dearth of structural redundancy in timber construction yields a lower inherent factor of safety. Through engineering, a structure's *actual* factor of safety can be tailored to the needs of the project.

Using CAD, a designer can compile a set of floor plans from a number of independent layers, each of which is associated with a specific discipline. Shop drawings are always for a *particular* trade and can be thought of as starting where the designer's layer for that trade leaves off. On a nonresidential building, shop drawings are prepared for much or most of the work. Thus, the shop drawings for a particular job, taken together, are nearly equivalent to its as-built drawings; they represent a much more accurate representation of the proposed building than the working drawings. This is not to suggest, however, that shop drawings supersede the working drawings on which they are based. On the contrary: They complement and expand upon them.

A good set of working drawings does not show *every* condition, just every *typical* condition. To call attention to slight variations, the designer will note that one condition is similar to another, and then describe the differences between them. Connections are shown schematically; the contractor is left to fill in the details, in accordance with the specifications and the standards referenced within them.

Working drawings depict certain aspects in great detail. These include the locations and sizes of horizontal and vertical openings, the spatial relationships between structural work and the nonstructural work hung from or attached to it, and the interfaces between different materials and systems. In general, working drawings emphasize the finished result, not the contribution each trade makes to achieving it. This is consistent with the contractor's responsibility for methods and means.

Whether or not the designer wished to review the shop drawings, they would still be prepared by the fabricator for his internal use. Fabricators need to know the quantities, qualities, shapes, and sizes of everything within their contracts. They need to choose among the options available to them regarding suppliers and manufacturers. They must determine the most cost-effective way of satisfying any performance specifications. And, they must show their own shop workers precisely what is to be done to each piece. (See Fig. 12.2.)

The timber shop drawings one fabricator might prepare for a given job are likely to be similar, but not identical, to those prepared by another. Any differences may be associated with local union rules, the relative cost of labor and materials, proprietary manufacturing techniques, or the limitations of the fabricator's equipment. On more complex projects, the differences may involve the types of connections and their relationship to the degree of shop and field assembly. Regardless of the specific differences, they are almost always driven by economics.

Integration of Architectural and Structural Information

Timber shop drawings must integrate two distinct aspects of the contract documents: architectural and structural. In preparing their respective drawings, the architect is concerned primarily with form and function, the engineer with structural adequacy and economy. The degree to which their domains overlap is a function of the character of the building's structure. Think of the physical parts of a building—walls, floors, and roofs—as the figure, and the spaces between them—the inhabited spaces—as the ground. The engineer's primary concern is the figure; the architect's, the ground. The engineer is concerned with the structure itself, the architect, with the space the structure defines. The architect and engineer, in other words, have complementary images of the building.

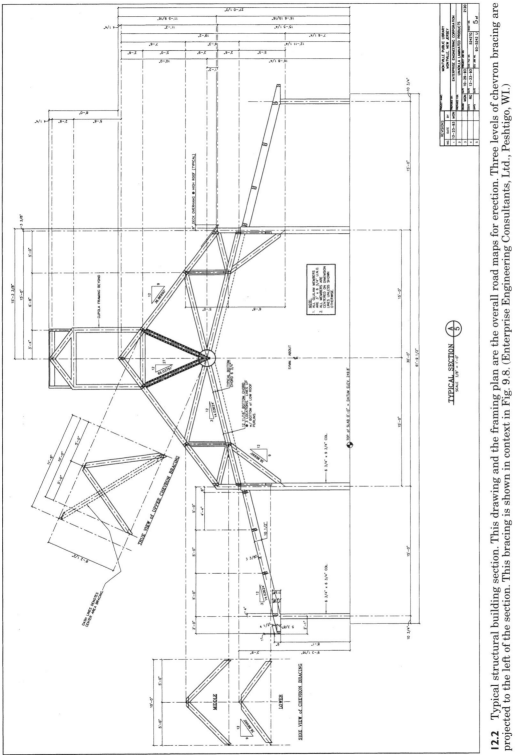

TYPICAL SECTION ⟨A⟩ ⟨5⟩
SCALE 3/8" = 1'-0"

NOTE:
1. ALL GLULAM MEMBERS ARE 5 1/8 x 9 3/4" U.N.O.
2. ALL MEMBERS ARE CENTERED ON DIMENSION LINES UNLESS SHOWN OTHERWISE.

MONTVALE PUBLIC LIBRARY
MONTVALE, NEW JERSEY
ENTERPRISE ENGINEERING CONSULTANTS
UNADILLA LAMINATED PRODUCTS

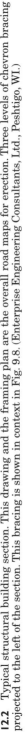

12.2 Typical structural building section. This drawing and the framing plan are the overall road maps for erection. Three levels of chevron bracing are projected to the left of the section. This bracing is shown in context in Fig. 9.8. (Enterprise Engineering Consultants, Ltd., Peshtigo, WI.)

In a building where the structure is hidden from view, the architect's concern may be limited to the locations of the columns and the thicknesses of floor and roof platforms. Where the structure is to be left exposed, however, there is potentially much more overlap between the concerns of the architect and engineer. The architect may wish to dictate the proportions and spacing of beams and joists, the types of connections, and the character of the deck. These need not be considered constraints on the structural engineering; with the structural work no longer limited by the dimensions of walls and floors, the engineer may find the range of structural options to have actually expanded. For the architect, however, "achieving an aesthetically acceptable and universally acclaimed building while working within the constraints of a structural system can require more care and planning than covering up structure with purely architectural materials."[1]

The timber detailer must find ways to bring these often different and sometimes conflicting visions of the project into harmony. In a sense, she must see the structure through the eyes of the architect. If the timber frame is to be hidden from view, it is as if the architect is wearing glasses with clear lenses; if the timber frame is to be exposed, the architect's lens lends a job-specific esthetic cast to the structure.

From the perspective of the designer, the measure of the detailer's success is how well the design intentions are translated into shop drawings. The review process—with its emphasis on structure within a particular architectural context—gives the designer the opportunity to check for concordance between his vision and that of the detailer. It is easy to see why drawings and specifications are not sufficient without a comprehensive point of view; they leave the detailer without direction. When this happens, the building ends up getting designed during the shop drawing process, rather than when it should have been, during the preparation of the working drawings.

Although the shop drawings are based on the working drawings, they are not simply more detailed versions of them. The concerns of the detailer, and therefore the emphasis of the shop drawings, are different from those of the designer, as the following examples will illustrate.

When an engineer initially analyzes a grid of beams and joists, his span calculations are based on the nominal spacing of the members, center to center. He may recommend that joist hangers be used. For esthetic reasons, the architect may prefer flush type joist hangers to more conventional ones. The engineer then sizes the members and labels his drawings accordingly, accompanied by a schematic illustration of the connections. What does the detailer do with this information?

Since the ultimate objective of the shop drawings is to guide the shop, the detailer's first priority is to translate the language of the designer into that of the fabricator. (See Fig. 12.3.) The fabricator doesn't care about nominal dimensions; he needs actual dimensions.

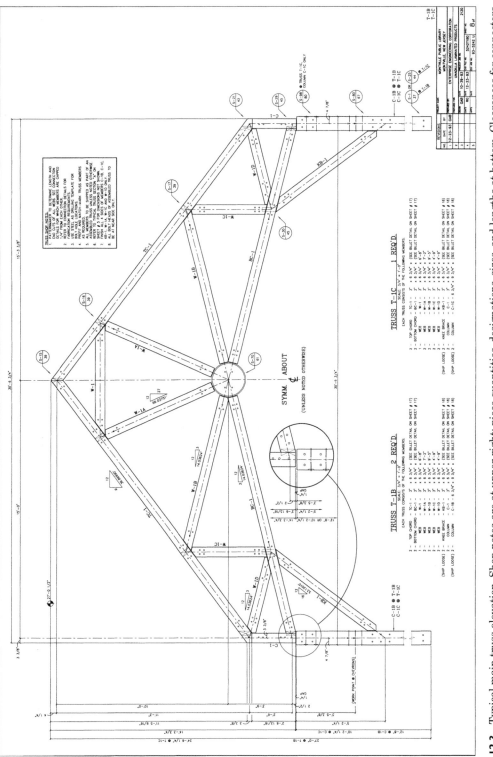

12.3 Typical main truss elevation. Shop notes are at upper right; piece quantities, designations, sizes, and lengths at bottom. Clearances for connectors are detailed. This truss is shown in context in Fig. 14.12. (Enterprise Engineering Consultants, Ltd., Peshtigo, WI.)

He wants to know that each of the 12-ft (nominal) joists is actually 11 ft 6⅞ in long, after subtracting the half width of the glulam beam at each end. Similarly, to him, the custom hanger is not simply a structural shape; it is an item from a catalog or a collection of steel pieces that have to be ordered, cut to size, welded, drilled, ground, and painted. The joist ends are not prepared for mortising simply by scribing the outline of the hanger on them; the shoulders of the mortise must be held back enough from the edges of the hanger to allow for moisture-related movement of the joist.

The Shop Drawing Review Process

Shop drawings are reviewed in a bounded cumulative iterative feedback process. The process is *bounded* by the scope of the contract documents; the contractor is not required to exceed it. It is *cumulative* in that the amount and accuracy of the content of the shop drawings increases with each review cycle. It is *iterative* in that it repeats until the designer finds the drawings approvable. And, it is a *feedback process* in that each successive version of the shop drawings reflects comments from previous versions.

Before generating the entire set of shop drawings, detailers will usually request confirmation that their basic understanding of the project is correct. This is accomplished through the submission of a small number of typical shop drawings or, in the case of particularly complex projects, just the framing or erection plans. Even though a partial early submission may be customary and appropriate, the detailer should not expect the designer to pass complete judgment on it, nor should the designer feel obligated to do so, until it can be reviewed in the context of either the complete set, or at the very least, of all related shop drawings within that set.

A partial submission is only entitled to a partial review. If, for example, the detailer is unsure of specific aspects of the building's overall geometry, the submission should be explicit in that regard, and the designer should respond accordingly. Suitable review comments might include: "Submission is in general accordance with the Contract Documents with respect to overall geometry only," or "Approval of this drawing is limited to its geometry, as noted," or "This shop drawing is conditionally authorized, subject to review of related shop drawings." The designer should limit feedback to the issues raised by the detailer, especially since the submission may be incomplete not only in the number of sheets, but also in the contents of each one.

Like an architect, a timber detailer assembles his drawings in layers, each of which follows from those that precede it. The framing plan, which often comprises the initial partial submission, represents the index, if you will, to the rest of the shop drawings. Its piece labels and section marks are pointers to distinct destinations elsewhere in the set.

Because she is most often operating under a fixed fee, the detailer has an interest in keeping the set of shop drawings as brief as possible. To achieve this objective, each piece drawing must apply to as many pieces as possible. Organizing the components of the structural frame into classes of timbers and connectors is therefore essential. Members are subdivided further into beams, purlins, joists, and columns. Truss members are detailed individually and as parts of assemblies. Within each class are members that are identical to one another or which are variations on a theme. Put simply, the organization of the timber shop drawings should follow from the organization of the timber frame.

It should be easy to navigate from the framing-plan-as-index to the details of the members and connectors. The numbering system should not only be complete, but also understandable. To take advantage of the power of object-oriented programming, there is a move afoot to integrate the specifications with the working drawings by labeling each object in the drawings with the number of the paragraph in which it is specified. While in favor of consistency in labeling, I am concerned that the worker in the field, normally without the benefit of the spec book, will be unable to decipher the code. I am also concerned that it will not be *obvious* whether a particular code number—other than its first digit, which corresponds to its spec division—is correct. Until the specifications for each CAD object inhere in it, errors will be easier to make and harder to catch with numerical labels than with textual labels.

The numbering of individual pieces is extremely important, not only during review of the timber shop drawings, but also during shipping, handling, shake-out, and erection of the timber frame. (See Fig. 12.4.) In fact, erection errors can often be attributed to unclear piece numbering systems in the shop drawings. Where appropriate, the detailer's numbering system should be tied to the building's structural grid, and the orientations of individual members should relate to the compass orientation of the building.

Stamping the Shop Drawings

Because designers review shop drawings solely for *general* conformance with the contract documents, there are only a couple of ways they can summarize their specific comments: "conforming," or "nonconforming."[2] By convention, each of these two options has two of its own suboptions. If the shop drawings are conforming, they are stamped either "Approved for manufacture," meaning the shop drawings are in general conformance with the contract documents and can be released as they are to the fabricator, or "Approved for manufacture, as noted," meaning that the fabricator must also account for the designer's specific written comments. If the shop drawings are nonconforming, they are stamped either "Revise and resubmit," meaning that the nonconforming aspects are too widespread or serious for the designer to approve the shop drawings in their current state, or "Rejected," meaning that the detailer missed the boat entirely, and must start again. Some firms have a fifth category, falling somewhere between conforming and nonconforming, called "Approved for manufacture, as noted. Revise and resubmit for record only." Contractors receiving this mark are free to release the products for manufacture, but are obligated to resubmit the shop drawings, to denote their understanding of the markups and to correlate the shop drawings with the actual construction.

It is not uncommon for a designer to stamp the individual sheets in a given set of shop drawings in a variety of different ways. Most of the piece drawings may be approvable, for example, even if the framing plan still requires minor revision. Unless all the shop drawings in a set deserve the same stamp, the best way to avoid confusion is for the designer to stamp each sheet individually, even if the set is quite lengthy. The detailer will appreciate knowing which sheets are "approved" or "approved as noted," since those sheets will not have to be resubmitted.

With the cost of preparing and revising shop drawings representing a significant percentage of the cost of a timber building, it is appropriate that there be a separate line for the shop drawings in the contractor's schedule of values. Like every other aspect of the job, payments for timber shop drawings should be made in proportion to the percentage complete, and the associated retainage should not be released until after substantial completion of the timber contract. Particularly if the designer has asked that the shop drawings be resubmitted purely for record purposes, the promise of retainage release is often the only incentive for the contractor to comply.

Perhaps the best advice to both the producer and reviewer of the shop drawings is "take nothing for granted." If there is even the slightest doubt as to whether a given detail applies to a particular condition, the detailer should request confirmation from the designer. Similarly, even if the details in the shop drawings correspond precisely with those in the working drawings, the designer's review must proceed systematically around the building, considering every timber condition and its relationship to the work of all of the other trades with which it will come in contact.

At the risk of oversimplifying, the architect is concerned primarily with geometric relationships and connections between different materials and systems, while the detailer is concerned primarily with those within his own trade. Given their different emphases, it should come as no surprise that the details in the working drawings often depict conditions different from those in the timber shop drawings. When the details in the shop drawings correspond exactly to those in the working drawings, it indicates either that the working drawings are overde-

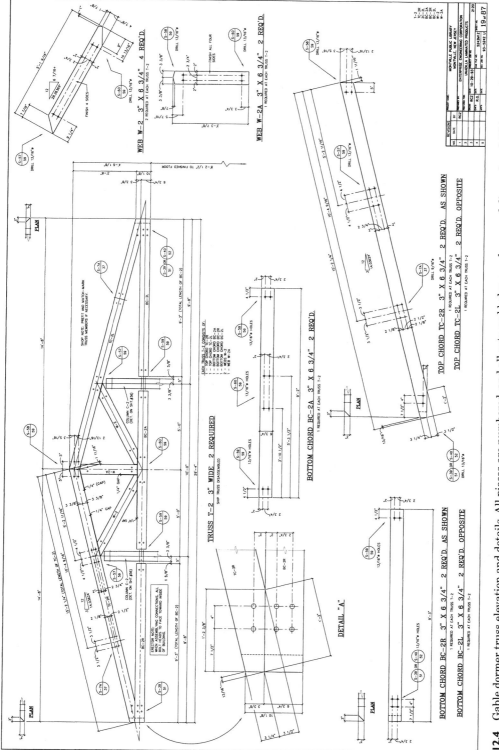

12.4 Gable dormer truss elevation and details. All pieces are numbered, and all cuts and holes are shown and dimensioned. This truss is shown in context in Figs. 8.1 and 14.10. (Enterprise Engineering Consultants, Ltd., Peshtigo, WI.)

tailed, or that the shop drawings are underdetailed. Do not hesitate to ask for additional information, including isometric drawings for complex three-dimensional intersections, for areas detailed inadequately.

Although the responsibilities of the designer and detailer complement one another, the detailer alone is responsible for bringing deviations to the attention of the architect and engineer: "Approval of shop drawings does not authorize any deviation from the contract documents unless the contractor gives specific notice of the variance and receives express permission to proceed accordingly."[3]

Deviations from the Contract Documents

Deviations come in many forms and should not necessarily be considered cause for rejection.

- Deviations may arise because specified products have been discontinued; in that case, the detailer has no choice but to deviate, and the designer has no choice but to determine the equivalence of the proposed alternatives.

- Deviations may be necessary because the indicated details may not be constructible; in that case, the designer must rethink them, making sure that the modified details are structurally and architecturally satisfactory.

- Deviations may arise while the timber shop drawings are being coordinated with timber fabrication processes or the constraints of other trades; in that case, the acceptability of the deviations is a function of their *relative* conformance: Assuming they are adequate, how closely do they capture the spirit of the contract documents? (See Fig. 12.5.)

- Deviations may relate to quality. Although this type of deviation may appear to fall into one of the other categories or may not even be identified in the shop drawings as a deviation, it is the type encountered most frequently.

Every quality-related deviation has cost implications. Imagine you are the architect for a timber building. Here are a few scenarios, seen from your perspective:

- You've called for the underside of all timber decking to be stained; the detailer proposes that this not apply to areas in which the decking will be hidden from view by a suspended ceiling. Since the proposed change does not affect the quality of the job, your response should be to seek a credit for the difference between the specified and proposed labor and material.

- You've specified that those column faces to be exposed to view in the completed building be stained; the detailer asks if she can stain all

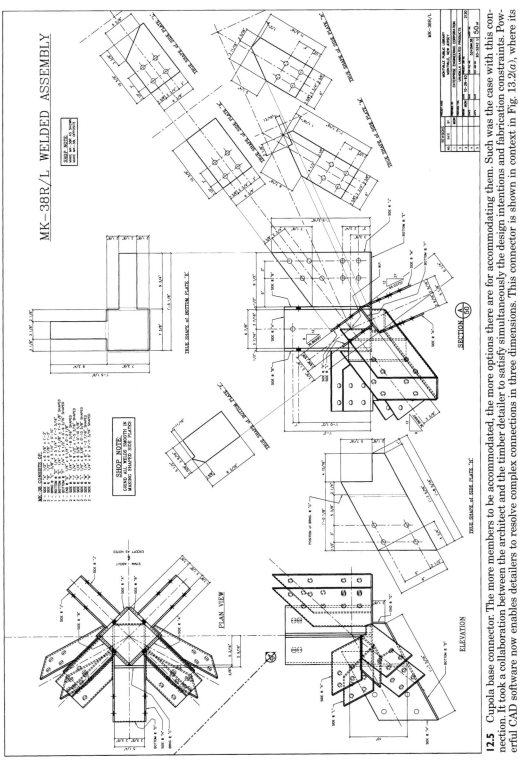

MK-38R/L WELDED ASSEMBLY

SECTION $\frac{A}{50}$

12.5 Cupola base connector. The more members to be accommodated, the more options there are for accommodating them. Such was the case with this connection. It took a collaboration between the architect and the timber detailer to satisfy simultaneously the design intentions and fabrication constraints. Powerful CAD software now enables detailers to resolve complex connections in three dimensions. This connector is shown in context in Fig. 13.2(a), where its piece mark is visible. (Enterprise Engineering Consultants, Ltd., Peshtigo, WI.)

faces, so that she doesn't have to mask those that will be hidden. Your response should be to allow it, but with no extra for any increase in labor and material, since the change is solely related to the contractor's methods and means.

- You've stipulated that the custom steel connectors be given two coats of paint—one in the shop and one in the field; the detailer proposes to apply both coats in the field, after the building is enclosed. You should reject this approach, since the connectors will otherwise be without corrosion protection while exposed to the weather. (You should also reject the idea of applying both coats in the shop, since abrasions resulting from the construction process may not otherwise get touched-up under the contract, and even if they are, they may not match the shop coat.)

As long as he has received the designer's explicit approval of deviations *from the contract documents,* the fabricator can commence fabrication with confidence. The contents of the approved shop drawings, however, are binding. Therefore, deviations *from the approved shop drawings* are absolutely forbidden.

Circumstances Where Field Work Is Preferred

Although shop work is typically performed under more controlled conditions than field work, higher overall quality can sometimes be achieved if certain fabrication activities are performed in the field. Generally, these activities relate to the natural variations among like wood products. For example, even though all the timber columns along an exterior wall may have been fabricated at the same time from stock of the same species and moisture content, none will be perfectly straight, and each will be imperfect in its own way. These slight differences may not be noticeable inside the building, but could present an unacceptable appearance on the outside, in a fascia for example. To minimize the collective effect of defects that individually are within acceptable tolerances, it would be wise, in this instance, to have all the rafters fabricated a little long. After erection, a line may be snapped at the proper overhang. Sawing to this line in the field will produce a straight fascia, rather than a crooked one composed of straight segments. The distribution of work between the shop and the field is normally considered part of the contractor's methods and means. Nevertheless, to the extent that that distribution will affect the quality of the final result, it is fair game for review by the architect in the timber shop drawings. The aspects of timber construction most commonly split between the shop and the field are assembly and finishing.

The default setting for timber fabrication is "in the shop." Therefore, whether field cutting is specified by the designer or deemed appropriate by the detailer, it is imperative that its precise extent be identified as such in the shop drawings. In fact, field fabrication should be prohibited except to the extent that is indicated in the shop drawings and approved. That said, conditions do arise in the field which must be resolved in the field to avoid delay. These usually occur at the interfaces between different trades, which the following trade is usually left to resolve. Such is the case with misaligned anchor bolts. Even if the preerection survey reveals that a few bolts are slightly out of line, there is often no time to remove and replace them. It would be helpful if the timber shop drawings included a protocol for such field modifications. It could be as simple as a general note requiring the timber engineer's written authorization for any field work not explicitly called for in the approved timber shop drawings. In anticipation of these inevitabilities, it behooves the designer to request such a procedure in the specifications or during review of the shop drawings.

Problems in Basing Shop Drawings on Working Drawings

Much has been written recently about the legal risks of incorporating copies of structural working drawings into structural shop drawings. In fact, we and many of our consultants have posted prominent warnings in our drawings and specifications, stating that such copying is prohibited. We fear that the detailer will rely on our working drawings to a greater extent than is appropriate. Even though the contractor is ultimately responsible for the fit of all members, we fear that failure to achieve that objective may become our problem, if the shop drawings are based too literally on our working drawings. Yet, as the offices of architects, engineers, and fabricators become more automated, and their clients demand faster and more accurate results at lower cost, this prohibition appears increasingly counterproductive and unnecessarily time-consuming.

Some believe the liabilities of designers and contractors can be kept separate even if they share a common database: "If the architect's interests are properly protected, sharing digital data can contribute to the successful execution of a project and can help the architect communicate design intent to those responsible for construction of a project."[4] I now believe that the current prohibition is arbitrary, in that it does not prevent a detailer from recreating the exact contents of the working drawings by tracing, digitizing, or scanning them. Regardless of your viewpoint, it is easy to see that ". . . the design professional's position in the Construction Process—principal professional adviser of the owner, interpreter of the documents, and judge of performance as well as the principal conduit between the owner and prime contractor—can appear to make him responsible if anything goes wrong.

Such an outcome does not take into account the primary responsibility of the prime contractor for selection of subcontractors, the fabrication processes, or construction methods or other activities within its control."[5]

Designers get in trouble when they exceed the limits of their authority. They are not responsible for reviewing quantities, even when such information is placed—as it should be—on the shop drawings." As a general rule, don't review shop drawings or other submittals concerning the proposed implementation of means, methods, procedures, sequences or techniques, or other temporary aspects of the construction process. Those are the responsibility of the contractor, and review of these submittals could subject you to responsibility not normally undertaken by a design professional."[6]

In spite of these warnings, the reviewer should make some random checks of the detailer's quantities as part of the process of verifying general contract compliance. More important, the sooner discrepancies are identified, the lower the cost of resolving them, and the smaller their impact on project progress. Justifiably, reviewers fear that stating what they believe are the correct quantities will make them responsible in the event that their quantities are incorrect. To avoid this risk, consider leading the detailer to the relevant portions of the working drawings, so that the conclusion he or she comes to is the same as yours.

Essential Components of Timber Shop Drawings

The components of the timber shop drawings fall into two categories, procedural and substantive. Procedural aspects facilitate the tracking of the shop drawings and are usually placed in the title block. They include project number and name, submission number and date, subcontractor name, detailer name, address, phone, fax, and e-mail address, and a place for the designer's stamp. If engineering is within the contractor's scope, then the same categories of information must be furnished for the contractor's engineer, as well as his or her license number in the project's state, and the name and edition of each code on which the engineering design is based.

Substantive aspects describe the building's geometries, materials, and systems. They include the drawings and their contents. Within the set of shop drawings, expect to find framing plans, piece drawings, and details; within each drawing expect to find dimensional information, columns lines, connectors, fasteners, openings, quantities, and allowances for other trades. (See Fig. 12.6.) At the beginning of the set, look for general notes, references to applicable codes and manufacturing standards, legends, grades or properties of materials, species, and timber finishing and treatment schedules.

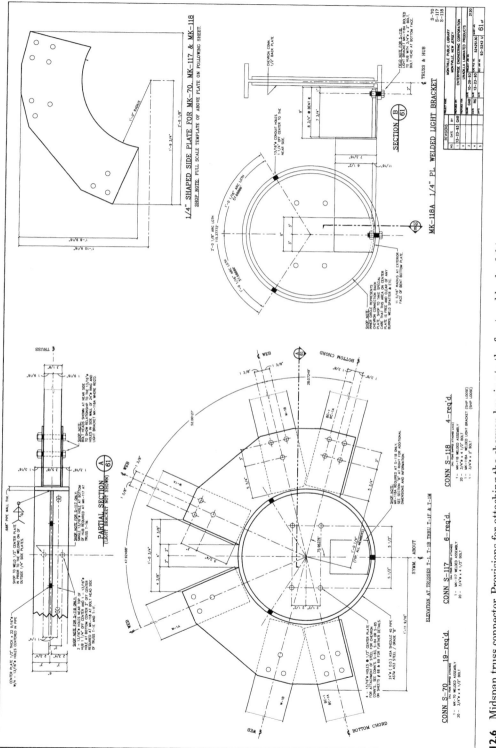

12.6 Midspan truss connector. Provisions for attaching the chevron bracing to the front and back of this connector are shown in a different shop drawing, but referenced in the section at the lower right of this sheet. That same detail refers to the brackets required to support lighting fixtures, provided by others, at certain of these connectors. This connector is shown in context in Fig. 14.9. (Enterprise Engineering Consultants, Ltd., Peshtigo, WI.)

Every submittal should be accompanied by a transmittal from the contractor. In addition, the contractor should sign each sheet certifying that it is in general conformance with the contract documents. The purpose of the contractor's review is to coordinate the timber shop drawings with the rest of the project. Until receipt of the contractor's certification, the designer's review should not proceed. Otherwise, the designer may end up doing the contractor's work for him.

It is crucial that the submissions be numbered and dated by the detailer. Knowing the current version enables the reviewer to identify and refer to comments on previous versions. Prior to approval, the reviewer must verify that all comments from previous versions have been addressed. After approval, that number identifies uniquely the version that will be the basis for fabrication. The intervals between the dates represent the turnaround times of the various parties. Where contractors or designers are obligated to submit or review shop drawings within a given time, these intervals are a measure of contract compliance.

Detailers' comments and questions are also a common feature of shop drawings. When they are unsure of something, they can bring it to the attention of the reviewers in several ways. The most common is for them to write their questions on the appropriate drawing, pointing to the area in question and "clouding" it. To stand out, the style of the note should be different from the style of the drawing's other text. If there are a number of questions, detailers will sometimes put them on a separate sheet attached to the drawings. The third approach is for the contractor to submit a Request for Information (RFI) to the architect prior to submitting the shop drawings. While this may seem like a reasonable approach—the detailer, after all, may be unable to proceed without certain information—it is usually unjustified. Rarely are the working drawings so vague, incomplete, or inscrutable that experienced detailers can't get started on their shop drawings. In my experience, RFIs full of questions more appropriately asked on the shop drawings are usually an indication that contractors are either trying to buy time, hoping to cast doubt on the completeness of the working drawings, or too lazy to look for the answers themselves: ". . . some contractors generate unnecessary RFIs at the drop of a hard hat, often when a simple review of the construction documents and other available data would reveal the required information."[7]

Organization of Shop Drawings

With the increased use of digital details furnished by product manufacturers, working drawings are becoming increasingly precise and proprietary, thus more like shop drawings. As long as the designer takes steps to maintain a competitive bidding environment, this trend is healthy and should reduce the cost and expedite the preparation and processing of shop drawings. However, even in trades, such as timber construction, where proprietary differences are few, there is much the designer can do to make shop drawing production and review as painless as possible.

The most important rule, I believe, is to dimension the working drawings and their details in a manner that is consistent with a conventional construction sequence. For example, consider the roof overhang. Many architects apparently think that finishes will only end up in the proper relationships to one another if they are dimensioned to each other. Consequently, they measure the overhang from the fascia to the face of the exterior wall finish. However, before "getting materials into the same pew, make sure they get into the same church." Recognize that the relationships between finish materials attached directly to structure are controlled primarily by the geometries of the structural components and systems to which they are attached. If the armature supporting those finishes is in the right place, the finishes will be too. Therefore, dimension the overhang from the outside face of the timber subfascia to the outside face or centerline of the perimeter columns. Use a similar approach when dimensioning cantilevers, floor and roof elevations, eaves and rakes, framed openings, and other conditions involving materials and assemblies attached to timber framing. This approach will not only enable the timber detailers to lay out the frame without having to derive its boundary conditions, but will also enable you to avoid having to repeat those calculations to check their work.

While there are no hard and fast rules for organizing timber shop drawings, the following aspects and those in Fig. 12.7 characterize a complete and well-organized set:

- Like architectural working drawings, information is most general at the beginning of the set, and most specific at the end.

- Framing plans are followed by cross sections which are followed by piece drawings.

- For large projects, key plans indicate the area covered by each framing plan.

- Similar parts have similar part prefixes (B- for beams, T- for trusses, etc.).

- Identical pieces receive identical piece numbers.

- Fastener orientation is indicated in the framing plans, wherever it is important.

- Pieces are dimensioned with reference to column lines and finish floor elevations.

- Complex connectors are shown in multiple views and at large scale (even full size).

Overall objectives
  Completeness
    In terms of contract
    In two and three dimensions
  Coherence
    For review and approval
    For fabrication and erection
  Conformance
    With drawings and specifications
    With codes and standards
  Coordination
    In contract vs. not in contract
    Field work vs. shop work
  Considerations
    For other trades
    For moisture-related movement
  Codification
    Of materials and sizes
    Of pieces and assemblies
  Chronologies
    For shop drawings and fabrication
    For delivery and erection
  Connectivity
    Within timber and with other trades
    Between plans, sections, details
  Constructability
    Within timber contract
    Within context of overall project
  Control
    General contractor vs. subcontractor
    Subcontractor vs. vendors/suppliers
  Correlation
    Typical conditions
    Atypical conditions
  Compilation
    Quantification of required elements
    Normalization of project details
  Contrast
    Concealed vs. exposed
    Interior vs. exterior work
  Complementary
    Within timber contract
    Between timber and other contracts
  Consistency
    Within timber contract
    Between timber and other contracts
  Conventions
    Defaults and deviations
    Plans, sections, and details
  Configuration
    Overall building form
    Arrangement of timber components

Procedural information
  Project information
    Project name and number
    Applicable code(s)
  Design team
    Name of architect
    Name of structural engineer
  Construction team
    General contractor and subcontractor
    Fabricator, detailer, and timber engineer
  Shop drawing information
    Date, version, and drafter
    Number of sheet and number of sheets

Substantive information
  General
    Dimensions and elevations
    Quantities and designations
    Slopes and angles
    Arrangements and spacings
    Changes in plane and orientation
    Clearances and alignments
    Openings and penetrations
    Offsets and discontinuities
  Framing
    Straight and curved
    Species and grades
    Sawn, glulam, or reinforced glulam
    Bearings and supports
    Firecuts and plumb cuts
    Bevels and notches
    Cambers and radii
    Bird's mouths and overhangs
    Fireblocking and draftstopping
  Decking
    Sawn, laminated, or panel type
    Jointed, patterned, or textured
    Species and grades
    Thicknesses and edge spacing
    Coursing and nailing
    Starting and ending conditions
  Connections
    Fasteners and connectors
    Setbacks and spacings
    Recessed vs. surface-mounted
    Fillets and mortises
    Welds and gauges
    Additive vs. subtractive
  Finishes
    Preparation and application
    Treatments, paints, and coatings

12.7   Checklist for timber shop drawings, for use in conjunction with a project's contract documents.

Advice for the Reviewer

If shop drawings are new to you (and they will be if you are coming from a residential construction background), heed the following advice: Regardless of the clarity of the shop drawings, allow ample time to complete your review; few individuals can comprehend and evaluate the contents of dozens of drawings in a single sitting. Because the interval between the completion of the working drawings and the receipt of the shop drawings can be anywhere between a month and a year, take time to reacquaint yourself with the project. Shop drawing review requires great concentration; work in a quiet place away from distractions such as the phone. And finally, allow time to let their *implications* sink in; some of your best thinking may take place when you're away from the office.

A thorough and unhurried review of the shop drawings is in everyone's interest. "A great many persons are charged with checking the drawings, and it is essential that each of them pass through as many hands as possible and be carefully examined by all. In this way the entire chain of construction command is made aware of the impact that a given material will have on the job."[8] While the project architect's familiarity with the entire project will enable identification of conflicts between the fabricator's proposed approach and the rest of the work, the specific individuals responsible for the contents of the timber specifications and working drawings are in the best position to determine, in general, whether the fabricator's proposed approach conforms to them.

Because shop drawing review is so time consuming, be careful not to perform more of it than you have to. If the contractor submits shop drawings for aspects other than those specified or required, stamp them "Not Required for Review," and return them at once.[9] Accept no resubmittals of shop drawings previously stamped "Approved" or "Approved as Noted." Respond to shop drawing–related RFIs by asking that their contents be put on the shop drawings, for review with them. Refuse to work from incomplete information. When reviewing aspects at the interface with other shop-fabricated systems, insist that the shop drawings for both trades be in hand before reviewing either one. Reject both if they have not been coordinated with each other. Learn about the review process from someone who is familiar with it. And, no matter how thorough the working drawings, don't be surprised if conditions turn up that you did not anticipate; fortunately, if you've done your homework, they'll be minor.

Don't abuse the process; the reviewer's job is to review, not to finish designing the building depicted in the working drawings. Once having perused the contract documents, the experienced detailer will be on the lookout for review comments that are inconsistent with them. Any that are found will elicit requests for clarification or additional compensa-

12

tion. Since contracts are normally construed against their authors, ambiguities in the contract documents are usually construed against their architects and engineers. The detailer is more likely to act like a collaborator rather than an adversary if you are upfront about problems rather than attempting to make changes in the guise of legitimate review comments. If you must make a change, recognize that the cost of modifying the shop drawings is an integral part of its overall cost (and may be the only cost, if the change has no effect on construction labor and material). Make sure there are procedures in place to keep each trade apprised of changes in the timber work which might affect them.

Summary

Shop drawings are the primary medium through which the timber fabricator and the architect and engineer communicate. Using conventions specific to the industry, the detailer indicates how the schematic intentions illustrated in the working drawings will be realized. In conjunction with other types of submittals, shop drawings help assure that the quality of the constructed facility will be as specified.

The sizes of their members and the complexity of their connections make shop fabrication mandatory for heavy-timber buildings. Shop fabrication mandates shop drawings. Shop drawings mandate a detailer, an intermediary between designer and constructor. Each additional player makes communication more cumbersome and increases the likelihood of misunderstanding. At the same time, the lack of structural redundancy typical of heavy-timber construction amplifies the penalties for miscommunication. It is in this context that timber shop drawings must be prepared and reviewed.

Timber shop drawings integrate architectural and structural information. The timber frame is supported by or gives support to nearly every other trade. As such, its shop drawings are the road map for much of the subsequent work. To the extent that the finished frame will be exposed to view in the completed building, *it can be thought of as a system of rough carpentry acting as a finish material.*

The review process commences with an initial submission—often consisting solely of a framing plan—and proceeds iteratively through the submission and, if required, resubmission, of increasingly complete shop drawings. The reviewer is entitled to additional information reasonably required to illustrate and explain fully the timber frame; the contractor is entitled to additional compensation for aspects not shown or not clearly shown in the contract documents. The tasks associated with reviewing, fabricating, shaking out, and erecting the timber frame are all made easier when the numbering system for the components of the timber frame is clear and unambiguous, and related to the organization of the working drawings and the geometry of the building.

Because of natural variations among timbers, the appearance of the overall project may be enhanced by performing certain tasks in the field rather than in the shop. In the timber shop drawings, the affected pieces should be identified, and any associated dimensional adjustments so indicated. Distinctions between shopwork and field work should be as explicit as those between the work of the timber contractor and that of other trades.

A complete set of timber shop drawings contains all the necessary procedural and substantive information necessary to get the job done. In depicting the various aspects of the timber work, the objectives of the shop drawings are differentiation, orientation, identification, dimension, and connection. The best timber shop drawings are exemplars of graphical excellence—they give the viewer "the greatest number of ideas in the shortest time with the least ink in the smallest space."[10]

Because the timber shop drawings must be completed and reviewed prior to the fabrication of the timber frame, and because most project work cannot commence until after the frame is completed, the general contractor has a keen interest in their progress. The contractor will monitor the review process with a submittal log, and integrate that information into the overall project schedule. In all likelihood, the shop drawings, fabrication, and erection of the timber frame will be on the critical path from the start of project construction.

12

Shop Drawings

Notes

1. Thornton, p. 4.
2. There is widespread disagreement regarding the legal implications of the stamps, including the use of the word "approve" and its variants. Whether the stamper or the stampee, you should consult your trade or professional association or legal counsel for guidance on their use and interpretation.
3. Clough, p. 62.
4. Sanders, p. 368.
5. Sweet, p. 288.
6. Dixon, p. 277.
7. Boston Society of Architects, p. 1.
8. Liebing, p. 72.
9. Dixon, p. 275.
10. Tufte, *Visual Display,* p. 51.

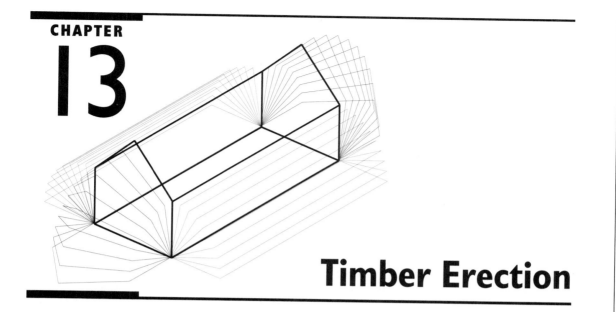

Timber Erection

Skill is concentrated more in the conception, setup, and beginning of work than in each successive detail of its execution.

EDWARD TENNER, *Why Things Bite Back*

Introduction

Timber is erected more like steel than like lightweight wood framing. Its members are too heavy to be lifted by hand, are primarily shop fabricated, and are installed with specific lengthwise and cross-sectional orientations. Unlike steel, however, timber is delicate, is likely to be left exposed in the completed building, and usually arrives at the job site with its connectors unattached.

Planning for the erection of a timber frame should not begin when the first truckload of materials arrives from the fabricator. On the contrary, to assure a smooth and efficient raising of the timber frame, erection planning should start at the time of bidding. Neither the architect nor the engineer cares how the design is produced, so long as the intended result is achieved. Keeping this in mind, the intelligent builder will have thought through the entire production process and reviewed it with the fabricator *before* establishing a price for the job. The interval between contract award and material delivery is the time to work out the logistics, "the planning, implementation, and coordination of the details"[1] of the erection operation.

The details of the erection process are many and varied (and all have design implications). They include site conditions and access, construction sequences and schedules, equipment availability and positioning, crew size and composition, safety and protection, and coordination

with other trades. Consider all relevant issues in anticipation of erection, and make contingency plans around critical aspects. Because the profitability of a job can disappear quickly if things don't go as planned, plan under the assumption that things *will* go wrong. While having a spare carpenter on call or ordering some extra decking might add slightly to your costs, such precautions should be thought of as a cheap form of insurance.

The Critical Interface

The critical interface is that between the fabricator and the erector. While each normally wishes to complete its work as quickly as possible, the constraints and procedures of one may be in conflict with those of the other. For example, consider the matter of erection sequencing. Except for buildings with small footprints or cranes with large reaches, erection of the entire building is usually not practical from a single crane location. Crane relocation being costly and time consuming, most erectors prefer to minimize it. Furthermore, they prefer to top out each area of the frame before starting on the next. Unless there is enough room on-site to store all the components for the entire frame, this approach mandates that materials be delivered and stacked near their final locations in plan.

The fabricator, on the other hand, is likely to prefer producing members by type—column, beam, joist, or rafter—rather than by location. This helps maximize not only the efficiency of the manufacturing process, but also the density of each truckload. After all, it is easier to stack similar members than those with different sizes and shapes. Another rule for loading the trucks is LIFO—Last In, First Out. "Load beams of a floor at the bottom of a truck bed and columns at the top. Columns are erected first and beams afterward."[2] When fabrication can be completed before the start of erection, the fabricator can ship the frame in whatever order the erector prefers, but may charge extra for any double handling or short loads which result.

Double handling again becomes an issue once the materials reach the site. Each time a member is handled, time is lost and the likelihood of damage is increased. Therefore, the mantra of time managers—*handle each item but once*—applies even more in the field than it does in the office. Ideally, each column and beam would be delivered and set down right next to its ultimate location in the building. Erection would then consist simply of, well, erection. Instead, the erector usually ends up with the responsibility of "shaking out" the job, or situating each member close to the location to which it has to be lifted. "Many are the cases where disputes ensued between erector and fabricator due to poor shop-loading arrangements, which resulted in unusually lengthy shake-outs. The fabricator sometimes pays for that time."[3]

While it may cost a little more and may not always be practical, I recommend delegating shake-out to the fabricator whenever possible.

Because the fabricator wants to minimize the turnaround time of its trucks, shake-out responsibility will provide an incentive for efficient loading and unloading. And, to the extent that handling by the erector is reduced, so will be the incidence of erection damage.

Regardless of who performs the shake-out, the results must conform to the erection strategy. Allowances must be made for movement of personnel and equipment, distribution, orientation, and protection of members, and handling of connectors and fasteners. To give them an initial feel for the job and guarantee that things are arranged the way they want them, many erectors prefer to do the shake-out themselves.

The timber shop drawings are the road maps that guide erection. It should not be surprising, therefore, that their framing plans are also known as *erection* plans. Since every member and custom connector should be labeled during fabrication or prior to shipping to correspond to the numbering system used in these drawings, it should be easy to distribute all components properly in the field. Once a timber is in the right place, however, it must still be given the proper orientation. There are two axes of orientation: lengthwise and cross-sectional:

- A member has the proper lengthwise orientation when its ends are pointing in the right directions. Prior to raising, the bottom of a column should be adjacent to its base plate. Assuming a small crane, the ends of a beam, joist, or rafter should be just below the connectors that will receive them. To save time in the field, the fabricator should establish in the shop drawings conventions for lengthwise orientation.

- A member has the proper cross-sectional orientation when a given side faces the correct direction in plan (if a column) or section (if other than a column). At joists, beams, girders, rafters, and purlins, the usual way of distinguishing top from bottom is to write TOP on their top surfaces. Since the ends of columns may be their only concealed surfaces, it is desirable to label them there. Even if it appears that a member would work with either its top or bottom surface facing up, follow the fabricator's marks. Intentional asymmetries are common: in the layup of certain glulams, in the finishing of prefinished members, and in the predrilling of holes for timber connectors.

Crew Size

When the amount of time available to erect the timber frame is fixed, the minimum crew size is given as:

$$\text{total output} = (\text{workers})(\text{output per day per worker})(\text{available days})$$

$$\frac{\text{total output}}{\text{available days}} = (\text{workers})(\text{output per day per worker})$$

If daily output is insufficient to complete the job on time, it can be increased by increasing the number of workers or increasing their output (by increasing output per hour or the number of work hours). The optimum crew size is a function of the size and complexity of the job, the available equipment, the quality of the workforce, and the experience the crew has had working together. Labor shortages may limit the quantity or quality of workers, forcing the erector's regular crew to work harder or longer than would otherwise be necessary. Unfortunately, extended periods of overtime tend to reduce productivity.

Erection cannot necessarily be performed with maximum efficiency on all of the days allocated to it; allowances must be made for inclement weather, learning curves, and contingencies. Furthermore, erection is not the only aspect for which the timber erector is typically responsible: Prior to starting erection, materials must be unloaded and shaken out; following substantial completion, punchlist items must be repaired or replaced.

Erection Schedule

The primary goal in scheduling timber erection (or any on-site activity, for that matter) should be to minimize the potential impact of activities beyond your control, such as material deliveries and preparations by others. Encourage the vendors and subcontractors responsible for those aspects to complete them far enough in advance of the scheduled start of your work to allow for verification of completeness and accuracy and correction of any deficiencies. Their failure to do so will make their problems your problems. Their activities are always on your critical path; therefore, any delay in their completion will result in an equivalent delay in your commencement.

A proactive stance should also be taken with respect to the weather. Rather than hoping for the best, plan for the worst (or at least for the average). Organize the work to get under cover as quickly as possible. Distinguish between those activities that can and those that cannot be performed on a rainy day. Review the general conditions of the construction contract for provisions related to the weather. That of the standard form of the American Institute of Architects is typical: "If adverse weather conditions are the basis for a Claim for additional time, such Claim shall be documented by data substantiating that weather conditions were abnormal for the period of time and could not have been reasonably anticipated, and that weather conditions had an adverse effect on the scheduled construction."[4]

While a weather-related claim can only be made after erection is complete, evidence of its impact on construction can be derived from the schedules you generate before it starts and in the records you maintain while it is underway.[5] The National Weather Service can substantiate the abnormality of the weather conditions during the erection period. Bear in mind that the adversity of the weather is relative not only to historical records, but also to the time of year. Recognize and plan for the fact that overall productivity will be lower if erection takes place in the winter or during the rainy season than at other times of year.

Another goal of scheduling should be to minimize the overall cost of crane rental. This is not necessarily the same as minimizing its duration. Match the equipment to the job and to the crew; the higher potential productivity of a more powerful crane will only be realized if the job is sufficiently large, the members sufficiently heavy, and the crew of sufficient size and experience. Accounting for the time required to secure and release slings, the most cost-effective way of placing some members may be by hand. If a second crane will already be on-site, consider using it to provide temporary support during the framing of three-dimensional forms. Compare its costs and benefits with those of scaffolding.

Preparations by Others

Before erection can start, various preparations must be made by other trades. Some involve permanent construction, while others fall under the category of general conditions. Irrespective of who is responsible for each preparation, do not proceed with erection until all are complete. If you proceed with the knowledge that anchor bolts are improperly located or that the required temporary electric service is not yet available, you will be presumed to have accepted responsibility for accommodating those conditions, even if they cause you, and those who follow you, additional effort or expense. It is always more economical to build correctly the first time than to demolish and rebuild aspects that are initially built incorrectly.

Perhaps the most obvious permanent preparation is the construction of the foundations, including the piers for the timber columns or bents. (See Fig. 13.1a and b.) As the locations of the anchor bolts are critical to the subsequent erection process, I recommend that the erector either set the bolts itself or survey the setting work performed by others before and after the pour. In the event that the general conditions assign responsibility for the foundation survey to others, make sure to review it thoroughly before starting erection.

While the use of accurate templates assures that the anchor bolts in each cluster are positioned properly with respect to each other, it does not assure that the cluster itself is properly located or oriented. Therefore, even though the foundation survey is usually plotted from the centers of clusters of anchor bolts, it must also indicate the degree to which the base plates are mounted out of square. The slight "play" typical in the pockets of column bases can accommodate slight misorientations in the anchor bolts. Prior to the start of construction, it is a good idea to

(a)

(b)

13.1   Steel base for timber column. (*a*) Column base anchored to interior concrete footing and surrounded by formwork for pier. (*b*) Column base cast into pier. Bearing plate is above top of pier to prevent contact between timber and concrete. Piers are oriented on the diagonal to avoid reentrant corners at corners of slabs. Shear-type control joints for slab radiate from corners of pier. Pier is wrapped with bituminous filler in preparation for pouring of slabs.

establish the tolerances you can accept and communicate that information to the general contractor. If the as-built conditions exceed those tolerances, insist that the defective work be corrected. If time constraints make modifying the timber column the only alternative, make sure that the work is done pursuant to specific directions from the fabricator's engineer and that the offending trade foots the entire bill.

Most contractors prefer to pour concrete under cover. If, as is usually the case with steel, the structural frame will eventually be concealed, there is every reason to complete the superstructure before pouring the slab-on-grade. In the case of exposed timber framing, however, there are a number of reasons why the slab should be completed *before* erection begins. First, if the pour is later, there is the risk that the timber will get splattered or that the lower ends of the columns will pick up moisture during the pour. Second, there is the possibility that concrete hoses and power trowels will damage the columns. Third, it may be beneficial to lock any otherwise freestanding piers into the slab before they are subjected to erection loads. And finally, the relatively light weight of timber componentry increases the likelihood that some shake-out and erection will be performed by rubber-tired equipment, for which the smooth, hard, and level surface of a concrete slab is desirable.

Of course, not all timber frames bear on concrete foundations. In pole construction, for example, pressure-treated timber poles may bear on compacted earth and gravel. Also, if it is high enough above the floor, timber roof framing may sometimes be permitted in concrete- or steel-framed buildings.[6] Whatever trades interface with the timber, be sure they have a complete understanding of their tasks and their scheduling and coordination responsibilities.

Preassembly

The controlled conditions of the shop nearly always result in higher quality work than can be achieved in the field. Furthermore, shop work is usually not susceptible to weather delays and is often performed by workers paid less than their field counterparts, especially where the shop is in a rural area and the site is in or near a city. Nonetheless, determining the appropriate level of preassembly is far from straightforward.

Even the simplest timber frame occupies much more volume assembled than disassembled. Thus, the costs of wrapping and shipping are higher for an assembly than for its components. In the case of preassembled frameworks—whether two- or three-dimensional—shipping costs can be enormous. For oversize loads, special trucks and off-hour trucking and delivery may be required. A larger crane may be needed on the site. Overall erection time, on the other hand, will be shortened, reducing site labor and the risk of weather-related delays. With proper provisions, it may be possible to install trusses and the like directly from the truck, minimizing shake-out.

(a)

(b)

(c)

13.2 Preassembly and erection of a cupola, an example of a complex substructure. (*a*) Detail of connection to the rest of the roof structure. The piece marks for the timber and steel components, visible when the cupola was on the floor, were out of sight once it was raised. (*b*) The entire cupola, ready to be lifted. (*c*) The cupola in final position. Scaffolding was needed to provide support and stability until all of the surrounding framing was tied into it.

Preassembly need not be limited to the shop; it can also take place in the field, on the ground, prior to erection. On a recent project, the timber erector concluded that preassembling two large cupolas on the ground slab would be easier, quicker, and less dangerous than assembling them way up in the air. (See Fig. 13.2*a, b,* and *c.*) While shop preassembly was an option, the proportions of the preassembled cupolas would have made shipping prohibitive. The same conclusion was reached by the erector of a pyramidal timber and steel truss covering an area 36 ft (11 m) square. (See Fig. 13.3*a* and *b.*)

Another issue to consider is whether preassembly will reduce the number of cranes or supports needed during the erection of complex structures. Frames of domes, pyramids, and other forms that derive their strength and stability from their three-dimensional geometries, often require multiple temporary supports until complete. Framed in the proper sequence, some such forms may be self-stable at intermediate erection stages, but most are not.

In certain special cases, a preassembled timber structure may even be partially collapsed for shipping. To accommodate highway height restrictions, some manufacturers of prefabricated housing modules, for example, ship them with their rafters resting on their ceiling joists. In the shop, the lower end of each rafter is bolted to a connector. In the field, the erector simply rotates opposing rafters upward and connects them at the ridge.

Parallel-chord trusses present unique opportunities for compact shipment. Since triangulation is what resists truss deformation, the removal of triangulation will turn them into parallelograms, capable of being collapsed. This transformation will only be easy if:

- Each connection consists of a single bolt or shear connector, so it can pivot.

- The remaining web members are all parallel to one another.

- The layers occupied by the web members alternate with those of the chords.

- The diagonals can be removed and reinstalled quickly and easily.

Issues related to preassembly are also driven by site constraints. The economics of erecting long-span timber trusses are very different when they form the roof of a one-story warehouse than when they comprise the frame of a bridge over a deep ravine. In the latter, if the required reach exceeds that of the available cranes, the erector may have to consider framing from both banks toward the middle. In that case, the appropriate degree of preassembly will be a function of the capacity of the structure and of its cranes at each stage of the process. The magnitudes and senses of the internal forces will be very different in two smaller unjoined cantilevered trusses than in the single larger simple-

(a)

(b)

13.3 Pyramidal glulam and steel truss. (*a*) The partially-assembled truss is stabilized by a sling hung from a crane. Steel boots transfer the substantial outward thrusts from the top chords into the bottom chords. (*b*) The fully assembled truss is lowered into position over the framing for a continuous band of clerestory windows. Perimeter beams act both as window headers and as tensile ties to minimize truss deflections. Kalafer Residence, Franklin Township, NJ. William Nastasi, general contractor. The Goldstein Partnership, architects. (William Nastasi.)

span truss of which they will become a part. The size, shape, and location of the staging area may also influence the economics of preassembly. Although it may be convenient to assemble on a floor slab, trusses there may interfere with the movement of workers and equipment.

Although the overall depth of a timber truss affects the forces in its members and the feasibility of shipping it completely preassembled, the critical portions of the top and bottom chords occur where the ratio of the bending moment to the truss depth is the highest (see Chap. 8). In a triangular truss, this ratio is highest at the ends. Therefore, although the overall height of a triangular truss is certainly important, it may not determine its overall capacity. If the trusses would otherwise be too tall to ship, consider preassembling and shipping the truncated trusses separately from their peaks.

Protection

Protection comes in many forms and spans a large portion of the construction process. It starts before the trucks leave the shop and doesn't end until the building is weathertight and the interior has reached equilibrium. It includes protection from water, impact, overloading, and the operations of other trades. The basic rule is to always employ protection methods that do no permanent damage. Therefore, whether specified or not:

- To protect the exposed edges of timber columns, use nonstaining wood bumpers continuous from just above the floor to a reasonable distance above the highest likely point of impact. Secure them with corrosion-resistant strapping clamped to itself, *not nailed to the columns*.

- To hide nail holes, fasten temporary braces to surfaces that will be concealed in the completed construction.

- To protect timbers prior to erection, keep them up off the ground on blocks, and cover with nonstaining waterproof tarps.

- To protect timbers while driving them into place, use wooden driving blocks; *never strike timber faces directly*.

- To protect wrapped members, keep the wrapping intact as long as possible. Once partially opened or damaged, remove it completely.

Unless the job is in the desert, the timber frame will get wet sometime during construction. In smaller members, wetting of discrete faces may cause minor temporary warpage, but this should abate once equilibrium throughout the piece is reestablished with the site environment. The biggest culprit in moisture gain is, of course, unsealed end grain, whether at cut ends or inside drilled holes. Sealing of end grain exposed prior to delivery is the responsibility of the fabricator; sealing of end grain exposed by field operations is the responsibility of the erector.

Limitations on Carpentry Field Operations

Even though timber construction has long been under the jurisdiction of carpenters,[7] it is appropriate on all but the smallest timber project to establish the limits of their authority and, if appropriate, to limit the exercise of their craft in the field. Timber frames are often largely shop fabricated; particularly complex conditions may even be shop assembled (and then disassembled) prior to shipping. In this context, fitting problems encountered in the field are at least as likely to be the result of misreading of the erection drawings as they are of misfabrication of the members. Under the mistaken assumption that the problem before them is someone's else fault and that field refabrication is the only appropriate course of action, field workers engaged in brief "corrective" carpentry may unnecessarily delay the project.

With occasional exceptions, the fabricator is typically responsible for all of the components of the frame, including the timbers, the connectors, and the fasteners. Therefore, when a problem with fabricated items arises in the field, the first response should be to contact the fabricator. From the perspective of the erector, the work of the timber fabricator is just another type of preparation work by another trade. Unauthorized field modification of such work shifts liability from the fabricator to the erector.

Each step in the fabrication process—from lay up (in the case of glulams), to drilling, from staining to wrapping, from shop loading to job site delivery—takes time and adds value to the job. Unless required to do so by the specifications, the fabricator will not normally furnish "spare stock," with the exception perhaps of some extra fasteners. For that reason, it behooves the erector to track the contents of each delivery and immediately to report any shortfalls to the fabricator. The field crew must treat each component of the frame as a unique item, the replacement of which will take time and cost money.

Even minor field work should be prohibited without the written direction of the fabricator. Parameters of corrective work affecting structural strength or stability, such as reaming out undersized bolt holes, should be established by the project's structural engineer, via the architect. Guidance regarding finished surfaces should come directly from the architect. The contractor's correction methods must be tailored to the job. If, for example, a shop-painted beam won't fit into the pocket of its hanger, field planing must be confined to those portions of the beam which will ultimately be concealed.

Temporary Loads and Bracing

Most connections within timber frames *act* like hinges, even if they don't necessarily *look* like hinges. Whereas steel and reinforced concrete construction permit connections capable of resisting rotation, such is not the case in timber construction. Consequently, stability of a completed timber frame is almost exclusively a function of the geometry of the frame or of the distribution of shear walls and floor diaphragms.

Stability of a timber frame that is under construction can be another matter, depending on the erection sequence. If the bracing of the frame is an integral part of the frame, and each component of it is installed with the members it braces, the need for temporary bracing can be minimized, if not entirely eliminated. On the other hand, if permanent bracing is not installed until after the frame is erected, temporary bracing will be required, both for the safety of the workers and for the protection of the work. (See Fig. 13.4*a, b* and *c.*) Safety regulations virtually assure that the timber erector will have the building footprint to himself for the duration of his work. Therefore, if installation of permanent bracing is under the jurisdiction of other trades, it is not likely to occur until the completion of framing.

Because the need for and arrangement of temporary bracing is a consequence of the construction method and sequence rather than the building design, its engineering is the responsibility of the timber contractor or erector rather than the project's architect or structural engineer. The erector must therefore review the proposed erection process—from unloading through demobilization—with the erection engineer. That individual should have extensive experience dealing with the differences in quantity, distribution, and direction between the loads timber members and assemblies encounter during construction, and those they will encounter in service.

Differences in temporary and permanent loads have many potential causes and effects. The most obvious differences in load direction are those that occur when members are stored on their sides. In that case, gravity loads act on the weak axes of the members. Differences in load paths between the incomplete and the completed structure can affect load quantity, distribution, and direction. Improper arrangements of lifting anchors can cause load reversals—of particular concern in timber trusses—or can induce bending or combined loads—of particular concern in members sized exclusively for axial loads. Load reversals can also lead to failure of connections not designed for them.

Concern for construction loads should not be limited to the timber erection period. On the contrary, loads arising from subsequent activities by other trades may exceed those of erection itself: A tall enough stack of gypsum board on the floor or of shingles on the roof could conceivably overload the structure. Although timber's capacity to accommodate high short-term loads is substantial, it is not unlimited. For this reason, the construction superintendent must be advised of and enforce all load limits. If construction loads will overstress any members, then the contractor must provide the necessary shoring at his expense. Under no circumstances may member sizes be changed without written authorization from the architect or engineer.

(a)

(b)

13.4 Rafter, purlin, and arch framing with temporary and permanent bracing. (*a*) Workers on the floor and wall lift a purlin toward another one at the roof, for seating in previously attached steel hangers. The knee braces visible on the left stabilize the framing in the longitudinal direction. (*b*) Purlins are stacked on the floor to the left, ready for installation. Temporary braces facenailed into columns and rafters stabilize the frame in the transverse direction. Acoustic ceilings in this area will eventually conceal the nail holes.

13.4 (*Continued*) (*c*) The steel shoes, in conjunction with the temporary wood nailers across adjacent arches, may be sufficient to stabilize this structure temporarily, but a few diagonals would make it much more rigid.

Although erection issues are many and varied, they need not be discouraging. All that is needed is a rational and well-organized erection plan integrating constructability issues and engineering constraints. Regardless of the specific plan, however, the following guidelines should be followed wherever possible, for safety and economy:

- Orient members in storage as they will be oriented in use.
- Load members during lifting as they will be loaded in use.
- Never lift bundles by their straps or bands.
- Reject all damaged members and assemblies.
- Space blocking more closely under members stored on their sides than those stored on their edges.
- Endeavor to keep construction loads below permanent loads.

Erecting the Timber Frame

Erection of medium- or long-span framing systems is similar, whether in timber or steel. Each crew is headed by a crew leader or foreman. The foreman directs shake-out, finalizes sequences (within the parameters established by the erection engineer), and is authorized to handle problems of fit.[8] The foreman is responsible for quality control and coordination. While crew members will be responsible for their own tools, the foreman will supply the equipment needed to keep the job on schedule and maintain a safe working environment.

The differences between steel and timber frames do affect their erection. Economical spans in timber are shorter than those in steel. For given cross-sectional dimensions, timber is lighter than steel. The allowable building height for timber is far lower than for protected steel. Compared to steel, then, timber members tend to be shorter and lighter, and often need not be lifted so high. Consequently, the majority of timber buildings are erected with relatively small cranes. (See Fig. 13.5.) In fact, lightweight rolling cranes capable of being supported by the ground slab are often sufficient. In conjunction with good crane operating practice, adhere to these guidelines:

13.5   A small mobile crane assisting in erection. Using the man-lift in the background, workers were raised up to the truss joints to bolt them up. A ladder provides access to the roof; temporary wooden rungs enable movement up and down the roof and its trusses.

- Maintain a safe distance from power lines, trees, and existing structures.[9]
- Plan the work to minimize crane relocations.
- Verify that the materials under the crane are adequate to support it, fully loaded.
- Stay well below the crane's maximum load and reach.
- Minimize tripping hazards in work areas.
- Secure load hooks whenever there is a risk of high winds.
- Provide a ground for electric hand tools.
- Do not start a lift until the workers are properly supplied.
- Use common sense and adhere to all safety regulations.[10]
- Staff the job amply throughout the erection.
- Designate as the signal person an individual experienced in that task.[11]

Both the detailer and the erector spend most of their time worrying about the connections, the detailer during preparation of the shop drawings, the erector during construction of the timber frame. Efficient member alignment is the key to maximizing the productivity of a crew charged with installing hundreds or thousands of bolts. To bring into line the holes in adjacent members, workers may use driftpins—short tapered round rods,[12] placed temporarily—or bolts, placed permanently. Check the erection drawings for indications of bolt orientation: Where the bolts will be left exposed, the designer may have called for a consistent orientation. To satisfy this requirement, experienced crew members will orient themselves and load the pockets of their pouches accordingly. For example, some carpenters may prefer to insert all bolts with their left hands, and place nuts and washers with their right hands.

Bolt orientation may also relate to the work of subsequent trades. To prevent weak-axis bending of timbers at curtain walls, mullion attachments should be made only at timber connections. Failure to orient threaded ends correctly for such subsequent work can result in costly rebolting by the timber erector. To further limit callback work, insist that other trades set their own clips and angles. Leave all connections structurally complete; properly detailed, the work of others should add to yours. Under no circumstances should other trades have to loosen or otherwise modify what you have done.

Inasmuch as a consistent interface is required between the unthreaded portion of each bolt and the inner surface of the hole bored for it (see Chap. 6), it is best if all drilling is done under the controlled conditions of the shop. If the fieldworker's bolt won't fit in a hole, it is at least as likely that the bolt is larger than the intended size as it is that

13

the hole is too small. Even if the latter, the solution is never to force the bolt into the hole. Not only will this compromise the integrity of the interface, but it could also result in cracks around the bolt, if the timber's moisture content is significantly above equilibrium at the time of bolting. For these reasons, all mismatches between holes and fasteners should immediately be brought to the attention of the fabricator for resolution.

Since timber is constantly shrinking and swelling, its connections never rely on friction. Therefore, nuts need only be snug. Overtightening is not only unnecessary, but undesirable. If the members are extremely dry at time of installation, room must be left for expansion; if they are green, subsequent shrinkage may necessitate resnugging when equilibrium is reached. Where an assembly will be subject to vibration, use lock washers.

Bolts must not only be of the proper diameter, but of the proper length. If too short, the threads of the nut will not fully engage those of the bolt. If too long, the productivity of the workers will decline, the resulting installation may look unsightly, and the projecting bolt shafts may interfere with subsequent operations. At column bases, projecting bolt shafts can also be a hazard to passers-by.

The erection sequence must take account of the structural roles of each of the members. For example, if the columns are sized to help resist lateral loads induced by gravity in the floors and roof, the knee braces should be attached to them before those surfaces are loaded. If they are sized only to resist lateral loads from winds and earthquakes, attachment should be delayed until the floors and roof have deflected under gravity loads.

Due to timber's natural variations, it is almost inevitable that certain aspects of the completed frame will appear to be out of line. Waviness is particularly noticeable to the viewer who is positioned to sight along edges of eaves and rakes. Have the fabricator make the timber rafters and lookouts a few inches long, and snap a line across them in the field to guide cutting. Although the carpenters responsible for the fascias and soffits could easily do this work, don't let them; only the erector should work the timbers. And, while any waviness could be eliminated by shimming the joint between the fascia and subfascia, that will take even more time than the above method, and will be visible if the soffit is left open.

Installing Timber Decking

Whether you are using sawn or glulam heavy-timber decking, there are basically two different lay ups: simple span and random length continuous. Simple span installations require that the lengths of the deckboards or planks correspond to the spacing of the framing. Planks are run from the midpoint of one support to the midpoint of the next. If supports are spaced uniformly, the lengths of all planks will be the same. This makes the erector's job easy, both in ordering and in installation. Whereas the absence of decking end joints between supports may make simple spans attractive to the architect, the inefficiencies inherent in them make them less attractive to the structural engineer.

Random length continuous decking installations (similar to controlled random lay up[13]) exploit the benefits of cantilevering. (See Fig. 13.6.) Whereas each simple span deckboard is supported at each end, the typical random length continuous deckboard need only bear on a single support. While all timber decking must have tongue-and-groove edges, only 2-in nominal (5-cm nominal) decking installed with continuous spans must have tongue-and-groove ends (also known as end-matching[14]) or employ metal end-splines for load transfer.[15] In general, joints in adjacent courses must be staggered a minimum number of feet, which

13.6 Random length decking, prior to cutoff. To comply with bearing and staggering requirements within the field of the floor or roof, both ends of random length decking courses are usually left ragged. Upon completion of decking, the carpenters trim off the excess by following a chalk line. As seen at lower right, midspan connectors for the purlins were attached prior to lifting.

approximates the nominal thickness of the decking in inches. End joints within 6 in (15 cm) of aligning must be separated by no fewer than two courses. And to provide continuous lateral restraint for the supports, occasional deckboards may be required to bear on two of them.[16]

Because it enables more complete utilization of timber resources, random length decking tends to be more economical than the uniform length material needed for simple spans. In timber construction, however, random length does not mean random distribution of lengths; the presence of too many short pieces will make it impossible to achieve consistently the required bearings. Therefore, just as it is customary, to satisfy applicable structural objectives, for soils engineers to specify the distribution of particle sizes for engineered fill, it is customary for the timber designer to specify the distribution of board lengths for timber decking. While the rules vary slightly with thickness,[17] a typical distribution would require that 90 percent of the pieces be at least 10 ft (3.0 m) long, that 50 percent be at least 16 ft (4.8 m) long, and that the minimum length of the longest 40 percent of the pieces, plus 2 ft (0.6 m), equal or exceed the typical spans.[18] Because of the strong relationship between the lay up and the distribution of decking lengths, it is wise to verify that lengths are ordered and delivered in the proper proportions. Otherwise, you may find yourself at an impasse, unable to complete the decking without violating the specifications.

Structural continuity makes random length decking substantially more efficient than simple span decking. For a given plank thickness, the deflection of a random length continuous installation will be about 40 percent less and the maximum bending moment about 16 percent less than one with simple spans.[19] Due to the significant structural implications of each of the possible lay ups, it is customary for the designer to specify which is desired. The challenge for the installer is to organize the crew and the materials for maximum productivity. Match the crew to the job: *Simple* spans are *simple* to install, whereas *continuous* spans require *continuous* vigilance on the part of the workers to assure proper layout.

Certain general guidelines apply to both types of lay up:

- Before starting decking of a particular floor or roof plane, determine how many courses will fit. Adjust the edge of the first course with respect to the edge of the area to be decked, so that neither the first or last course will be less than half a course in width, after final trimming.

- Orient the boards so that their tongues point in the direction you are moving as you install them: upward on a sloped roof where the decking is supported on rafters, outward where it is supported on purlins.

- Unless otherwise indicated in the contract documents, lay decking so that the finished or pattern face will be exposed to view in the completed building.

- Using nails of the sizes, types, and spacings recommended by the manufacturer of the decking, face nail and toenail each piece into its supports. At 3- and 4-in decking, spike courses to each other.[20]

- To avoid damage, drive boards into place using driving blocks straddling the tongues.

- To provide full bearing *on* the deck, bevel the ends and edges of deckboards at changes of roof plane orientation. To provide full bearing *for* the deck, the designer or detailer should have made provisions in the top surfaces of the supports. If they are absent, immediately notify the designer. Field modification may involve shaping the bearing surfaces by cutting away material or by adding beveled shims.

- To minimize staining from exposure to the weather, use nails with corrosion-resistant coatings.

- Know the ultimate extent of each decking installation. Use boards of full width, even if they project beyond the edge of the decked area. On completion of decking, snap chalk lines at the limit lines, and rip off any overhangs using a circular saw set to the thickness of the decking.

- At hips and valleys that are symmetrical—where the hip or valley rafters or beams bisect the angle described by the lower edges of the intersecting roofs—align the courses on each roof plane.

- Reject any boards that are damaged or warped to a degree that cannot be eliminated during installation.

- To minimize splitting of narrow supports where the shanks of the decking nails are large, drill pilot holes to receive them.

- Where surfaces to be decked are tightly curved about an axis parallel to the planks, mock-up a small area using scraps, to confirm that the tongues and grooves can accommodate the curve. If they cannot, mill the tongues or grooves of the planks as required.

- Nail deckboards from one end to the other. This will maximize the leverage available to straighten out minor warpage and will provide better control of the tightness of joints.

- Set the widths of edge and end joints in relation to the anticipated moisture-driven movement of the decking. If, at the time of installation, the decking is drier than it is likely to be in the completed building, leave joints looser than if it is green.

- If, to achieve diaphragm action, gluing is required—between deckboards or between the decking and its supports—apply glue in moderation. Remember that glue that squeezes out of a joint has neither structural nor esthetic value.

13

Timber Erection

Other guidelines are lay-up specific:

- At simple span lay outs, direct the work by snapping chalk lines down the middle of the top of each supporting member. Where the member is arched or severely cambered, use a tri-square to help draw the midline.

- At random length continuous lay outs, let the planks in the last bay overhang the end of the area, "ragged" fashion. On completion of decking, snap a line across the decking at the limit line, and cut off the excess in one pass. This will be quicker and neater than measuring and cutting each piece as it is installed.

Installing Structural Panel Decking

Structural panels such as plywood and oriented strand board can be used alone in 1.125 in (2.90 cm) thickness as decking in roofs, and as the required covering in 0.50 in (1.25 cm) thickness for timber decking in floors.[21] Virtually all of the general installation guidelines listed above for timber decking apply to structural panels used as roof decking. For structural panels used as a covering for timber floor decking, conform strictly to the installation requirements of the specifications or of the decking manufacturer.

While there are certainly numerous installation concerns common to planks and panels, there are also several important differences:

- Unlike planks, panels come in a small selection of standard sizes— usually 4 ft × 8 ft (1.2 × 2.4 m)—and are nearly always drier when delivered than they will be until long after the building is enclosed. Therefore, dimensional control is critical and, to avoid buckling of the joints as the panels expand, maintenance of proper spacing is essential.

- Because they are much thinner than planks, panels can usually span no farther than 48 in (1.2 m) between supports. Therefore, while nails for panels are shorter and more slender than those for planks, there are many more of them within a given area.

- Although roof panels in heavy-timber buildings are far thicker than in residential construction, their tongues and grooves are still very fragile. Therefore, care should be taken in unloading, storing, and handling the panels, to avoid the frustration and delay associated with repairing damaged edges while on the roof.

Protection and Seasoning of the Timber Structure

It is the rare timber structure that is intended to be left exposed to the elements. More commonly, the timber erector is eager to have the roofer cover the structure as quickly as practicable. If possible, the roofing should follow immediately after the installation of the roof decking in each building area. (See Fig. 13.7.) This will only work, however, if roofing operations do not overtake decking installation. This is a function of the size and production of the respective crews, and their construction sequences.

Protection from the elements—even if only temporary—is better than no protection at all. To avoid any delays associated with coordinating the installation of this protection, it is customary to assign it to the timber contractor, through a provision similar to the following:

> *Temporary protection for completed areas of decking:* At the end of each workday (or each work period, if the workday is interrupted by inclement weather), provide appropriate temporary protection from moisture, such as a layer of roofing paper, over all completed areas of roof decking. Maintain such protection until all roof decking is substantially complete, as adjudged by the Architect. (Temporary protection shall be removed and replaced with permanent roofing system under "Shingle Roofing & Sheet Metal".)[22]

To realize the full potential of this approach, a complementary paragraph should be included in the roofing section of the specifications, making the roofer responsible for installing the permanent coverings immediately upon removal of the temporary ones.

Though roofing is good at protecting decking, the exterior walls and windows must be in place to protect the perimeter of the structural

13.7 Roof decking with and without temporary protection from the weather. Ideally, installation of temporary roofing paper should follow immediately after decking. Closely spaced wood nailers minimize wind damage to roofing paper.

frame. Once the building is weathertight and, as a result of heating or cooling, interior conditions begin to approximate those of the completed facility, seasoning of the frame will commence. Although this process typically starts long after the timber contractor has vacated the site, it can have a great impact on the ultimate appearance of that contractor's work.

The objective is to minimize seasoning defects, such as checking. Since the rate of moisture movement in wood is largely a function of the relative humidity of the surrounding air, the temperature of the wood, and the moisture gradient from the surface to the core,[23] this objective can only be achieved if the building's temperature is raised and its relative humidity lowered very gradually.[24] If the contract documents are silent on the need for *gradual* heating and drying of the building interior, it would be prudent to bring it to the attention of the general contractor as early in construction as possible. If you meet with resistance, ask the architect to help champion your cause. It would be a shame if carelessness by others in the final weeks of the job resulted in permanent scarring of the work so carefully fabricated and erected by the timber contractor.

Once the timber contractor has left the site and the other trades have returned, protection of the work becomes everyone's responsibility. I recently worked on the design of a replacement for an historic timber railroad station that had burned to the ground as a result of carelessness by roofers involved in its restoration. The lesson here is simple: The slow-burning characteristics of heavy timber should not lull the roofer, the plumber, or anyone else who works with an open flame into thinking that a timber building is noncombustible. Just in case, the general contractor should distribute fire extinguishers around the building and verify the reliability and adequacy of the water supply. Furthermore, construction debris should be removed daily and kept far from the structure in dumpsters. Sawdust and small timber cutoffs are particularly dangerous. Finally, smoking should be prohibited on the entire job site.

Although recycling of timber construction debris may not save much money, it is the responsible thing to do. Prior to starting fieldwork, find out who in the area collects such materials and what they do with them. Using them for fuel may be preferable to landfilling them. Make sure to put aside any treated lumber for separate disposal.

While the assembly and disassembly of timber formwork are beyond the scope of this book, it is appropriate to bear in mind that formwork represents just another type of timber structure (albeit a temporary one). (See Fig. 13.8.)

Punchlist and Warranty Work

The timber work is normally substantially complete well in advance of the other trades. To avoid being held responsible for repairing damage to the timber work by trades that follow, request that the designer inspect

I3.8 Timber in formwork for concrete construction. On large repetitive projects, timber formwork may be reused numerous times. With their emphasis on adjustability and quick disassembly, fasteners and connectors for temporary construction are fundamentally different from those for permanent construction.

it before you leave the site. Ask that the timber punchlist be issued, even if it is mutually agreed that certain corrective items thereon should not be addressed until after the *entire* project is substantially complete. (Consider requesting that the warranty period for the timber construction commence when it, rather than the entire building, is substantially complete, but don't expect the architect's agreement.)

Common items on timber punchlists include missing or loose fasteners, stained or damaged members, grade stamps or other markings that are exposed to view, and missing or poorly applied prefinishes. You know the designer will be carrying a checklist during his or her inspection; use Fig. 13.9 as your guide, and correct any problems you find before the

Timbers
  Workmanship
    Compliant
    Consistent
  Provisions
    For specified future work
    For moisture-related movement
    For attaching connectors
    For work of others
  Framing
    Complete
    Properly aligned and spaced
  Surfaces
    Clean and undamaged
    Even and well-matched
    With eased edges
    Free of loose splinters
  Members
    Free of excessive warpage
    Free of excessive distortion
    Treated and sealed, for exterior use
  Finishes
    Of proper color and sheen
    Uniform, within acceptable tolerances
  Bearing surfaces
    Smoothed and shaped for full bearing
    Separated from masonry and concrete
    Bearing on seats, where provided
  Blocking and bracing
    Properly located
    Snugly fitted
  Temporary protection
    Of vulnerable corners, with bumpers
    Of roof decking, with roofing paper
  Temporary construction
    Fastened in concealed locations
  Grade and inspection marks
    Present, but hidden from view

Decking
  Layout
    With required bearings and offsets
    Detailed correctly at ends and edges
    Aligned across hips and valleys
  Deckboards
    Spaced correctly at ends and edges
    With specified pattern and/or texture
    With pattern properly oriented
    Installed to produce diaphragms, where
      required
  Fasteners
    Spaced correctly
    Driven the proper amount
    Installed to avoid splitting

Connectors
  General
    Oriented properly
    With unobstructed weep holes
    With grouted baseplates
    With tightened anchor bolts
    Prepared for future work
    Prepared for work by others
  Surfaces
    Protected from corrosion
    Painted uniformly in proper color(s)
    Clean and undamaged
    Free of tool marks and open joints
    With eased edges
    With welds ground smooth, if required
  Fasteners
    Installed where required
    Tightened snugly
    With washers under bolt heads and nuts
    Oriented properly
    Trimmed to reasonable length
    Upset to prevent removal

Work by others
  Holes and notches
    Of authorized sizes and shapes
    In authorized locations
    Cut to minimize stress concentrations
  Clearances
    From concrete and masonry
    From chimneys and appliances
    For finishes

Closeout
  Submittals
    Record drawings
    Bolt testing report
    Plumb and level survey
    Certificates of conformance
    Releases of liens and claims
  Removals
    Temporary bracing
    Construction equipment and tools
    Staging and scaffolding
    Visible grade stamps
    Construction debris
  Repair or replacement
    Of nonconforming work
    Of damaged or abraded surfaces
    Of warped or distorted members
    Of timber work damaged by others
    Of unacceptable finishes
  Conditioning
    Gradual, to minimize checking

13.9 Timber inspection checklist, for use in conjunction with contract documents.

designer arrives. In the unlikely event that there are any defective components that cannot be repaired, but must in fact be replaced, immediately order the parts, notify the general contractor of their anticipated delivery date, and request that nothing be done in the affected area which would interfere with the replacement work.

Summary

Erection of a timber structure can be a straightforward process, but only if it is planned in the context of the entire project. An appropriate balance must be found between shop and field assembly. Preparations by others for the timber work and by the timber contractor for the work of others must be coordinated. Allowances must be made for aspects beyond the erector's control.

When coordinating the erection process with the detailer, fabricator, and shipper, recognize that each maximizes its profit in a different way. The fabricator does so through mass production of similar components, the shipper by moving the fewest number of full loads, and the erector by efficiently orchestrating all of the related on-site activities. The likelihood of cooperation and success will be greatly increased if the overall strategy has something in it for everyone.

The erector decides where to start the work and how to proceed, and communicates his intentions to all of the affected parties. Materials, manpower, and equipment must be deployed to support the intended erection sequence. The relative stability of the structure must be envisioned during each stage of erection; take appropriate temporary measures to maintain stability until the completion of permanent bracing. The relative exposure of the structure to the weather must also be assessed for each stage of erection; take appropriate temporary measures to protect it until building enclosure is complete.

In short, every hour and dollar spent in advance planning, scheduling, coordination, and communication will pay enormous dividends during field operations. Prepare your workers for the worst, so they can perform at their best.

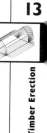

13

Timber Erection

Notes

1. Flexner, p. 1131.
2. Bunéa, p. 29.
3. Bunéa, p. 29.
4. AIA, General Conditions, p. 12.
5. Such records can also become part of the database upon which future timber construction bids are based.
6. BOCA, 1996, p. 62.
7. "Innovative Design . . . ," p. 136.
8. Bunéa, p. 133.
9. Korman, p. 14. "The most common cause of crane-related death is electrocution from contact with a power line. . . ."
10. Bunéa, p. 135.
11. Arnold, p. 54.
12. Harris, Cyril, p. 166.
13. TCM, p. 7-742.
14. Harris, Cyril, p. 183.
15. TCM, p. 7-737.
16. TCM, p. 7-742.
17. AITC 112-93, p. 3.
18. CWC, *Wood Reference Handbook,* p. 211.
19. Filler King, p. 4.
20. TCM. p. 7-741.
21. BOCA, 1996, p. 254.
22. The Goldstein Partnership, Montville Specs., p. 6B-6.
23. Panshin, p. 330.
24. AITC, 111-79, p. 4.

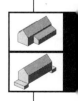

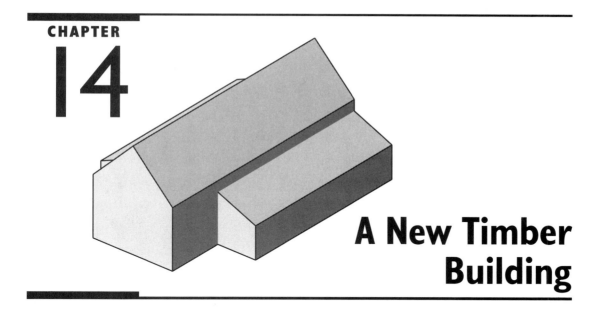

A New Timber Building

... beauty in a building is not merely a matter of decoration,
something to be added at will, but is inherent in the lines and masses
of the structure itself.

GUSTAV STICKLEY[1]

Background

Before starting graduate school, I spent two summers building custom wood-frame houses in Massachusetts. During the second of these, I was privileged to work with one of my professors, Ed Allen, on the construction of his own house, which he had designed. Then, during graduate school, I took his course on timber design and engineering. These experiences taught me a number of invaluable lessons about the building process, and how it dovetails with the design process (pun intended).

My first postgraduate job was with a small design/build firm in northern Vermont. The principal needed someone to design a huge wood-framed dairy barn, and my background was perfect. I started design work in early July, but because of the short building season, construction had to start immediately. There were days in which we would work out a detail in the morning, and it would be built that afternoon. To minimize cost, we framed the floor of the hayloft with a series of 16×16 in creosoted timbers, salvaged from a nearby railroad trestle, which had just been dismantled. The roof framing consisted of extremely lightweight plate-connected 2x trusses, with closely spaced lines of bracing along their bottom chords. They were so long and slender that they twisted in the wind as they were hoisted into place. Even though the ridge was left open, the updraft from the warm cows was so great that precipitation rarely entered the building.

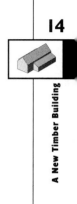

The footprint of the barn was almost half an acre (100 ft wide × 200 ft long). With its glulam columns and knee braces, massive hayloft beams and decking, and dense roof framing and bracing, the complexity of the framing was overwhelming, and presented a marked contrast to the large simple forms of the exterior (see Fig. 9.6).

On every project since then, I have thought about leaving the roof structure exposed to view, to recapture some of the excitement and surprise I felt every time I entered that barn. It wasn't until more than 12 years later, when I began work on a new library for Montville Township, NJ, that I had the chance to do it. This chapter is a case study of that building and is included in this book to illustrate how the issues covered in previous chapters were integrated into the design and construction of a real project.

Building Program and Architectural Context

Montville is a township large in area, but, until recently, small in population. Although it is being overrun by housing developments, old family farms can still be found. The site designated for the library was a west-facing meadow adjacent to the high school, on a prominent hillside underlain by bedrock. The program called for a building of 18,000 GSF. From our extensive experience designing public libraries, we knew that such a facility would operate best if it were all on one floor. Unfortunately, a plateau large enough for the building would require an enormous amount of fill.

After my initial design concept was deemed too modern, I asked for more guidance as to the character the Board was looking for. I was pointed to a barn visible from the site. It was being used as a community center by the township, and was reported to be a favorite of town residents. With that, I decided to pursue the possibility of housing the Montville Library in a high-tech barn.

Design Evolution

As soon as the Board approved the *idea* of a timber library, I started poring over books about barns and timber frames. The chapels of Fay Jones, an architect from Arkansas, were particularly inspirational, due to their delicate frames and largely transparent skins. Here, I thought, was a master of structure. Although each of these buildings contained a single space, the lessons seemed applicable to buildings with more complex programs.

- *Building program:* The library was to contain several book collections, including adult, children's, and reference. Associated with each collection were to be a number of seats for readers. Provisions were to be made for future expansion of the building. Offices were required for the librarians responsible for the collections, as well as for the library director. Technical services were to be centrally located, and be adjacent to the circulation desk. A separate wing, containing large and small meeting rooms, was also desired.

- *Form and design:* Early on, I decided that the layout of the library wing should be comprehensible the moment one walked in the front door. I placed the service desks and the terminals that accessed the computerized catalog in the main lobby, with the three collections (and their seats) in three building wings radiating from that core. All spaces requiring acoustic and/or visual privacy were distributed along the building perimeter. The components of the meeting wing were organized in a similar manner. (See Fig. 14.1.) Rather than express this wing as a separate form, I decided to let the main gable of the library extend over it. A gallery, to be left open in good weather, would pass between the two building wings, and would have views into each.[2]

- *Daylighting:* In contrast to traditional barns, with their limited windows, I wanted the interior of the library to be as open and airy as possible. My objective was to bring enough sunlight into the building that supplementary artificial lighting would be unnecessary during daylight hours. The offices and support spaces around the building perimeter would, of course, have windows. The problem was how to get daylight to the bookstacks and reading areas, located 15 to 30 ft from the exterior walls. Having already designated the functions to occupy the perimeter and interior zones, I knew that there would normally be a wall between them. Rather than terminate this wall at the underside of a roof spanning both zones, I decided to break the roof above this wall, creating a clerestory along the edge of the interior zone, to permit borrowed daylight.

Truss Form

The bottom chords in a series of parallel trusses create a virtual ceiling plane. Depending on its height above the floor and on the height of the real ceiling, this virtual ceiling can be more important than the real one in establishing the scale of a room. For Montville, I was not interested in horizontal bottom chords, for they would create a visually static environment. Instead, I decided to align the chords with the sloping roofs of the adjacent side-bays. From the exterior, this would make the side-bays appear to be extensions of, rather than attachments to, the central bay. We went so far as exploring the feasibility of having the roof trusses span the full width of the building, with the side-bay portions cantilevered beyond the interior columns, but rejected this approach on the basis of cost.

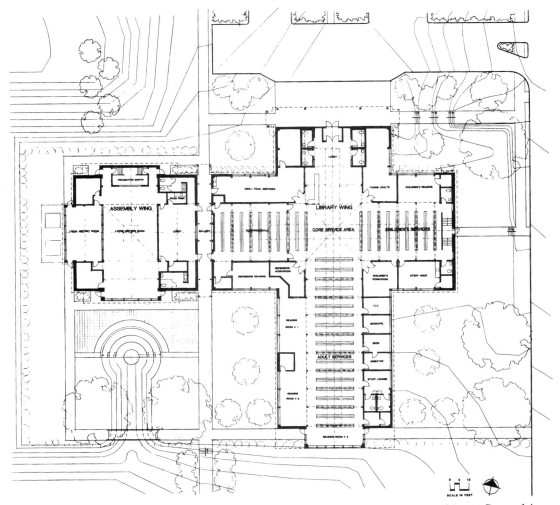

14.1 Floor plan of the new public library, Montville, NJ. The Goldstein Partnership, architects. Severud Associates, structural engineers. Timber engineer and detailer: Enterprise Engineering Consultants, Ltd., Peshtigo, WI. Timber fabricator, Unadilla Laminated Products, Unadilla, NY. Timber contractor, Dajon Associates, Hackensack, NJ. General contractor, The Conklin Corp., Franklin Lakes, NJ.

Establishing an acceptable arrangement for the truss' web members was a matter of trial and error. The final scheme has the lower webs extending the lines of the bottom chords, and the upper webs defining a zone for the ridge skylights. We then built a crude study model, showing one of the transepts, the roofs all around, and the cupola on top. (See Fig. 14.2*a* and *b.*) Fred Severud, PE, our structural engineer, did his preliminary member sizing, and Unadilla, a timber fabricator with whom we had worked on previous projects, prepared a preliminary estimate of the timber contract.

(a)

(b)

14.2 Conception and reality: similar views of related artifacts. (*a*) Structural study model cut through a typical building wing. The side bay roofs align with the bottom chords of the trusses. The ridges are undecked to accommodate skylights. The cupola evolved from an octagon to a square. (*b*) Timber structure nearing completion. At indicated elevation, central bay projects further forward than side bays.

14

A New Timber Building

Interior Partitions

Maximizing the visibility of the timber frame in the building interior was a high priority. In an open barn or other unpartitioned space, this is no problem. In a public library, on the other hand, not only are there typically lots of partitions, but also ducts, air-conditioning equipment, and, for sound control, acoustic ceilings, all with the potential to interfere with the view of the structure. It was obvious that hiding these systems, by running ductwork within the floor slab, for example, was cost prohibitive. Therefore, I was committed to integrating these systems into the architecture of the interior.

With the offices and other support spaces in the side bays, partitions would be needed along the edges of the central bay. For structural reasons, the interior columns in these areas would be much wider than the partitions running between them. The lateral bracing had to meet the columns at roughly middepth, but would be much narrower.

The presence of these interior partitions made it inevitable that the structure of the side bays would be hidden, when viewed from the central bays. To what extent would the locations of the partitions, with respect to the interior columns, affect the architectonic qualities of the central bays themselves?

Since the column alignment and spacing had evolved from simple and straightforward planning modules, the most obvious approach would have been to run the partitions from column to column, aligned with the column centerlines. This would have left the inner column faces visible from within the central bay and the outer ones visible in the side bays. The lateral column bracing would have been concealed by the upper partitions. As a practical matter, framing of the upper partitions would have been complicated by the presence of the lateral bracing and its connectors.

Using cardboard study models, we looked at the architectural implications of this approach, as well as two others: running the partitions continuously past the inside faces of the columns, and continuously past their outside faces. (See Fig. 14.3a and b.) Because it avoided the whole issue of framing between the braces, the former was simple to build, but it left only the roof structure of the central bay visible. I found the result uninspiring, primarily because it made the roof structure look like it was resting on walls, rather than integrated with a system of braced columns. The latter had the same construction advantages as the former but, by leaving the interior columns and bracing visible, emphasized the three-dimensional nature of the structure, and achieved the spatial richness I was after. Furthermore, it enabled the partitions between offices to frame into the outside faces of these longitudinal partitions, simplifying wiring. The design approach finally seemed to be falling into place.

It was at about this time, in the fall and early winter of 1992, that lumber prices jumped to record levels. Our project was in jeopardy. To save it, we decided to redesign the structural frame.

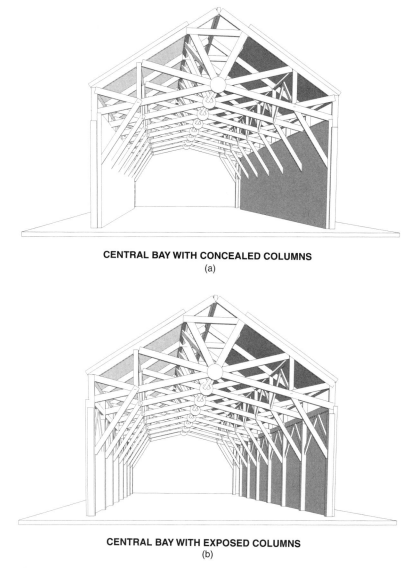

CENTRAL BAY WITH CONCEALED COLUMNS
(a)

CENTRAL BAY WITH EXPOSED COLUMNS
(b)

14.3 Computer model of final structural frame. (*a*) Concealed columns. (*b*) Exposed columns. Note the difference in architectural character.

Material-Driven Versus Labor-Driven Costs

On hearing of our efforts to increase the efficiency of the superstructure, several friends from outside the construction industry have asked why we didn't design the building that way in the first place. Why, they ask, were you intentionally wasting material in the original design?

The construction industry is unique in many ways, one of which is that materials have usually represented a smaller portion of the cost of a building than the labor needed to fabricate and install them. Under such circumstances, the most economical way of building something is often that which involves the least labor, even if it means using more material. As the cost of materials rises, however, this equation changes. Conventional solutions are no longer appropriate.

Another aspect that comes into play here is the effect that improved engineering and fabrication techniques have had on construction. Unlike the master builders of medieval Europe, for whom confirmation of the structural adequacy of their cathedrals came from whether they collapsed, today's architects and engineers rely on fundamental scientific principles in their structural work. The evolution of this understanding has manifested itself in the form of buildings of greater and greater efficiency. This increase in efficiency means that today's structures are lighter in weight and less redundant than their predecessors. Combined with an increasing degree of prefabrication and the rise of other advanced quality-control procedures, today's structures are also more predictable than those fabricated primarily with field labor.

Given this background, it should not be surprising that innovations in the use of construction materials have often been driven by increases in their costs. The development and proliferation of wood I-beams and laminated veneer lumber are directly related to the decreasing availability (and increasing price) of high-quality dimension lumber. These products start with previously undesirable raw material and end up yielding products of superior performance.

The impetus for our redesign in Montville was an unprecedented short-term escalation in the cost of lumber, resulting from the imposition of severe constraints on logging in the Pacific Northwest. With rising costs for our primary building material, but a fixed project budget, we had only a few options:

- Switch to a more economical framing material, such as steel (rejected for esthetic reasons)

- Use a more efficient engineered wood product (none found, with an acceptable appearance)

- Develop a more efficient framing system

Obviously, this third option was the one we eventually selected. Our objective, then, was to reduce the *quantity* of lumber in the building sufficiently to offset the rise in its *unit price*.

Equivalent Thickness

One way of comparing wood-framing schemes is in terms of their *equivalent thickness*. (See also Chap. 5.) In a one-story building such as ours, this is calculated by dividing the total volume of lumber in the building by the building area. Conceptually, it is as if the wood members were made of a frozen liquid which, when melted, would flow to a uniform thickness throughout the building footprint.

Take, for example, a residential floor system, consisting of 2×10s at 24 in on center, with ¾-in subflooring. In any 10 × 10 ft area would be five 10-ft lengths of 2×10s and 100 sq ft of ¾-in plywood. At 1.5 × 9.25 in, each 2×10 has a cross-sectional area of 13.875 sq in. Thus, the average volume of the joists is:

$$[5 \times (10 \text{ ft} \times 12 \text{ in/ft}) \times (13.875 \text{ in}^2)]/100 \text{ sf}$$

Or

$$(5 \times 120 \times 13.875)/100$$

Or

$$83.25 \text{ in}^3/\text{sf}$$

Or

$$0.58 \text{ in}^3/\text{in}^2$$

To this must be added the subflooring, yielding a total of 0.58 + 0.75 or an equivalent thickness of 1.33 in for this framing system. To be fair, the calculations should also include girders, blocking, and, perhaps, even columns and lateral bracing. That is why the above total seems so low.

In a building, of course, programmatic constraints typically limit the architect's freedom in the sizing of structural bays. Such is certainly the case in libraries, the planning of which is driven largely by the lengths of bookstack ranges (standard 3-ft increments, with 21 ft practical maximum) and their center-to-center spacing (standard 5-ft increments for accessibility). Consequently, to maximize the capacity of the bookstacks, the width of the building's central bay was set at 30 ft (21-ft bookstack ranges, with aisles at each end), and the spacing of roof trusses was set at 10 ft (the largest multiple of 5 ft which could be spanned reasonably by timber decking).

When lumber prices began their meteoric rise, we were too far along in the design to start rethinking the 10-ft spacing of the *primary* framing, the roof trusses. Thus, our equivalent thickness calculations did not examine the implications of changing the truss spacing to 15 or 20 ft, even though such a change might have been beneficial. Instead, we took the 10-ft spacing as a given, and looked only at the *secondary* framing, the purlins, lateral braces, and roof decking. This much we knew:

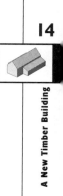

- The bottom chords of the trusses needed to be braced, and this was normally done with horizontal members running from truss to truss.
- The spacing of the top chord intersections or *panel points* of each truss was tighter than the spacing of the trusses themselves (approx. 7 vs. 10 ft).

The question I asked of my structural engineer was whether the equivalent thickness of the roof would decrease if we ran the decking downslope, rather than across the trusses, supporting it on purlins more closely spaced than the trusses. His calculations showed a significant decrease, not only because the decking got much thinner, but also because the top truss chords, which had been combined members, would now only be in axial compression. I next asked him to determine the net impact, if any, of changing the bottom chord braces to chevron braces, supporting each purlin at midspan.

Although the number of bracing members doubled (from one *horizontal* to two *diagonals* per span), and their total length increased (since the horizontals took the shortest route between trusses, and the chevrons did not), there was a significant net increase in efficiency, for the following reasons:

- Although the total length of braces increased, the length of each brace decreased. The cross-sectional dimensions of axially loaded wood members are usually controlled by the tendency of such members to buckle in compression. Shorter members, in other words, are more resistant to buckling than longer ones. Therefore, shorter braces could have smaller cross sections than longer ones. Since the tendency to buckle is a function of the square of the length of the brace, the decrease in brace cross section more than offset the increase in overall brace length.
- In a structure with conventional horizontal bottom chord truss bracing, the purlins would be spanning 10 ft between the top chords. In a structure with chevron bracing, the braces would support the purlins at midspan, cutting their free length in half. The depth of wood bending members is usually controlled by their clear span. Shorter members are more resistant to bending than longer ones of similar cross section. Therefore, shorter purlins could be shallower than longer ones. Because the tendency to bend is a function of the square of the length of the purlin, the decrease in purlin depth more than offset the increase in chevron brace cross section resulting from the purlin load it was now carrying.

Interestingly, we found that the new member sizes we needed were slightly smaller than what the building code permitted for a heavy-timber building. Consequently, the efficiency we actually achieved was slightly less than that which was theoretically possible, purely from a structural perspective. Unfortunately, it was too late in the design process to reexamine the truss spacing, to see whether a change from 10 to 15 ft, for example, would align the theoretical member sizes more closely with the minimum sizes mandated by the code.

The net effect of these structural transformations was to reduce the equivalent thickness of the roof system from 4.5 to 3.5 in, a reduction of 22 percent. This translated into a savings of 2000 ft^3 of lumber, with a total weight of approximately 35 tons, and a total installed value of about $100,000. However, although we had reduced the overall weight of the system, we had increased its complexity. We had doubled the number of braces, and increased the number of brace connections by 50 percent from 2 per brace per span, to 3. To what extent would these costs offset the benefits associated with our new scheme?

To answer this question, we considered these facts:

- Chevron braces would be more numerous, but much shorter and lighter in weight than straight ones, thus reducing labor per brace.

- The purlins would have midspan connections, but would be much lighter in weight than if they had not, thus reducing purlin material.

- The area of roof decking would not change, but the decking would be thinner and lighter in weight, this reducing decking material.

- All truss members, purlins, and chevron braces would now be the same size: GL (glulam) 3×7, thus enabling manufacturing and installation efficiencies.

On balance, we determined that the cost of fabricating and erecting the additional connections and braces would approximately equal the savings that would be achieved:

- In fabrication, using members of consistent size
- In erection, using members of lighter weight
- In decking installation, using shorter and more slender nails

Only then, having completed an evaluation of the overall impact of our proposed changes, was I prepared to say that the changes were worth making. (See Fig. 14.4.) We now had a workable roof structure, but still had to find a way of resisting lateral loads generated by winds and earthquakes.

Axially Loaded Versus Bending Members

The greater load-carrying capacity of columns, compared to beams, is obvious from their relative sizes on a typical construction site. Whereas a timber girder might be a 6 × 12 in, the column supporting it might only need to be 4 × 4 in. Members subject to a *combination* of loads,

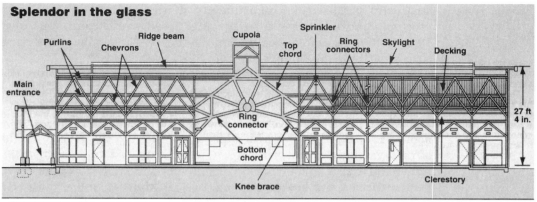

Redesign added purlins, chevrons and switched decking direction but maintained initial concept, shown in model of original cupola area (top).

14.4 Longitudinal building section. (Reprinted from *Engineering News-Record,* February 6, 1995. Copyright, The McGraw-Hill Companies, Inc., all rights reserved.)

some parallel to their long dimension (i.e., axial) and the rest perpendicular thereto, are called *combined* members. The axial forces acting on such members can be compressive or tensile.

A long column loaded too heavily in compression will fail by buckling. It is not surprising, therefore, that bending of a slender compressively loaded member drastically reduces its load-carrying capacity. To resist such forces, the member must look more like a beam than a column, with its larger cross-sectional dimension parallel to the bending load.

To maximize efficiency on this project, I initially tried to conceive a frame in which all of the primary structural elements were only loaded axially. Unfortunately, this was easier said than done.

First, I ran into trouble at the roof truss. Heavy-timber decking could readily span the 10-ft truss spacing contemplated. This meant that the top chord of each truss would be a combined member, carrying compressive truss loads and bending decking loads. This was not all bad, for it eliminated the need for secondary framing to support the deck. However, once we decided to add purlins and run the decking across them, this issue became moot.

Second, I ran into trouble at the typical interior column. Since interior shear walls or diagonal braces between the floor and the underside of the roof structure would have limited the effectiveness of future partition rearrangements, we rejected their use. By default, the only means of laterally bracing the frame involved making the main columns large enough to resist the combined loads to which they would be subjected (compression, from gravity acting on the roof structure, and bending, from lateral loads transmitted through knee braces). But even this did not exact much of a penalty. The code required a minimum size of 6 × 8 in, but to simplify turning corners, we opted for a square column 8 × 8 in. To accommodate bending, we only had to go up one column size, to 10 × 10 in (nominal). The fact that it was square was ideal for resisting wind forces along both axes.

Special Conditions

In general, I had wanted to house connector plates in mortises at the ends of the truss members. However, since the members were already of minimum thickness, the code prohibited further reduction. Accordingly, we tried to use standard beam hangers and column caps everywhere, but found that our frame required more than its share of custom connectors. These were concentrated around the transepts and ridges.

Valley framing was inevitable, where the sloping roofs changed direction at the corners of the transepts. I was uncomfortable with the idea that the top edge of the valley rafter would not be parallel to the undersides of the two planes of roof decking it was to support. Therefore, I insisted that that surface be beveled inward, to provide full bearing for both deck surfaces meeting there. (See Fig. 14.5.)

In addition to continuous clerestories between the lower and upper roofs, I had proposed the use of linear translucent ridge skylights. The uppermost purlin was a natural line of support for the lower edge of the skylight; no support was needed at the ridge.

14.5 Valley rafter, showing typical inward beveling to support roof decking at intersecting planes.

Calling upon my barn design experience, I thought a cupola atop each transept roof would help emphasize the special nature of the space below it. In a moment-resisting steel frame, framing an opening in such a location would not have affected the rigidity of the roof frame. However, in wood, with its "pin" connections, the presence of such an opening destabilizes the frame, allowing the four converging valley rafters to rotate. Stability can only be restored by triangulating the corners of the opening. We achieved this by carrying into each transept a variation of the midspan chevron bracing used elsewhere. To assure that the transept framing acted as a unit in resisting wind loads, Fred, our engineer, recommended that the purlins be extended along all four sides of each cupola opening. (See Fig. 14.6.)

This meant that the upper end of each valley rafter was braced by two perpendicular struts from nearby trusses, and that one leg and two inclined wall supports, plus four purlins, all converged at each cupola

14.6   Opening for cupola at roof of assembly wing. A valley rafter, two purlins, and two chevron braces converge at each corner of the cupola.

corner, for a total of ten members, each one at a different angle and/or orientation. This was the most complex connection in the building.

Another interesting condition was to occur at the top of each transept's corner columns. Each of these had to support the ends of two perpendicular trusses. Although the trusses might have been simpler to erect if each included its own vertical end member, this approach would have resulted in clerestory conditions substantially different from those elsewhere around the building. Therefore, we chose, instead, to let these trusses share their end members.

After investing all that time and energy to make the structure elegant and efficient, the last thing we wanted to do was crowd it with sprinkler piping.[3] We decided to integrate the layout of the piping with that of the structure. Mains would run just above the clerestory windows, with branches running upslope between each pair of trusses. But somehow, the piping around each corner of the transept had to connect to every other corner, and had to continue down into the ceilings of the side bays beyond.

Much time was spent searching for a way to hide these piping links. In the end, we concluded that exposing them would be fine, as long as they were coordinated with major framing members. As illustrated, we hung the mains from the undersides of the valley rafters, and had a custom-welded piping loop fabricated for the bottom of each cupola opening. (See Fig. 14.7.)

Intending to run the link down to the side bays along the same diagonal as the link up to the cupola, I found the way was blocked by the

14.7 Custom sprinkler loop under cupola at library wing. Piping is connected to each corner and descends along the underside of each valley rafter.

shared corner post of the trusses. Although, within a typical steel column cap, the plate resting on top of the column is often the same piece of steel as the plate on which the beam or truss bears, neither Fred nor I saw any reason why this had to be the case. For this job, we developed a connector with two bearing plates, separated vertically by enough space for the sprinkler pipe to pass through, on the diagonal. (See Fig. 14.8.)

The midspan truss connector, although the most prominent one in the building, turned out to be relatively easy to work out. With 10 members converging there, a spherical connector, like a space frame node, seemed like a reasonable point of departure, until we looked into the cost. On our second try, we decided to accommodate the 6 coplanar truss members using plates projecting from a cylinder. The cylinder consisted of a short length of heavy-gauge 24-in-diameter steel pipe. The four chevron braces would be secured to tabs welded to the faces of a plate inside the cylinder. (See Fig. 14.9.) (I would have preferred to have used a hollow cylinder, but it would have had to have been much deeper, to pick up the braces on its outside face.)

14.8   Corner column of transept, under construction. The connector opening at the middle of the photo allows the sprinkler piping to pass through the column. Bolts on adjacent column faces are offset vertically to avoid interference.

14.9 Midspan truss connector. The plate to which the chevron brace tabs are welded is bolted to a separate plate at the center of the ring. See Fig. 12.6 for shop drawing detail.

There are several locations, including the front entrance overhang, where the roofs of the central bay step down to align with those of the side bays. The bottom chord of the end truss thus became a combined member, picking up the purlins of the lower roof. We opted to deepen this member by lowering its bottom edge, rather than raising its top, to leave some room to properly flash the base of the curtain wall at this gable end.

There are also several locations where the shed roofs typical of the side bays are interrupted by gable dormers. In each, a spandrel beam was added under the clerestory, to pick up the inboard end of the dormer ridge. The outboard end was supported by the peak of a trussed frame. (See Fig. 14.10.) These spaces were created to take advantage of dramatic views of a landscaped courtyard, with the hills beyond.

We required that the timber connectors at the ends of the building be capable of simple field conversion into interior connectors, when building expansion occurs. There was insufficient room between the outside face of the last bent and the inside face of the curtain wall for the steel plates needed for a future bay, but there was enough room for bolts with which to attach such plates.

Although the design and engineering of our custom frame was now just about complete, we knew that it had to be understandable to the bidders, if we were to achieve our budgetary objectives, and get the project built. As we were preparing the contract documents, we had a number of conversations with Unadilla, the fabricator. Among the

14.10   Gable dormer truss. Combined loading of the top chord was avoided by framing the purlins into the panel points of the truss.

issues discussed were whether a fully assembled roof truss could feasibly be transported over the road (see Fig. 14.11), what sort of temporary covering they could provide to protect the roof deck until the roofers arrived, and the availability and pricing of various decking products. Their input assisted us in preparing the timber specifications and in detailing the frame. It also led to our decision to specify galvanized decking nails, to minimize the likelihood of rust stains during construction. Finally, they convinced us that a "plumb and level survey" of the completed frame, a standard procedure in steel buildings, was of little value in a diagonally braced timber one. They said the pieces would only fit if the frame were square, and that, even if the final result were *not* perfect, there was little you could do about it at that point.

The Timber Frame: Integration and Finishing

It was apparent that the library's site was very windy. A deep overhang was needed at the entrance. This would not only protect those using the front doors, but would also provide an opportunity to bring the structure outside, giving a hint of what was to be revealed within. For economy, we switched from timber to plywood decking for parts of the entrance canopy, and from architectural to structural grade glulam material for side-bay roofing framing, which would be hidden from view.

14.11 Shop-assembled truss. For protection during shipping and construction, the connectors were shop primed, and the timbers were shop wrapped.

This brought to five the number of different conditions to be finished:

1. Exposed interior structure, stained as light as possible, to maximize reflectance of artificial uplighting
2. Exposed exterior plywood soffits, painted with a heavy-bodied paint, to hide defects
3. Exposed exterior structure, treated to resist moisture
4. Hidden interior structure, left unfinished, for economy
5. Exposed steel connectors, painted where visible inside, for appearance and, coated, where outside, for corrosion resistance

It was important to me that the timber structure be visible, especially at night, from the major street passing in front of the building. To achieve this, the gable ends had to be glazed; to reinforce the illusion that the side bays were extensions of the central bay, the upper portion of the curtain wall (that which was outboard of the roof truss), needed to be appear separate from the lower portion. Ultimately, we decided that the mullions of the upper portion should echo the truss members behind it.

The panel points of the truss were laterally braced by the system of chevrons we had developed. But since the truss members themselves were only 3 in wide, they were unable to resist lateral loads at midspan. It was critical, therefore, that the curtain wall manufacturer put his wind load connections at the panel points of the truss, and nowhere else. (See Fig. 14.12.) Furthermore, the truss had to be able to deflect under roof loads, without taking the curtain wall with it. A detailed description of how this was accomplished would fill another chapter, but suffice it to say that small structural steel sections were placed inside some of the mullions, to transfer loads to the panel points, and some connections were configured to allow independent vertical movement of the curtain wall and truss.

At the interface between each building wing and the gallery, we placed interior windows within the plane of roof truss. To avoid cracking of the glass when the truss deflected and distorted under load, we called for slip joints in the head framing of each opening.

The central bay was narrow enough to be conditioned from each side. To maximize comfort and control, our mechanical/electrical engineer, Sid Garner, PE, suggested positioning small fan coil units at 10 ft on center along the lengths of the side bays, above their acoustic ceilings. Air would be supplied, via short flex-ducts, from the ceilings down into the offices and through the drywall fascias across into the central bay. There would be no need for return ductwork; the central bay itself would act as a return plenum. The transepts would be supplied with ventilation by pairs of supply registers at their four corners. Makeup air entered through soffit vents. To limit stratification within the high central bays, paddle fans were hung at regular intervals from the

14.12   Curtain wall framing at bottom chord of main truss. To prevent weak axis bending and combined loading of the truss members, connections between the curtain wall and truss were permitted only at panel points.

underside of the ridge beam. Its wiring was run out of sight on top of that beam, and fed down through bored holes. Large condensing units were placed outside the building and, in conjunction with modular boilers, circulated water of appropriate temperature through a supply loop that fed each fan coil unit.

To get the structure to glow at night meant that some of the lighting would have to be aimed at it. If stained a light color and illuminated by uplights, the framing and the underside of the roof deck would help distribute artificial light uniformly throughout the central bay. The uplights would be mounted on the fascias along both sides of the central bay.

Task lighting for the readers and bookstacks would be supplied by high-performance linear fluorescent downlights spaced 5 ft on center along the central bay. Sid, an amateur pilot as well as an engineer, came up with a way to hang these using aircraft cable screwed into the purlins above. The ⅛-in-diameter cables are so slender that they are nearly invisible, making the stack lights appear to hang in midair. Where the stack lights face into the transepts, their vertical faces are used for signage. (See Fig. 14.13.)

14.13 Corner column of transept, nearing completion, supporting two perpendicular trusses. Sprinkler branch piping passes through truss openings. Bookstack lighting hangs by aircraft cable from purlins. Uplights are mounted *on* and supply registers are mounted *in* the side walls of the central bay.

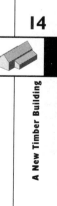

The transepts were perhaps the most difficult areas to light, what with their great height, and our desire not to clutter them with fixtures. And, since each transept was surrounded by open trusses rather than solid walls, it was not clear where its uplights could be mounted. Ultimately, we decided to attach the uplights for the transepts to the ring connectors at the midspans of their trusses. Eight large cylindrical downlights were hung from the roof deck, around the cupola. Their wiring was concealed in conduit, within the thickness of the roof insulation, and fed down through holes in the roof deck.

In traditional barns, the building skin, including the girts and siding, wraps tightly around the perimeter columns. In our case, by contrast, the heavily insulated portions of the exterior walls were thick enough to encase the 6×6 in perimeter columns. But with most of the lower walls and all of the clerestories to be glazed, the issue was whether the window system should fill the space between columns, or run continuously past their outside faces. Architecturally, the former would tend to emphasize the structural rhythm, while the latter would emphasize the horizontality of the glazing. Economically, the column covers associated with the former would make it more costly than the latter, both to fabricate and to install. Primarily for that reason, we eventually opted for the latter, with one exception.

At the cupolas, the windows were sized to fit within the framed openings, and aluminum trims wrapped the corner and peak framing. If I had a chance to do it over again, I would have framed each cupola face with a single perimeter frame extending its full width and height. By narrowing the trims, this approach would have lightened the finished appearance, particularly at the cupola peaks.

Construction

The glulams came to the job site marked in accordance with the shop drawings and wrapped for protection from the elements. The connectors came boxed or crated. The erectors distributed the thousands of pieces of this 3-D puzzle around the floor slab, unwrapped the timbers to be used that day, and then began erection. Heavy members were raised with the aid of a small rolling lift.

The trusses had come largely preassembled, enabling them to be set much sooner than would otherwise have been possible. Given the elevation at which the cupolas were to be set, it seemed to me that assembling them on the floor would have a lot of similar benefits. The erector apparently agreed. (See Fig. 13.2.)

Once a significant area of the roof was framed and braced, installation of the decking began. This proceeded smoothly until the valleys were reached. Since the decking would be seen from below, great care was needed to be sure that the boards on one side of the valley lined up with those on the other. (See Fig. 14.14.)

14.14 Underside of cupola. Symmetry was a high priority around the peak and, notwithstanding the asymmetry of the ridge framing, within the peak connector itself.

Postscript

Between the time of our initial design and building occupancy, the areas allocated to bookstacks and readers' seats were reversed. The owner eventually came to the realization that provisions for additional readers were more useful than empty shelves several years away from being filled. If the final proportions had been established earlier, the furniture distribution and, therefore, the feel of the building interior, might have been very different. Although I would have preferred to have dedicated the central bays exclusively to readers (as I had recently done in another new library), I had to place the bookstacks there too, because there were simply too many of them to go anywhere else. (Fig. 14.15.) Upon completion, it was apparent that my initial objective could have been achieved.

To the extent that building design involves problem solving, it shares with other such activities the tendency to be driven by the manner and sequence in which its problems are defined. Architects know that a single building program can yield a multitude of different plans, depending on how it is read and interpreted.

Although, in retrospect, I might have rearranged the library furniture, I am generally pleased with the timber structure. (See Fig. 14.16.) It is frustrating, however, to look in the timber connector catalogues and find virtually nothing appropriate for exposed architectural trusses. Fred and I had fun developing the custom connectors for the

14.15 Interior of completed library from just inside the front door.

14.16 Exterior of completed library from a reading area within one of the two gable dormers.

Montville Library, but it should not have been necessary. In Europe, especially Germany, connector design is far ahead of that in the United States. We still have a long way to go.

The winter of 1995–1996 was the first one for the completed library, and what a winter it was. The blizzard in early January dropped over 2 ft of snow, and the winds on this site produced enormous drifts on the roof. Shortly thereafter, the weight of the snowpack was dramatically increased by a soaking rain. I am happy to report that throughout it all, the timber frame performed admirably. As a result of deflections, doors located under the roof areas with the largest drifts (typically the northwest corners of the transepts) started binding in their frames, but this problem disappeared as soon as the drifts slid off the roof.

AUTHOR'S NOTE: This chapter was adapted from an article in the August/September/October 1996 issue of *Joiners' Quarterly*.

Notes

1. Quoted in Mitchell, p. 12.
2. This gallery was to extend north of the building as a barrier-free covered walkway, and run all the way to the high school, but for budgetary reasons, was never constructed.
3. Under the BOCA Code, sprinklers are required in A-3 uses, such as libraries, with areas in excess of 12,000 SF. (The Montville Library has an area of 18,000 SF.) Also, the concealed spaces above ceilings had to be sprinklered, since the structure was not considered "incombustible." I understand that the AFPA is developing formulas under which equivalent fire ratings will be calculable for timber decking, but code acceptance of this approach appears to be several years away.

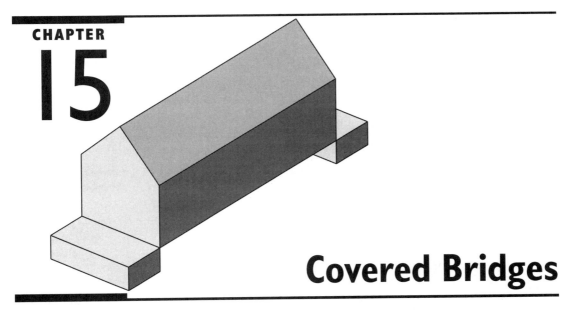

CHAPTER
15

Covered Bridges

Phillip C. Pierce, PE

What stories could these bridges tell
If they could only talk?
They'd tell us of the ones who rode
And those who had to walk,
The rich, the poor.....those in-between
Who used their planks to cross,
The soldiers, farmers, businessmen
In buggies, sleighs, by "hoss,"
Like sentinels these bridges stand
In spite of flood and fire,
Their rugged, stalwart strength remains
Our future to inspire.

AUTHOR UNKNOWN

Introduction

Covered bridges offer generic lessons about all types of timber structures, including buildings, other types of timber bridges, maritime timber structures such as wharves and piers, and timber shelters such as park structures and gazebos. Covered bridges contain unique combinations of proportions, spans, exposure, loads, age, and joining systems. While covered bridges are worthy of special attention in their own right, study of them can offer many useful lessons applicable to other types of timber structures.

Covered Bridges

For those of you unfamiliar with covered bridges, Figures 15.1 and 15.2 present exterior and interior views of classic examples. Figure 15.3 presents views of a typical bridge with nomenclature for the various skeletal components.

A covered bridge is basically a roofed timber structure spanning a roadway obstruction, such as a stream. The sides of the structure support it and the traffic crossing it. The roof protects the structure from the elements. Covered bridges and buildings both provide an enclosed area via use of roof and walls (although there are only two walls in a covered bridge). They have common requirements, such as the need for bracing to resist wind loads and the need for timber trusses or arches to span long distances. When the span of a structural component exceeds the capacity of a conventional *bending* member (say 30 ft [9 m] for timber members), then one must use another type of structural component. In timber, a *truss* is the next logical step beyond the beam. Timber trusses are an important feature in many larger buildings and have been an integral part of building construction for a long time. In the United States, timber trusses were quickly adapted for use in bridges, following the evolution of the truss concept. Trusses support almost all remaining covered bridges, although a few are supported by some sort of arch—usually "tied."

Clearly, without trusses, covered bridges would not exist. Therefore, the focus of this chapter is on the various facets of timber bridge trusses and the bracing required to make them function properly. While the

15.1   Classic example of a New England covered bridge. Upper Falls Bridge, Weathersfield, VT. (Phillip Pierce, PE.)

15.2 Interior view of Codding Hollow Covered Bridge, Waterville, VT. Queen Post configuration trusses. (Phillip Pierce, PE.)

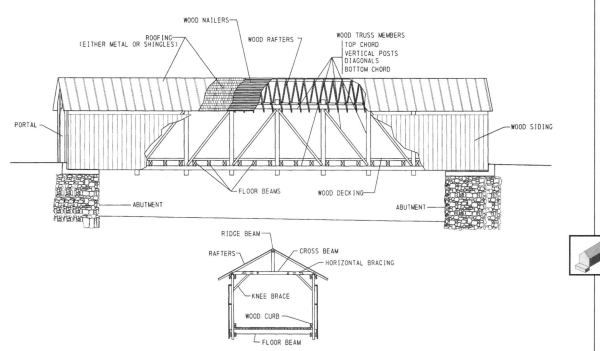

15.3 Typical bridge views. Multiple King Post configuration shown. *For a close-up of typical framing and bracing, see Fig. 1.4.* (Phillip Pierce, PE.)

remarks are aimed at covered bridges, they are presented in comparison to buildings. The intent is to provide further insight into the general behavior of timber buildings, drawing on the specific performance of timber bridges.

There are many types and configurations of timber trusses used in the support of covered bridges. Figure 15.4 depicts the more common types. For clarification, the use of "type" in this context denotes the configuration of the truss. It is not intended to differentiate highway from railway or pedestrian bridges, since the vast majority of covered bridges are highway bridges and will be so considered herein. Neither does type denote the number of spans in the bridge, an issue that will be more clearly discussed later.

As explained in Chapter 8, triangular shapes are the most stable geometric forms and are thus the building blocks for truss construction. Chapters 6 and 7 discussed the difficulties of connecting timber components. Timber bridge trusses must resist combinations of axial, shear, and flexural forces induced by the imposed loads. Each type of truss takes advantage of one or more features to efficiently support the loading over a given span. A thorough discussion of the nuances of the various truss types could fill another volume. Therefore, this chapter focuses on some general aspects of trusses, without delving into the unique features of each type.

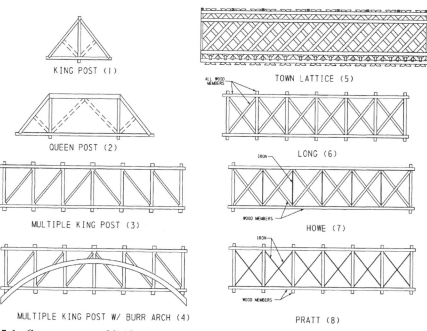

15.4   Common covered bridge truss configurations. (Phillip Pierce, PE.)

Local artisans built most of the original bridges, drawing on insights gained from examples built by others in an adjacent geographic area. Therefore, the bridges display strong "local flavor." A notable example of local flavor is the flat roof typical of the bridges of Madison County, Iowa (of book and film notoriety) compared with the more common gabled roof. Such evolution makes it hard to generalize about covered bridges.

The examples cited herein are based on firsthand experience with 75 bridges in Vermont. Involvement with other covered bridges in New England and middle Atlantic locations also influences my remarks. Vermont's diverse collection of bridges includes all of the major types found elsewhere in the United States. A review of any of the numerous "coffeetable" books on covered bridges reveals that while the portals (end views) or roof lines may be different, the structures remain basically the same throughout the country.

Timber trusses in larger and older buildings include many salient features originally identified with covered bridges, allowing several direct comparisons to be made between these types of buildings. Traditional and modern timber support systems, including trusses, are a primary focus of this book; therefore, relevant comparisons and contrasts will be made between these types of structures.

Although timber trusses can be built in multiple spans, most of the remaining covered bridges are single-span structures. Longer bridges were often located at crossings ultimately demanding greater capacity (in number of vehicular lanes or magnitude of vehicular loads) than this style of structure could provide and have long since been replaced with more modern structures. Consequently, the remaining single-span structures are usually located in rural settings over small streams, and have nearby detour structures, so that their limited capacity can be accepted. Similarly, the majority of timber trusses in buildings are simple spans. The idiosyncrasies associated with multiple spans are relevant to both structure types, but are beyond the scope of this book.

The remaining sections of this chapter offer:

- A history of the development of covered bridges in the United States
- Comparisons between bridges and buildings
- A review of important design issues
- Evaluation of existing bridges
- Repair and strengthening of existing bridges
- Actions for preservation
- An examination of issues related to the construction of new timber bridges

A closing introductory remark is in order: Since there are fewer than a thousand covered bridges remaining in the United States, few practicing engineers come in contact with them. Fewer still gain broad experience with their idiosyncrasies. While there are many sources of historical information, including *American Wooden Bridges,* published by the American Society of Civil Engineers (which includes J. P. Snow and Robert Fletcher's classic, "History of the Development of Wooden Bridges"), there have been few articles and still fewer books with technical information. In the following pages, I have attempted to convey engineering information about covered bridges while helping the reader develop an appreciation of the relationships between covered bridges and buildings.

A Brief History of Covered Bridges

Before discussing bridges in relation to truss development, let us compare them to other early timber structures, such as mills and churches. As new communities were being built in the United States, some of the first structures to be erected were mills for the production of required supplies (boards/timbers, grains, etc.). After conventional frame houses and storage structures were built, churches soon followed. These types of structures were generally built of post and beam construction. Surviving examples of mills and churches date from the 1700s. Bridges usually followed the construction of these initial important structures, after the local population became tired of dealing with fords or ferries.

The following discussion focuses on the relationship between the development of timber trusses and the evolution of covered bridges. Typical for such historical topics, there are many conflicting dates and claims. The following information is based on the references previously cited.

Andrea Palladio is credited as the first to publish information about timber built into the form of a truss in "A Treatise on Architecture," published in 1570. Distribution of the information was slow. It was nearly 200 years before large timber bridges were built in the United States using his principles.

Credit for the first timber truss bridge in the United States is not clear. Truss configurations were initially added to structures built on cantilever principles and were combined with arches. John Bliss, noted as "one of the most curious mechanics of the age," completed a crossing of the Shetuck River (124 ft [37.8 m] long and 28 ft [8.5 m] above the river) at Norwich Landing, Connecticut, in 1764. Others cite Enoch Hale for building the first timber truss highway bridge in the United States across the Connecticut River at Bellows Falls, Vermont, in 1785.

The first bridges supported by timber trusses were *not* covered. The exposure of the main support members to the detrimental effects of wetting and drying led to failures from serious deterioration after only a few years. The first truss covering is credited to Timothy Palmer for the construction of the "Permanent Bridge" over the Schuykill River at Philadelphia in 1805. The next few years saw the construction of several covered bridges at various locations in New England.

Apparently associated with the covering, Mr. Palmer secured a patent "for his improvement and construction of bridges" in 1796 (although there is seemingly no record of construction of a covered bridge until 1805). Theodore Burr obtained the first patent for a truss configuration in 1817 for his combination of truss and arch in the commonly recognized configuration known as a "Burr Arch." (Burr's earlier patent dated 1806 is missing.) Ithiel Town obtained a patent in 1820 for a lattice truss that arguably represents the most important development of covered bridges in the United States (more later). Lewis Wernwag may have been the first to patent covered bridge wooden trusses with iron *components*—cast-iron seats, bolts, and adjustable horizontal bracing (1812 patent missing; second patent, 1829). However, William Howe was the first to incorporate iron *members* into a timber truss (tensile verticals) and received a patent for this system in 1840.

While there is no reliable way to know the total number of covered bridges built in the United States, one finds estimates in excess of 10,000 by 1885. Once they became readily available, iron components were quickly adopted for use since they could be handled easily, could support heavier loads than timber, and were not as susceptible to environmental decay. By the 1920s, with the rise of iron and steel, timber trusses were losing favor with bridge owners and builders.

At the time of the 1987 edition of the *World Guide* (published by the National Society for the Preservation of Covered Bridges), 871 covered bridges remained in the United States, with 1517 in existence worldwide. The more common truss configurations (Fig. 15.4) of the remaining bridges include King Post, multiple King Post, Queen Post, Burr combination of multiple King Post with supplemental arch, Howe, Long, Pratt, and Town Lattice. Less common configurations included the unusual Haupt and Paddleford trusses. Many mongrel combinations of various configurations survived only because their weak initial constructions were strengthened over the years.

Figure 15.4 shows that King Post trusses represent the basic configuration used for the shortest bridges. Longer spans are supported by Queen Post or multiple King Post trusses. The Burr combination was intended to extend the span via the sharing of loads between truss and arch. However, in practice, the combination has many practical difficulties in achieving its goal (e.g., yielding arch supports, weak connections between truss and arch, and insufficient arch strength). Each type required construction skills possessed only by true timber artisans.

The Town Lattice type is unique in detail and design, although it still relies on triangular component assemblages. Town developed the lattice structure, in part, to eliminate complicated timber joinery. His truss system involved connections of planks with round timber pegs ("trunnels") so that relatively unskilled workers could construct a bridge quickly and efficiently.

A Comparison of Timber Bridges and Buildings

Geometrics

There are two primary differences between typical covered bridge trusses and typical building trusses. First, bridges are built with parallel-chord trusses, while building trusses often have a sloped top chord (e.g., Fink truss or Bowstring truss).

The second difference has to do with the number of trusses. Covered bridges are usually built with only enough trusses to support a single-lane travelway, one on each side. (However, there are a few "double barrel" bridges that use a central third truss that increases the capacity of the bridge enough to accommodate two separate travel lanes.) Buildings usually contain multiple trusses, potentially adding redundancy to the structure.

Another issue involving geometrics is the depth of the bridge truss as it relates to the vertical clearance for traffic. Often the clearance was minimal, reducing the amount of material needed to build the bridge. But, reduced height increases the material in the top and bottom chords, the members that do most of the structural work. Most surviving bridges are of a similar height; however, there are occasional examples of unusually shallow construction that are extremely overstressed and hard to rehabilitate authentically.

Related to geometrics of covered bridges, but unrelated to the comparison to buildings, most covered bridges are of similar width, regardless of length. Some are slightly wider than others; however, the width has been based more on convenience to vehicles than on design ramifications. Although true that the narrower the bridge, the more it tends to be weak against lateral wind loading, I am not familiar with any bridges with legitimate lateral bracing systems that have failed due to wind loading.

Materials

Original examples of covered bridges and older buildings were usually built of old-growth timber from nearby forests. In New England, many trusses were built of eastern white pine. Later, eastern spruce and northern pine species came into use. By the mid-to-late 1800s, southern yellow pine and, then, Douglas fir timbers were cut and shipped to the northern states for use in timber bridges.

More modern trusses are frequently built of Douglas fir or southern yellow pine. While the modern material may be as strong, there is anecdotal evidence that the old growth material was even stronger. The old growth forests represented an environment in which there was extreme competition for sunlight. The widely spaced trees grew taller and more slowly. Tall canopies meant long logs with few knots. The resulting wood had tightly spaced annual rings and was available in large diameter, enabling avoidance of the pithy heartwood. The resulting timber, as a material, was superior to any sawn lumber available today. (See Chap. 2.)

Original trusses which survive were nearly all built with timber from old growth forests. Weaker members failed long ago and were replaced. Failures are caused by many factors, but capacity of the material is the most important issue.

Investigation of old growth material has been limited. The Vermont Agency of Transportation recently sponsored such an evaluation in conjunction with the rehabilitation of two bridges. Although the investigation did not find substantial differences from current *National Design Specification for Wood Construction* (NDS) allowables, the statistical sample was small. However, the information provided interesting insights and is, therefore, considered an important development in this area.

This discussion is based on comparisons between old growth and contemporary sawn timbers. Glue laminated products (glulams) are often much stronger. For many years, glulam members have been used in covered bridge rehabilitation projects when sawn material is not sufficiently strong. Where substitution is required, many preservationists prefer glulams over metal components.

Another topic related to materials is the use of fresh-cut (green) material in most covered bridge construction. Some components were seasoned before installation, but most were not. An obvious disadvantage of using green wood was the inability to predict the locations of future shrinkage cracks and to react accordingly. Also, the slope of grain is difficult to detect in many green species. Therefore, you find occasional examples of inferior members with large shrinkage cracks and/or inclined grain. This situation should cause you to be especially careful in reviewing existing timber structures—whether bridges or buildings.

The use of green material also required details that enabled the structure to accommodate shrinkage. Mortise connections with "folding wedge" details in the top lateral system represent one method of retightening a system loosened by shrinkage. This system of opposing wedges penetrated a mortised connection and allowed the ends of the laterals to be adjusted, thereby maintaining a tight system.

Connections

The oldest covered bridges were constructed exclusively of timber. Most of their truss configurations involved timber joinery, some of which was very sophisticated. See Chapter 7 for a thorough discussion of traditional timber joinery.

Howe trusses employed iron components. Various tension splices were developed with iron parts. One particular joint detail developed in the late 1800s used a pair of metal rods, parallel to the sides of the timber component, with metal connecting bars through a mortise. The timber strength was developed via horizontal shear in the shear planes above and below the connecting bar. As previously mentioned, Town Lattice trusses used round timber dowels to fasten the planks together. By contrast, most of the larger building trusses that survive contain metal connectors, such as bolts, split-rings, shear plates, or steel splice plates.

Loading

There are important differences between covered bridges and buildings with respect to types of loading.

Vertical Loading

Dead load (self weight).   An important difference between the two types of truss-supported structures is that the bridge must contain a floor heavy enough to support very high concentrated loads. Also, the twin bridge trusses must support the entire weight of the floor, roof, and siding, whereas these loads are carried, in buildings, by multiple trusses. Therefore, self weight is a larger proportion of the loading for a covered bridge than for a building and often represents a major portion of the total load.

Live load (vehicular, pedestrian).   Although building trusses must often support forces associated with transient loads, the live loading of bridges is much more substantial and variable. In large part, vehicular loading of bridges is only limited by what kind of vehicle can fit through the opening. (One of the popular myths for explaining the covering was to neatly shape a load of hay traveling through the bridge!) Many vehicles with weight exceeding the posted limit of the bridge routinely use it anyway. Therefore, bridge trusses must provide a substantial reserve capacity for live loading, more substantial than trusses in buildings.

An understanding of the historical growth of vehicle weights is important. The earliest national bridge design specifications were published in 1931. Obviously, most of the surviving covered bridges were built long before there were such established codes. Accordingly, there were no "standard" weight vehicles for design purposes. A value of 15 tons (133 kN) has often been used as a reasonable weight for gasoline powered vehicles. Weights of horse-drawn wagons and skids are much more variable. Large "sleds" for winter time log hauling could easily be heavier than 15 tons. Recognize, however, that many covered bridges were built by trial and error without formal calculations of capacity. Modern structures, by comparison, are routinely designed for single vehicles weighing 25 tons (222 kN) and combination vehicles weighing 45 tons (400 kN).

Another ramification of vehicular loading of covered bridges is the potential for reversal of loading in the web members of the trusses near the center of the span. That is, a diagonal subject to axial tension when the vehicle is at one end of the bridge, can be subject to axial compression when the vehicle reaches the other. This quirk of some bridges requires the members to be designed for both situations. I am unaware of examples of structural failure related to this action. However, there are many examples where this behavior has caused loosening of the truss connections. Some members have even slipped off their moorings. The diagonals in the center of a bridge are usually only fully loaded under the weight of moving vehicles. Some of the original connections in bridges with multiple King Post trusses were bearing type, which relied solely on friction. Unless built with substantial initial camber, these connections tend to loosen when the bridge is empty. The lesson here is to provide means for holding the connection together regardless of loading, and to review regularly the structural behavior of the bridge.

There are a few remaining covered bridges that accommodate railway loading. Figure 15.5 shows such a bridge still in use in Vermont. Railway loadings are much heavier than those of automobiles. To accommodate them, bridge engineers either strengthened the single trusses along the sides of the bridge or doubled them up. However, since railway loading is not the norm, and since its nuances do not add appreciably to this chapter, the remaining discussion focuses on conventional automobile and truck loading.

15.5 Covered railway bridge. Known as the Fisher or Wolcott Bridge and framed with an unusual combination of Town Lattice and Pratt trusses, this is the last covered railway bridge still in use in Vermont, and one of only a handful still in use in the United States. Constructed in 1908, it crosses the Lamoille River near Hardwick, VT.

Snow load.   Although snow loading is not relevant to warmer climates, at northern sites it will affect both bridges and buildings. Often, the weight of snow can represent a significant load, sometimes exceeding that of vehicular traffic, and rivaling the dead load of the structure itself. However, an important difference involves the tendency of an active bridge to shed its load sooner than an unheated building. (Certainly a heated building without roof insulation will shed its snow load more quickly than an unheated structure.) The vibrations due to the passage of vehicular traffic often causes the snow to slide off the roof faster than it would without traffic. Interestingly, there are many examples of covered bridges that have experienced distress and failure while closed to vehicular traffic.

Horizontal Loading

Wind loading.   Wind forces can be significant; to resist them, structures require adequate bracing. (See Chapter 9 for an expanded discussion of lateral bracing.) Both bridges and buildings require bracing in the top. However, bridges must also resist wind on the bottom portion of the structure and must adequately transfer the forces along the bottom chord to the foundations. Many existing bridges contain a lateral bracing system beneath the flooring to resist wind loading. Bracing in the plane of the floor members is difficult to detail and install due to the close proximity of many structural components to one another. Occasionally, bracing is installed beneath the floor system; but this reduces clearance under the bridge. As an alternative, the floor can be used as a diaphragm to transfer wind forces to the foundations. To carry vehicular loads, the deck needs to be very substantial; so substantial, in fact, that the deck, by itself, can accommodate the lateral load imposed on it by the bottom chord of the trusses. Obviously, if the stability of the trusses is dependent on the diaphragm action of the floor, then the bridge will require temporary horizontal bracing during redecking operations.

Ice and flooding.   Except in coastal areas, buildings are not expected to accommodate flooding, but bridges are routinely subject to the relentless and considerable forces of flooding and ice. Bridges that remain are usually situated where water-related forces are no longer a threat, since previous floods carried bridges away that were too low or improperly founded. Bridges that contained other weaknesses, but which were destroyed in floods, are no longer around to teach us any lessons.

Seismic loading.   Concern for seismic loading is a current "hot topic"; however, I am unaware of covered bridges that have suffered serious distress due to earthquakes. In an earthquake, I would expect little damage to a covered bridge, because of the inherent flexibility and high impact absorption capacity of timber structures. The other loadings cited above are much more relevant to these types of structures.

Anchorage against horizontal loading. Bridges must be anchored against the action of the horizontal loadings previously described. These inordinately heavy structures are primarily "self-anchoring." However, they can be tied down in a variety of ways, primarily using steel cables or rods. The examples I have observed have usually been ineffective. While cables have often been used to stabilize structures distorted from unrepaired deterioration, they are less effective at anchoring against horizontal loading.

Floors

Bridges must have a floor to support vehicles as they cross an abyss. Buildings may or may not have elevated floors (the structure of a single-story building is usually confined to the roof and its supports). There are any number of floor support systems for buildings but relatively few for bridges. We address them next.

In general, bridge floors are comprised of timber floor beams (transverse members supported by the trusses) topped by longitudinal decking. The floor beams may be sawn timbers up to 8×16 in (200×400 mm); contemporary replacements may be glulams. The decking may consist of sawn planks up to 4 in (100 mm) thick or "laminated" from smaller-sized members with nails or glue, yielding a total thickness up to 8 in (200 mm).

In some configurations, longitudinal timber stringers may be positioned on top of the floor beams. Sometimes, solid sawn decking may be in two, or even three, layers crisscrossed for additional capacity. The presence of the deck (acting as a solid horizontal diaphragm) can be helpful for keeping the bridge square and for distributing lateral wind loading to the abutments.

Unique to covered bridges is the use of "running planks." These longitudinal planks rest on top of the regular deck to act as sacrificial members that can be replaced when worn out by vehicular traffic. They are usually found in two rows along the wheel tracks. By slowing traffic and guiding it along the middle of the bridge, running planks help to reduce loading variations between the trusses and the damage resulting from those variations. Centering the load within the width of the bridge enables it to be shared equally by the trusses.

Shared Weaknesses and Common Problems

All structures contain elements or details that are critical to their capacity. In new structures, components must be properly sized, and connections properly detailed. Existing structures are prone to deterioration from a number of different causes. We turn now to weaknesses and problems common to both buildings and bridges.

Compression Buckling

Both bridges and buildings require adequate bracing of compression members. Individual members may fail under heavy loads by "buckling" within the frame of the structure. In addition, the overall structure may become distorted due to improper or inadequate bracing. Out-of-plane bracing denotes that which is required to prevent individual members from moving sideways. A second type of bracing, in-plane or cross bracing, is required to keep the entire structure "square." Buildings and bridges differ in how they are braced.

Buildings, particularly those with flat roofs, sometimes use their roofs as diaphragms to provide out-of-plane buckling resistance. Additional out-of-plane bracing may be provided for erection but left in place afterward. The walls of the building may help keep the structure square against sideways movement, or additional bracing may be included internally. Bridges, by contrast, usually have sloped roofs that cannot contribute significant out-of-plane buckling resistance to the top chord. If you think of a covered bridge in plan as a long, shallow beam, it is obvious that a permanent and substantial lateral bracing system is required. Inherent structural redundancies enable many smaller bridges to survive without such a system, but it is better to provide an engineered system than to rely on such uncertain aspects.

The bigger issue, however, is the lack of ends on the bridge to keep the structure square in cross-section. A system of "knee braces," spaced at regular intervals along the length of the structure, is required to keep the bridge upright. Because the braces must be short so as not to interfere with vehicular clearance, this system is often relatively weak, allowing the structure to lean sideways. To hold them in proper position, many bridges have been retrofitted with external "guy wires" attached at the corners. Their presence therefore suggests a structure in distress that has not been properly rehabilitated.

Tension Failures

The proper handling of tension forces is the most important issue in timber truss structures. This is as true in bridges as in buildings. A careful inspection of the tension joints of a structure reveals a lot about how the structure is handling its loads.

Constructed, as they were, before the advent of metal connectors, early bridge trusses employed a variety of interesting ways to transfer tensile forces internally, including scarf joints, "bolt-of-lightning" joints, and shoulder joints. Each required skill to construct and was uniquely tailored to the stresses experienced in a particular type of joint. Nevertheless, it is difficult to estimate accurately the capacity of such joints. Building trusses, by contrast, more commonly use metal connectors, such as split rings and shear plates, to achieve tension splices. Such connections have been tested sufficiently to enable a relatively accurate estimate of their structural capacity to be made.

In both bridges and buildings, it is desirable for there to be multiple components in the tension member, for redundancy and concentricity. Reliance on a single tension member is dangerous. Double chords are common in truss construction, straddling the web members.

Component Deterioration

Bridges suffer from deterioration of components, as do buildings. Two of the more prevalent types are moisture-related deterioration from mold/fungi ("rot") and degradation from insect attack. (See Chap. 3 for a detailed discussion of these subjects.)

Bridges are subject to rot at any place along the truss where the roof has leaked. (See Fig. 15.6.) Occasionally there may be deterioration near open windows in the sides of the bridge. Bridges are also prone to deterioration near their ends from the unrestricted entrance of roadway drainage into the structure. Another source of rot is moisture retained in the dirt and debris tracked into the bridge by vehicles. Deterioration can also result from improper separation of primary timber components from foundation materials such as stone and concrete. (See Fig. 15.7.)

In addition to termites and carpenter ants, powder post beetles represent a significant source of deterioration. Beetles favor the hardwood species (oak, birch, maple, etc.) typically found in the trunnels of Town Lattice trusses. Additional deterioration results when woodpeckers attack the beetles. The resulting loss of section can be extreme.

15.6 Closeup view of the upper chords of a Town Lattice bridge that have obviously suffered from the ravages of a long-term leaking roof. The material was so punky as to allow removal by hand! It was a testament to the reserve capacities of such a bridge that it had not collapsed. (Phillip Pierce, PE.)

15.7 View of the underside of a bridge following recent authentic rehabilitation. Note the lighter color of the newer wood. Bottom chords have been selectively replaced using wooden peg connections. New bottom lateral bracing has been carefully mortised into the existing floor beams. The ends of the trusses have been supported on sacrificial timbers above the concrete. (Phillip Pierce, PE.)

Design and Analysis Issues

Material Properties

As indicated above, original timbers must be analyzed in light of their superiority to the contemporary materials on which current codes are based. Current specifications, used without thoughtful modification, will yield unnecessarily conservative results. On the other hand, if a quick response is required, and the cost penalties are not significant, current codes may be followed.

Modern specifications are based on statistical evaluations of the strengths of many test specimens. In timber design, the controlling value is the upper limit of the group made up of the weakest 5 percent of the samples. Under the so-called "95 percent Exclusion Rule," 95 percent of the members are stronger than the controlling value. While this approach may seem overly conservative with respect to stress-graded lumber, the industry has adopted it to address the wide variations in the quality of that lumber.

Similarly, one could argue that the use of the 95 percent Exclusion Rule is overly conservative with respect to the members of a covered bridge that has successfully served for more than a hundred years. Little is available to guide the selection of an alternate value. If desired, such an alternate value should be selected only by an engineer attuned to the subtleties of covered bridge design and behavior.

NDS Tension Penalty for Large Members

The 1971 edition of the NDS initiated the use of an adjustment factor in the allowable stresses for members exceeding 6 in (150 mm) in nominal width. The reduction was based on new results obtained by direct tensile tests. Previous values had been based on flexural results only, since the means were not available to test large members in tension without breakage at the grips. The new testing techniques indicated strengths significantly *lower*—up to 45 percent lower for members 10 in (250 mm) or wider—than previously identified.

At the time of the code change, I was involved in the examination of many World War II vintage military warehouse structures supported by timber trusses. Interestingly, the many failures observed in bottom chord members emphasized the need for such a change. While none of the buildings had yet collapsed, collapse was considered imminent and was averted only by strengthening of their trusses.

Use of the large member reduction factor during engineering evaluation of old timber trusses should therefore be considered, except in cases where the results of actual tests are available. Exercise caution in the consideration of that value.

Combined Stresses

One of the reasons that iron components were adopted for trusses is that they can accommodate axially loaded connections more easily than timber. It is difficult to connect the components of a large-timber truss for axial load only. Due to inherent construction inaccuracies and eccentric connections and loads, most timber truss components are subjected to combinations of axial, shear, and flexural forces. The degree of complexity in analyzing such components has necessitated the use of simplifying assumptions. Today's thoughtful practitioner must balance the benefits of analytic accuracy with the costs of achieving it.

I am currently involved in the analysis of a Town Lattice truss simulated by a sophisticated computer model containing over 14,000 elements. Although this analysis is one of the most intensive such efforts ever attempted, it still relies on several simplifications regarding the trunnel connections. While the behavior of a Town Lattice truss can generally be approximated by a "plate girder analogy" with respect to primary forces at midspan, this more advanced simulation enabled an evaluation of each of the individual components.

Obviously, the initial analyses performed by early truss designers were less rigorous. Nevertheless, the success of the remaining struc-

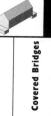

tures speaks for itself. These covered bridges have survived longer than any other type of bridge, except those comprised of masonry arches and slabs.

Design Specifications

No nationally recognized specifications have been published for the evaluation of covered bridges. One might consider following the guidelines of the *Standard Specifications for Highway Bridges* as published by the American Association of State Highway and Transportation Officials (AASHTO), but these are more appropriate for the engineering of new bridges than for the analysis of existing ones, especially ones as idiosyncratic as covered bridges. Examples of critical design issues not addressed by the standard specifications include:

Unit weight: One of the more important parameters for the examination of such bridges is the unit weight of the timber. AASHTO suggests the use of 50 pcf (801 kg/m^3) for timber materials, based on the expected use of pressure-treated hardwoods. This value is very conservative for old growth timber, which may weigh as little as 30 pcf (480 kg/m^3).

Snow load: Snow loading is not included in the AASHTO specifications since it is assumed that snow is removed by plows. Therefore, there is no concurrent full snow and vehicular loading, as could occur at a covered bridge with snow on its roof.

Load combinations: The above is but one example of the larger issue of combinations of dead, live, and snow loads. Selection of an appropriate probability of occurrence must accompany consideration of each load combination.

Load duration: The duration of load factor is critical to the establishment of timber stresses. For each load combination, there is an appropriate duration factor. Rural bridges having little traffic may have less critical requirements than bridges on more heavily traveled roadways.

While the *National Design Specification* has been available for a long time and is highly respected for its thoroughness and accuracy, it remains only a guideline. Specifications, like the AASHTO *Bridge Specifications* or the NDS, must be applied with engineering *judgment.* If the experiences of the engineer suggests the appropriateness of other values, they should be used. However, this position should only be taken by experienced engineers who have a detailed knowledge of the subject and an appreciation of the consequences of such decisions.

Evaluation of Existing Structures

Let me begin this discourse with counsel from a wise craftsman. To paraphrase what he volunteered: *Engineers recommend analytical evaluation regardless of the circumstances.* Often, however, a careful physical examination by an experienced person is sufficient to identify the source of structural distress. Many individuals have the technical ability to analyze a simple structure, but few can appreciate the complexities of a timber truss. Experience can often guide one to the critical elements of such a structure.

The evaluation of existing bridges involves careful consideration of several key issues. The following discussion addresses the most important ones. Although it focuses on bridges, it is just as relevant to buildings.

Field Examination

Before attempting an analytical evaluation of any structure, it is prudent to conduct a careful examination of it in its current condition. The critical issues for covered bridges include the horizontal and vertical alignment of the trusses, the degree to which the structure is out of square along any of its axes, and the condition of the tension splices.

Whereas the trusses in most of the remaining covered bridges are flat and have parallel chords, some have an appreciable vertical curvature. The "camber" of the truss is a measure of its "out-of-flatness." A truss with a slight upward camber is probably still able to support its loading without significant distress. However, this is not always the case. Conversely, a sag of the truss often indicates structural distress.

Bridges often develop angular distortions as a result of deterioration of various components, most commonly at the bearing areas. Such angular distortions make the structure resemble a parallelogram in plan, elevation, or cross-section. In some instances, steel cables are installed as a stop-gap measure to keep the bridge from collapsing.

The alignment of the truss along its axis is a measure of its load carrying capacity. Bows (horizontal displacement or curvature of the members) can result from buckling of overstressed compression members in the trusses. Misalignments of the truss chords (bows or kinks) can also indicate that the out-of-frame bracing of the compression members is inadequate. Since tension members often are the weak link in the structure, a careful examination of the splices is essential. Large gaps justify a more thorough investigation; tight splices inspire greater confidence.

The remainder of the investigation should follow standard practice, depending on the intent of the work. Many references offer useful guidance about the evaluation of timber structures in general, and the inspection of bridges in particular. (See Michael Ritter, *Timber Bridges*,

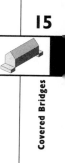

15

Design, Construction, Inspection, and Maintenance, USDA Forest Service, 1992.)

Species Identification

An evaluation of the bridge should include identification of the species of each of its critical members. Do not assume that one species was used throughout; the structure may have been built with different species for different member types (chords, diagonals, bracing, etc.). This may have been an attempt to provide greater strength where it was needed, or the result of circumstance (availability). Mixed species could also be the result of structural repairs or modifications involving the substitution of a new component for an existing one of another species.

Fortunately, there are trained specialists who can identify species from small samples—as small as a short pencil—at nominal cost (as little as $20 per sample). In the absence of such identification, one should be extremely careful in selecting allowable stresses, choosing appropriately conservative values.

Strength Ratio Concepts

Current timber specifications are based on extensive tests of numerous specimens. Various allowable stresses have been developed to account for the behavior of timber members having different growth characteristics, including width of annual rings, slope of grain, and size and distribution of knots. (See Chap. 2 and ASTM D245 and D2555). A properly trained individual can examine a structure and identify and record information relating strength parameters to standards. The resulting "strength ratios" can be used to generate allowable stresses. Obviously, the examination is limited to the visible portion of the structure; however, that usually includes an amount sufficient for a reasonable generalization. This type of special inspection of a complex covered bridge truss might cost a few thousand dollars or more.

Analytical Evaluations

Obviously, there are bridges where an analytical evaluation is required. Such an evaluation can involve manual or computerized techniques. For complex structures, a more sophisticated computerized analysis may be required, e.g., a finite-element model. As in all such evaluations, a complete understanding is necessary of the limitations of the software. Proper simulation of the boundary conditions is important. Timber trusses rarely behave strictly as pin-connected structures; similarly, they rarely behave strictly as fixed-joint frames. Distinctions between elastic and inelastic behavior must be incorporated into the model. Field verification of predicted behavior is a necessary follow-up to sophisticated computer simulations. An ideal evaluation would start with a carefully constructed model, followed by comparison of behavioral predictions with output from a reliable alternative. Next, the predictions would be compared with field measurements of pertinent displacements. If field behavior matches predictions, one can have confidence in the simulation. Discrepancies between predictions and behavior should result in careful reconsideration both of the model and of the accuracy of the field procedures. Further refinement of the model may be required to better match field performance.

Load Testing

Techniques for testing of structures are improving rapidly. However, the nonisotropic nature of timber makes it more difficult to test than more uniform materials such as steel and concrete. In metal structures, strains can be measured and readily equated to stresses; that opportunity does not exist to the same extent in timber structures. Nonetheless, it is possible to perform meaningful load testing of timber structures.

This topic could be the subject of an entire volume, but for purposes of this discussion, the critical relationship is that between loads and displacements. A recent example involved the installation of dial gauges at critical bridge locations. The measured displacements were then compared with computer predictions. (See Fig. 15.8.)

Nondestructive Testing

Much effort has been expended to develop reliable means of nondestructive mechanical evaluation of timber components. Current techniques include sound wave propagation and electrical resistance. I am not aware of any commonly used test method having strong general support. Many methods may be acceptable for certain situations, yet have serious weaknesses in others. While there are many *general* references regarding this matter, you should seek expert project-specific advice at the time testing is needed, to stay abreast of new developments.

Component Destructive Testing

In extreme situations, it may be possible or necessary to obtain information through destructive tests of structural components. A recent project included the testing of spruce bridge members approximately 3×12 in (75×300 mm) by 20 ft (6.1 m) long. The members were initially loaded to failure in tension (ASTM D198). The broken pieces were then used to prepare small-scale test specimens for compression, shear, and flexure behavior (ASTM D143). Be cautious in interpreting the results of such tests when the number of samples is small. Consider, too, the moisture content of the specimens at time of testing. Tests of this type are expensive and require special equipment. For that reason, such testing is unusual.

15

Covered Bridges

15.8 View of the underside of a bridge undergoing load testing. This is the Brown Bridge, crossing the Cold River in the town of Shrewsbury, Rutland County, VT. The two dial gauges were installed to obtain information relating the distortions and displacements of bridge components when under the influence of a known loading. Many other gauges were installed at other locations on the bridge during the testing program. (Phillip Pierce, PE.)

Hidden Defects and/or Deterioration

When evaluating existing structures, always keep timber's variability in mind. Although, in general, timber structures are quite open and easy to inspect, portions of them may not be. Concealed members or connections can contain nontypical defects and/or deterioration. Therefore, be very careful in generalizing about the condition of a structure from observations solely of its most visible aspects.

Town Lattice trusses are especially difficult to evaluate due to the large member areas in face-to-face contact. Moisture from a leaking roof collecting on concealed mating surfaces can result in advanced forms of wood deterioration without evidence at visible surfaces. In such structures, more deterioration is often found during rehabilitation than anticipated at the time of the initial evaluation.

Repairing and Strengthening Existing Structures

The repair and/or strengthening of an existing timber structure can be very difficult, often more challenging than design of a new one. The reasons include:

- Constraints associated with the existing structure and its component geometry
- Restrictions related to historic preservation
- Difficulties associated with matching the strength and/or flexibility of the existing structure

When repairing historic masonry, one of the problems is finding replacement mortar as "soft" as the original. Harder mortar will do more harm than good, by overstressing the stones when it expands. An analogous problem exists in the repair of timber trusses: Replacement of existing members with new ones of different strength can change the structure's behavior under load, by redistributing the stresses. The structural consequences can be serious.

Inadequate Repairs

Although it is possible to design, detail, and install a proper repair for a given weakness or problem, more often than we care to admit, repairs are not well done and do more harm than good.

Good repairs require precision each step of the way in accordance with the following guidelines:

Design: The repair must have adequate capacity for the given loading and must maintain the intended function of the component while matching the desired fixity of the connections.

Detailing: The repair must be detailed to enable implementation, must match the intent of the original design, and must be understood by the person performing the repair.

Installation: The workmanship must enable the member or connection to perform as intended.

Books have been devoted to proper—and improper—repairs. While a more detailed discussion of this subject is beyond the scope of this chapter, suffice it to say that repairs deserve careful consideration by experienced individuals.

Common Types of Strengthening

The strengthening of weak covered bridges requires many actions similar to those used in buildings. However, there are some actions unique to covered bridges.

One common form of strengthening involves the reinforcement of existing timber components with new ones. The new material must be connected in such a way as to enable the new and existing components to share the loading. An example would be the installation of extra chord members to an existing truss. This would best be done with the structure supported on falsework, to enable dismantling prior to the addition of the new material. Replacement of existing connectors with larger ones would be necessary. The holes in the new material would have to be drilled one side at a time using the existing holes as a pattern.

Another form of strengthening involves the installation of supplemental metal rods, cables, or shapes to an existing member. An example would be the addition of posttensioned chord members to relieve the loading of overstressed existing members.

A form of strengthening common in covered bridges, but unusual in timber buildings, is the installation of timber arches adjacent to the truss. Theodore Burr's system, patented in the early 1800s, incorporates an arch with a truss. Since then, for structural reasons, arches have been added to virtually all types of timber trusses.

The analytical evaluation of such combination arch/truss structures is challenging. There is no consensus on the intent of Mr. Burr's combination. Some believe that the trusses carry the dead load, while the arches carry the live load. Others disagree. The many variables involved result in differences between theoretical and actual behaviors with respect to yielding at supports, local buckling, and failure of connections. Often the arch is not large enough to support a significant portion of the load. In other cases, the arch supports most of the load. Be very careful in evaluating such a hybrid structure or in designing a new one or retrofitting an existing one.

Less-Authentic Actions

Due in part to the complexities of reinforcing covered bridges to carry current loads, it is common to replace the existing floor system with an independent structure. This could be comprised of longitudinal steel beams with a concrete or timber floor deck acting independently of the covered bridge shell. While some argue that this approach retains rather than destroys the bridge, others argue that the resulting bridge has been "gutted." This action can be cause for concern: Future safety inspections of the structure (mandated by federal law) will focus on the new floor (for support of vehicles), while the existing trusses and roof structure may be ignored. Without necessary repairs, the existing shell will deteriorate rapidly.

Frequently, engineers familiar with steel trusses attempt to repair timber trusses using similar details. Classic examples in covered bridges include the use of bolted "heel" plates at the ends of Queen Post trusses, the replacement of the lower ends of lattice members with bolted lap connections, or the use of bolted connections in bottom chord replacements. Often, these "repairs" are a poor substitute for compatible repairs installed by skilled artisans. In many instances, "repairs" further weaken the structure. The repair may reduce the net section of tension members, thereby causing a reduction in capacity. The introduction of a second material such as metal can lead to condensation at material interfaces, leading to rot.

Drawing a comparison with the situation in building repair is difficult, but it highlights the need for attention to the complete structure. No repair takes place in isolation; always consider its impact on the entire structure.

Historic Considerations with Existing Structures

Because most covered bridges are eligible for inclusion in the National Register of Historic Places (in accordance with the Department of the Interior Standards), their rehabilitation must often satisfy special regulatory requirements. In essence, this means that most work must employ "authentic" details and materials, like those previously discussed in "Common Types of Strengthening." In some instances, approval has been obtained to replace the floor system with an independent system, as already discussed in "Less-Authentic Actions." Before proceeding with rehabilitation of an older structure, it is wise to wait until all necessary reviews are conducted and approvals granted. (See Fig. 15.7.)

Preservation Actions for Covered Bridges

The remaining authentic covered bridges have survived in large part because of the care that they have received. Those that did not receive occasional attention collapsed long ago or were destroyed. Preventive maintenance, if performed regularly, need not be costly.

Sound Roof and Siding

Protection of the main truss members from deterioration caused by rot is one of the most effective ways of preserving a covered bridge. Proper maintenance of roofing and siding will lengthen substantially the useful life of a covered bridge.

When residing the bridge, maintain good air circulation around the truss members. If there is a space below the eaves and above the siding that enables air to circulate along the length of the bridge, preserve it. If there is no such space, provide it. Do not nail the siding directly to the truss members. Attach it with nailers to avoid direct contact with the main members. A direct contact area provides an ideal environment for development and concealment of decay.

Notwithstanding the above advice, recognize that historical preservationists often insist on repairs consistent with the bridge's original details. The area of coverage of the siding is often of paramount importance to them—if there was no gap at the top, they may prohibit its

15

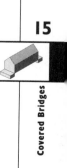

Covered Bridges

addition. However, the inclusion of nailers between the truss members and siding is often acceptable, because it improves the service life without visibly altering the look of the bridge. Focus on this issue when educating reviewers about project details.

Roadway Drainage

A second important action involves preventing roadway drainage from flowing into the bridge. An ideal situation would combine two features: The roadway would "ramp" *up* into the bridge, and excessive drainage would be collected at the curbs and piped down the side slopes to avoid erosion of the embankment at the abutments. If the approach roadway does not and cannot ramp up into the bridge, other measures should be considered. In some cases, the structure can be raised. In others, a trench drain can be installed across the roadway to collect runoff and direct it away from the bridge.

Most bridges have no special provisions for removing water blown or tracked inside. On decks with tight joints, water seeks a low spot to escape. On decks with loosely spaced planks, water falls through the floor system. While the structure would be better off without the moisture, its presence is inevitable. In general, intermittent wetting is not harmful, provided that ventilation is sufficient to allow the moisture to evaporate.

Proper Foundation Support Details

Covered bridges will deteriorate where they come into direct contact with foundation materials. Therefore, truss members and end floor beams should be supported on sacrificial timber blocking. This should be supported by concrete or stone foundations and backwalls (the vertical faces at each end of the bridge, separating the surfaces of the approach drives from the depressed bearing seats for the bridge structure), so that structural timbers do not contact the earth embankment.

Timber in contact with stone or concrete will eventually rot and decay. However, because the blocking can still last 20 to 40 years, its replacement does not represent a major investment over the life of the structure. While replacement requires raising the bridge from the foundation, it is usually not a problem for a competent contractor. The intent of such blocking is also to allow good air circulation around the main members. This should include an air space of at least 2 in (50 mm) between the end of the deck and the backwall, across the roadway.

Regular Cleaning

Traffic tracks a surprising amount of dirt into a bridge. Such accumulations retain moisture and contribute to timber deterioration. Therefore, the structure should be cleaned of such material each year. The best removal methods involve hand tools and air-blast cleaning. Water blast can also be very effective, but the introduction of additional water into the structure is not desirable.

While the deicing salts used in northern environments will be tracked in and will concentrate when dried, they do not pose the sort of serious hazard to timber bridges that they do to concrete or metal bridges.

Fire Protection

Arson continues to destroy many covered bridges. Fire retardance was not an issue when covered bridges were originally constructed. Spray-on fire retardants have recently been used in some situations. Since siding and roofing materials are more vulnerable than the thicker components of the trusses, they have occasionally been replaced with products treated with fire retardants. Some bridges have been fitted with sprinkler systems, others with alarm systems. There is no simple solution to this problem. A site-specific evaluation is necessary, and the resulting recommendations must reflect the needs of the owner and the details of the bridge.

New Construction

Covered bridges were losing favor with bridge owners and builders by the early part of the twentieth century. For many decades, no new examples were built; and many existing structures were destroyed to make way for more modern structures.

Fortunately, a few committed believers remained. Arguably, the most renowned covered bridge restoration/construction specialist in the second half of the twentieth century was Milton Graton (1908–1994). Mr. Graton worked on dozens of existing bridges and built several new examples from the 1950s until his death. His family continues the tradition and involvement. Over the last 20 years, other skilled artisans have responded to the interest expressed by covered bridge preservationists.

Occasionally, new examples of authentic construction are now built to duplicate a lost structure or provide a new one where one had not previously stood. Often, a Town Lattice configuration is adopted due to its redundancy and strength. (See Fig. 15.9.)

Federal funding legislation now mandates the consideration of life-cycle costs in the selection of bridge construction systems. When the total cost of initial construction and all anticipated maintenance and repair costs is divided by the expected life of the structure, covered bridges are very competitive, especially for carefully selected sites where they are adequate for the anticipated traffic. In those situations, new covered bridges can be cost effective.

15

Covered Bridges

15.9 Interior view of the recently built replacement of the Schofield Ford/Twining Ford Bridge. It crosses the Neshaminy Creek in Tyler State Park, Bucks County, PA. The two-span bridge is supported by Town Lattice style trusses and includes inclined web diagonal braces modeled after those of the original 1874 bridge, which burned in 1991. The bridge also includes rather unusual wooden peg connections. (Phillip Pierce, PE.)

This chapter has intentionally avoided the sizing of new bridges or their components. Covered bridges are usually single-span structures enclosing a single lane. Clearances for the passage of vehicles determine the minimum width and height of the structure. There are no standard rules of thumb for proportioning a covered bridge. The required strength is a function of the desired vehicular capacity. Based on national bridge design guidelines, a common single-vehicle capacity is 25 tons (222 kN). The member sizes are a function of the type of truss employed.

The members one might encounter in covered bridge trusses could be up to 40 ft (12 m) long with cross-sections up to 4×12 in (100×300 mm) for a Town Lattice chord. Other types of trusses might have chords even larger than that. Such large members would normally be cut from Douglas fir or southern pine to provide sufficient strength and workability.

New covered bridges have been constructed in several different states, including Vermont, New Hampshire, and North Carolina, as well as in several countries (see Fig. 15.10). For those wishing to visit such a structure, see the *World Guide to Covered Bridges,* published by the National Society for the Preservation of Covered Bridges, Inc., David Wright, Pres., P.O. Box 171, Westminster, VT.

15.10 Speed River Bridge: This 145-ft (43.5-m) Town Lattice covered bridge spans the Speed River, in Guelph, Ontario. The bridge has a clear span of 120 ft (36 m) and could carry just about any vehicle that would fit through it. Four hundred members of the Timber Framers of North America Guild built the bridge in 1992. That week-long paroxysm of effort coincided with their annual conference. The construction of this bridge was the subject of a fascinating video, entitled *Covered Bridges: 400 Timber Framers Build a Bridge,* see "References." (Benson Woodworking Co., Inc.)

Summary

This chapter has attempted to relate timber covered bridges to timber buildings. There are many similarities and differences between the two types of structures. With their long spans, heavy dynamic loads, and harsh service conditions, covered bridges represent a type of timber construction pushed to the limit, and from which, therefore, there are many lessons to be learned.

The fact that covered bridges are surviving active structures over 175 years old is astounding, especially given the service life of 60 to 75 years now standard in the bridge design industry for bridges of this size.

The information contained herein has touched only on issues relevant to this book. Many other topics have been omitted intentionally because they would not contribute to an understanding of the relationships between bridges and buildings. If I have been successful, you now have a better general appreciation of covered bridges and will look at timber buildings, in the future, from a slightly different perspective.

CONTRIBUTOR'S NOTE: I wish to recognize and extend my appreciation to the Vermont Agency of Transportation for its authorization to use the

15

Covered Bridges

photographs contained herein. Much of the material in this chapter was accumulated during a Long-Range Planning Study of covered bridges owned by individual towns in Vermont. I served as Project Manager of this study, which was sponsored by that agency and which was the largest, most extensive study of covered bridges ever undertaken.

List of Acronyms

Building design and construction are highly regulated. This book makes reference to a number of the organizations and agencies involved. To assist the reader, the acronyms associated with those groups, and with a number of related terms of art, are defined in this appendix.

AASHTO	American Association of State Highway and Transportation Officials
ACSA	Association of Collegiate Schools of Architecture
ACEC	American Consulting Engineers Council
AFPA	American Forest and Paper Association
AIA	American Institute of Architects
AITC	American Institute of Timber Construction
ANSI	American National Standards Institute
APA	American Plywood Association
ASCE	American Society of Civil Engineers
ASHRAE	American Society of Heating, Refrigerating, and Air Conditioning Engineers
ASTM	American Society for Testing and Materials
ASD	Allowable Stress Design
AWI	Architectural Woodwork Institute
AWPA	American Wood-Preservers' Association
AWS	American Wood Systems
BM	Board measure
BOCA	Building Officials and Code Administrators International, Inc.
CBC	Canadian Building Code
CWC	Canadian Wood Council
CSI	Construction Specifications Institute

DPIC	Design Professionals Insurance Company
ELFP	Equivalent Lateral Force Procedure
EMC	Equilibrium moisture content
FPL	Forest Products Laboratory (USDA, Forest Service)
FPS	Forest Products Society
IBC	International Building Code
ICBO	International Conference of Building Officials
ICC	International Code Council
LRFD	Load & Resistance Factor Design
MAP	Modal Analysis Procedure
MBF	Thousand board feet
NDS	National Design Specification
NFPA	National Fire Protection Association
NAHB	National Association of Home Builders
NIBS	National Institute of Building Science
OSB	Oriented-strand board
RFI	Request(s) for Information
SBC	Southern Building Code
TCM	Timber Construction Manual
UBC	Uniform Building Code
USDA	United States Department of Agriculture
WPIRC	Wood-Products Information and Research Center
WWPA	Western Wood Products Association

References

AIA/ACSA Council on Architectural Research. 1992. *Buildings at Risk: Seismic Design Basics for Practicing Architects.* Washington, D.C.: AIA/ACSA.

Allen, Edward. 1990. *Fundamentals of Building Construction: Materials & Methods, Second Edition.* New York: John Wiley & Sons.

———. 1980. *How Buildings Work: The Natural Order of Architecture.* New York: Oxford University Press.

Allen, Edward, and Joseph Iano. 1989. *The Architect's Studio Companion: Technical Guidelines for Preliminary Design.* New York: John Wiley & Sons.

Ambrose, James E. 1994. *Design of Building Trusses.* New York: John Wiley & Sons.

American Association of State Highway and Transportation Officials. 1996. *Standard Specifications for Highway Bridges, Sixteenth Edition.* AASHTO.

American Forest & Paper Association. 1991. *Design Values for Wood Construction.* Washington, D.C.: AFPA.

———. 1997. *Design Values for Wood Construction.* Washington, D.C.: AFPA.

———. 1991. *National Design Specification for Wood Construction.* Washington, D.C.: AFPA.

———. 1997. *National Design Specification for Wood Construction.* Washington, D.C.: AFPA.

American Institute of Architects. 1987. *General Conditions of the Contract for Construction, Fourteenth Edition.* Washington, D.C.: AIA.

American Institute of Architects and American Consulting Engineers Council. 1993. *A Project Partnering Guide for Design Professionals.* Washington, D.C.: AIA/ACEC.

American Institute of Timber Construction. 1979. *AITC 111-79: Recommended Practice for Protection of Structural Glued Laminated Timber during Transit, Storage and Erection.* Englewood, CO: AITC.

———. 1993. *AITC 112-93: Standard for Tongue-and-Groove Heavy Timber Roof Decking.* Englewood, CO: AITC.

———. 1992. *Inspection Manual AITC 200-92 for Structural Glued Laminated Timber.* Englewood, CO: AITC.

———. 1994. *Laminated Timber Design Guide.* Englewood, CO: AITC.

———. 1996. *Technical Note 7: Calculation of Fire Resistance of Glued Laminated Timbers.* Englewood, CO: AITC.

———. 1985. *Technical Note 11: Checking in Glue Laminated Timber.* Englewood, CO: AITC.

———. 1997. *Technical Note 19: Guidelines for Drilling or Notching of Structural Glued Laminated Timber Beams.* Englewood, CO: AITC.

————. 1985. *Timber Construction Manual, Third Edition*. New York: John Wiley & Sons.

American National Standards Institute, Inc. 1982. *Minimum Design Loads for Buildings and Other Structures. Standard A58.1-1982*. New York: ANSI.

American Plywood Association. 1981. *Nonresidential Roof Systems*. Tacoma, WA: APA.

American Society for Testing and Materials. 1997. *Annual Book of ASTM Standards*. West Conshohocken, PA: ASTM.

American Society of Civil Engineers. 1975. *Evaluation, Maintenance and Upgrading of Wood Structures*. New York: ASCE.

————. 1975. *Wood Structures: A Design Guide and Commentary*. New York: ASCE.

American Society of Civil Engineers History and Heritage Committee. 1976. *American Wooden Bridges*. New York: ASCE.

American Wood-Preservers' Association. 1995. *Book of Standards*. Granbury, TX: AWPA.

American Wood Systems. 1995. *Glulams: Product and Application Guide*. Tacoma, WA: AWS.

Angell, W., and W. Olson. 1988. *Moisture Sources Associated with Potential Damage in Cold Climate Housing. CD-FS-3405*. St. Paul, MN: Cold Climate Housing Information Center.

Apple Computer, Inc. 1993. *HyperCard Script Language Guide: The HyperTalk Language*. Cupertino, CA: Apple Computer, Inc.

Architectural Woodwork Institute. 1977. *Guide to Wood Species*. Arlington, VA: AWI.

Arnold, Rick, and Mike Guertin. 1995. "Raising Roof Trusses." *Fine Homebuilding* 99: 50–54.

Bachelard, Gaston. 1964. *The Poetics of Space*. Boston: Beacon Press.

Bealer, Alex W. 1972. *Old Ways of Working Wood*. Barre, MA: Barre Publishing Co.

Benson, Tedd. 1980. *Building the Timber Frame House: The Revival of a Forgotten Craft*. New York: Charles Scribner's Sons.

————. 1988. *The Timber-Frame Home: Design•Construction•Finishing*. Newtown, CT: The Taunton Press, Inc.

Benson Woodworking Co., Inc. 1994. Company brochure. Alstead Center, NH: Benson Woodworking Company, Inc.

Blandford, Percy W. 1984. *The Illustrated Handbook of Woodworking Joints*. Blue Ridge Summit, PA: TAB Books, Inc.

Bodig, Jozsef, and Benjamin A. Jayne. 1982. *Mechanics of Wood and Wood Composites*. New York: Van Nostrand Reinhold.

Boston Society of Architects/AIA. Undated brochure. "What Every Owner Needs to Know About RFIs." Monterey, CA: Design Professionals Insurance Company.

Botsai, Elmer E., et al. 1975. *Architects and Earthquakes*. Washington, D.C.: AIA Research Corporation.

Brannigan, Francis L. 1995. *Building Construction for the Fire Service, Third Edition*. Quincy, MA: NFPA.

Breyer, Donald E. 1993. *Design of Wood Structures, Third Edition*. New York: McGraw-Hill.

Brungraber, Robert L. 1992. "Engineered Tension Joinery." *Timber Framing* (March): 10–12.

————. 1985. *Traditional Timber Joinery: A Modern Analysis*. Ph.D. Dissertation (unpublished), Stanford University.

Brunskill, R. W. 1985. *Timber Building in Britain*. London: Victor Gollancz, Ltd.

Buchanan, Robert, Jason Pitcole, and Steve Tyler. 1997. "Specifying Firestopping to Save Lives." *The Construction Specifier* 50(6): 40–46.

Building Officials and Code Administrators International, Inc. 1996. *The BOCA National Building Code, Thirteenth Edition*. Country Club Hills, IL: BOCA.

————. 1993. *The BOCA National Building Code, Twelfth Edition*. Country Club Hills, IL: BOCA.

Bunéa, S. Paul. 1987. *Means Structural Steel Estimating*. Kingston, MA: R. S. Means Company.

Canadian Society for Civil Engineering Technical Committee on Wood Structures. 1995. "Reinforced Glulam Beams Using Advanced Composites," *Engineered Timber Structure News* 2. (winter).

Canadian Wood Council. 1997. *Comparing the Environmental Effects of Building Systems: A Case Study*. Ottawa, Ontario, Canada: Canadian Wood Council.

————. 1995. *Wood Reference Handbook: A Guide to the Architectural Use of Wood in Building Construction, Second Edition*, Ottawa, Ontario, Canada: Canadian Wood Council.

Cheung, Kevin. 1998. "Shrinkage in Multi-Story Wood-Frame Construction, Part 1: Shrinkage Calculation and Detailing." *Wood Design & Building* 3. (spring): 35–37.

Cleveland Steel Specialty Company. 1997. *Design Manual for Teco Timber Connectors.* Cleveland, OH: CSSC.

———. 1998. *Engineered Steel Connectors.* Cleveland, OH: CSSC.

Clough, Richard Hudson. 1975. *Construction Contracting. Third Edition.* New York: John Wiley & Sons.

Committee on History and Heritage of American Civil Engineering, ed. 1976. *American Wooden Bridges.* New York: ASCE.

Cook, Paul J. 1989. *Quantity Takeoff for Contractors: How to Get Accurate Material Counts.* Kingston, MA: R. S. Means Co., Inc.

Covered Bridge: 400 Timber Framers Build a Bridge (videocassette). 1992. Ontario, Canada: Alexander & McCormick Communications.

Cowan, Henry J. 1966. *An Historical Outline of Architectural Science.* New York: Elsevier.

———. 1976. *Architectural Structures: An Introduction to Structural Mechanics.* New York: Elsevier.

Crawley, Stanley W., and Delbert B. Ward. 1990. *Seismic and Wind Loads in Architectural Design: An Architect's Study Guide.* Washington, D.C.: AIA.

Demkin, Joseph A., ed. 1997. *Environmental Resource Guide.* New York: John Wiley & Sons.

Diesenhouse, Susan. 1998. "The Craftsmanship of Timber Framing." *The New York Times* sec. 11. (24 May): 1, 6.

Dietz, Albert G. H. 1974. *Dwelling House Construction, Fourth Edition.* Cambridge, MA: The MIT Press.

Dixon, Sheila A., and Richard D. Crowell. 1993. *The Contract Guide: DPIC's Risk Management Handbook for Architect and Engineers.* Monterey, CA: DPIC Companies, Inc.

Duggar, John Frederick III. 1984. *Checking and Coordinating Architectural and Engineering Working Drawings.* New York: McGraw-Hill.

Edlin, Herbert L. 1969. *What Wood is That? A Manual of Wood Identification.* New York: Viking.

Endersby, Elric, Alexander Greenwood, and David Larkin. 1992. *Barn: The Art of a Working Building.* Boston: Houghton Mifflin Company.

Faherty, Keith F., and Thomas G. Williamson, eds. 1995. *Wood Engineering and Construction Handbook, Second Edition.* New York: McGraw-Hill.

Feist, William. 1997. "The Challenges of Selecting Finishes for Exterior Wood." *Forest Products Journal* 47(5). (May): 16–20.

Ferguson, Eugene S. 1992. *Engineering and the Mind's Eye.* Cambridge, MA: The MIT Press.

Filler King Company. 1995. *Laminated Wood Roof Decking.* Homedale, ID: Filler King.

Goldstein, Eliot W. 1988. "Geometry in the Green Mountains." *Fine Homebuilding* 44. (February/March): 76–81.

———. 1996. "The Montville Public Library." *Joiners' Quarterly* 32. (fall): 12–18.

———. 1993. "Space and Structure." *The Military Engineer* 85(554). (January/February): 35–38.

———. 1991–1992. "Steel-And-Wood Trusses." *Fine Homebuilding* 71. (December/January): 74–76.

———. 1980. *The Summit House: Forces and Forms Cooperating in an Extreme Environment.* Master of Architecture Thesis. Cambridge, MA: MIT.

Gordon, J. E. 1978. *Structures; or, Why Things Don't Fall Down.* New York: Da Capo Press.

Götz, Karl-Heinz, et al. 1989. *Timber Design and Construction Sourcebook: A Comprehensive Guide to Methods and Practice.* New York: McGraw-Hill.

Graham, Frank D. 1947. *Audels Carpenters and Builders Guide #1.* New York: Theo. Audel & Co.

Graton, Milton S. 1978. *The Last of the Covered Bridge Builders.* Plymouth, NH: Clifford-Nicol, Inc.

Green Mountain Precision Frames. *Introducing the Timberbek™ Joinery System.* Windsor, VT: GMPF.

Groak, Steven. 1992. *The Idea of Building: Thought and Action in the Design and Production of Buildings.* London: Chapman & Hall.

Grogan, Tim. 1997. "Glulam: Surviving High Timber Prices." *Engineering News-Record* 238(26). (30 June): 29.

Grosse, Larry, and Fred Malven. 1996. *Fire Safety in Buildings*. Washington, D.C.: National Council of Architectural Registration Boards.

Guthrie, Pat. 1995. *The Architect's Portable Handbook*. New York: McGraw-Hill.

Harris, Cyril M. 1975. *Dictionary of Architecture and Construction*. New York: McGraw-Hill.

Harris, Richard. 1978. *Discovering Timber-Framed Buildings*. Bucks, U.K.: Shire Publishing, Ltd.

Hewitt, Cecil A. 1997. *English Historic Carpentry*. Fresno, CA: Linden Publishing.

Hodgson, Fred T. 1905. *Light and Heavy Timber Framing Made Easy*. Chicago: Frederick J. Drake & Co.

Holan, Jerri. 1990. *Norwegian Wood: A Tradition of Building*. New York: Rizzoli.

Holzbau-Praxis, Hinweise für die Ausführung nach DIN 1052. 1991. Karlsruhe: Bund Deutscher Zimmermeister, Bruderverlag.

Hool, George A., and W. S. Kinne, eds. 1942. *Steel and Timber Structures, Second Edition*. New York: McGraw-Hill.

Howe, Malverd A. 1912. *The Design of Simple Roof-Trusses in Wood and Steel, Third Edition*. New York: John Wiley & Sons.

Hubka, Thomas C. 1984. *Big House, Little House, Back House, Barn: The Connected Farm Buildings of New England*. Hanover, NH: University Press of New England.

Huntington, Whitney Clark, and Robert E. Mickadeit. 1975. *Building Construction: Materials and Types of Construction, Fourth Edition*. New York: John Wiley & Sons.

ICBO Evaluation Services, Inc. 1995. *Evaluation Report Number 5100: Fiber-Reinforced Plastic (FiRP™) Reinforced Glued-Laminated Wood Beams* (1 September).

"Innovative Design, Know-How Build Timber-Frame Library." 1995 *Carpenter* (July/August): 29.

International Code Council. 1997. *International Building Code 2000, May 1997 Working Draft*. Birmingham, AL: ICC.

Ivy, Robert Adams, Jr. 1992. *Fay Jones: The Architecture of E. Fay Jones, FAIA*. Washington, D.C.: AIA.

Jackson, Donald C. 1988. *Great American Bridges and Dams*. Washington, D.C.: The Preservation Press.

James, J. G. 1992. "The Evolution of Wooden Bridge Trusses." *Journal of the Institute of Wood Science (U. K.)*. (June and December): 168–193.

Jozsa, L. A., and G. R. Middleton. 1994. *A Discussion of Wood Quality Attributes and Their Practical Implications* (December). Vancouver, British Columbia, Canada: Forintek Canada Corporation.

Kamiyama, Yukihiro, ed. 1995. *Zusetsu Mokuzo Kenchiku Jiten (Encyclopedia of Wood Architecture), First Edition*. Kyoto, Japan: Gakugei Shupan Sha.

Ketchum, Milo S. 1956. *Handbook of Standard Structural Details for Buildings*. Englewood Cliffs, NJ: Prentice-Hall, Inc.

Kidder, Frank. 1899. *The Architect's and Builder's Pocket-Book, Thirteenth Edition*. New York: John Wiley & Sons.

Kidder, Frank, and Harry Parker. 1931. *The Architect's and Builder's Handbook, Eighteenth Edition*. New York: John Wiley & Sons.

King, Edward G., Jr. 1993. *Commentary on the National Design Specification for Wood Construction*. Washington, D.C.: AFPA.

Kitt, Jamie. 1994. "An Old-Fashioned Barn-Raising." *New Jersey Country Roads* (fall): 45–53.

Korman, Richard, and Sherie Winston. 1997. "Count of Crane Mishaps Remains a Mystery without One Database." *Engineering News-Record* 239(11). (15 September): 14.

Krieger, Alex, ed. 1988. *The Architecture of Kallmann McKinnell & Wood*. Cambridge, MA: Harvard University Graduate School of Design.

Kroloff, Reed. 1996. "New Life for Old Wood." *Architecture* 85(10). (October): 163–169.

Lathrop, James K., ed. 1988. *Life Safety Code Handbook, Fourth Edition*. Quincy, MA: NFPA.

Leichti, Robert J., Paul C. Gilham, and Daniel A. Tingley. 1994. "Partially Reinforced Glulam Girders Used in a Light-Commercial Structure." *Wood Design Focus* 5(4). (winter): 3–6.

———. 1993. "The Taylor Lake Bridge: A Reinforced-Glulam Structure." *Wood Design Focus* 4(1). (summer): 3–4.

Levy, Matthys, and Mario Salvadori. 1992. *Why Buildings Fall Down.* New York: W. W. Norton & Company.

Lewandoski, Jan. 1995. "The Erection of Church Steeples." *Timber Framing* 36. (June): 6–7.

Liebing, Ralph W., and Mimi Ford Paul. 1977. *Architectural Working Drawings.* New York: John Wiley & Sons.

Lstiburek, J., and J. Carmody. 1993. *Moisture Control Handbook.* New York: Van Nostrand Reinhold.

Mafell Carpentry Machines. 1996–1997 catalog. Oberndorf, Germany: Mafell.

Mainstone, Rowland J. 1975. *Developments in Structural Form.* Cambridge, MA: The MIT Press.

Mark, Robert, ed. 1993. *Architectural Technology up to the Scientific Revolution: The Art and Structure of Large-Scale Buildings.* Cambridge, MA: The MIT Press.

McCoy, Esther. 1975. *Five California Architects.* New York: Praeger.

McLeod, John A., III, and Joseph R. Loferski. 1993–1994. "All About Nails." *Fine Homebuilding* 85. (December/January): 42–47.

Miller, Warren. 1978. *Crosscut Saw Manual* (June). Missoula, MT: USDA Forest Service.

Mitchell, James. 1997. *The Craft of Modular Post & Beam: Building Log & Timber Homes Affordably.* Point Roberts, WA: Hartley & Marks.

"Montville Public Library." 1995. *Wood / le bois* (spring): 3–4.

National Association of Home Builders. 1978. *Basement Water Leakage . . . Causes, Prevention, and Correction.* Washington, D.C.: NAHB.

National Fire Protection Association. 1981. *Fire Protection Handbook, Fifteenth Edition.* Quincy, MA: NFPA.

———. *National Fire Codes.* Quincy, MA: NFPA.

Newman, Morton. 1995. *Design and Construction of Wood-Framed Buildings.* New York: McGraw-Hill.

Nigro, William T. 1987. *Redicheck Interdisciplinary Coordination.* Stone Mountain, GA: William T. Nigro.

Norris, Charles Head, and John Benson Wilbur. 1960. *Elementary Structural Analysis, Second Edition.* New York, McGraw-Hill.

Nunberg, Geoffrey, ed. 1996. *The Future of the Book.* Los Angeles: University of California Press.

Olin, H., J. Schmidt, and W. Lewis. 1990. "Moisture Control." pp. 104-1–104-24 in *Construction Principles, Materials & Methods.* New York: Van Nostrand Reinhold.

Ozelton, E. C., and J. A. Baird. 1976. *Timber Designers' Manual.* London: Crosby Lockwood Staples.

Packard, Robert J., ed. 1981. *Architectural Graphic Standards, Seventh Edition.* American Institute of Architects. New York: John Wiley & Sons, Inc.

Panshin, A. J., and Carl de Zeeuw. 1980. *Textbook of Wood Technology, Fourth Edition.* New York: McGraw-Hill.

Parker, Harry, and James Ambrose. 1994. *Simplified Design of Wood Structures, Fifth Edition.* New York: John Wiley & Sons.

Peurifoy, Robert L., and Garold D. Oberlender. 1996. *Formwork for Concrete Structures, Third Edition.* New York: McGraw-Hill.

Pooley, Bruce D. 1996. "Reinforced Glued Laminated Timber." *Civil Engineering.* (September): 50–53.

Post, Nadine M. 1997. "Annoying Floors: Help Coming to Keep Floors from Being Too Flexible for Comfort." *Engineering News-Record* 238(20). (5 May): 28–33.

———. 1995. "Library's Framing Is an Open Look." *Engineering News-Record* 234(5). (6 February): 34–36.

Quirk, J. Thomas, and Frank Freese. 1976. "Effect of Mechanical Stress on Growth and Anatomical Structure of Red Pine: Compression Stress." *Canadian Journal of Forest Research* 6(2): 195–202.

Redi-Lam II Glue Laminated Stock Beams. 1994. Product catalog, Louisiana-Pacific. Portland, OR.

Resch, Helmuth. 1987. *Dimensional Stability of Western Lumber Products, Second Edition.* Portland, OR: WWPA.

Rhude, A. J. 1996. "Structural Glued Laminated Timber: History of Its Origins and Early Development." *Forest Products Journal* 46(1): 15–22.

Ritter, Michael. 1992. *Timber Bridges: Design, Construction, Inspection, and Maintenance.* Washington, D.C.: USDA Forest Service.

Rogan, Sharon L. 1985. *United We Build: The Legacy of 100 Years.* Madison, NJ: New Jersey State Council of Carpenters.

Rosenfeld, Walter. 1985. *The Practical Specifier: A Manual of Construction Documentation for Architects.* New York: McGraw-Hill.

Rosengren, Maureen. 1995. "Wood Works in Distinctive Applications." *Building Design & Construction* 36(12). (December): 42–45.

Salvadori, Mario. 1980. *Why Buildings Stand Up: The Strength of Architecture.* New York: McGraw-Hill.

Sanders, Ken. 1996. *The Digital Architect: A Common-Sense Guide to using Computer Technology in Design Practice.* New York: John Wiley & Sons.

Scheffer, T. 1971. "A Climate Index for Estimating Potential for Decay in Wood Structures Above Ground." *Forest Products Journal* 21(10): 25–31.

Scheffer, T., and A. Verrall. 1973. "Principles for Protecting Wood Buildings from Decay." *USDA Forest Service Research Paper FPL 190.* Madison, WI: Forest Products Laboratory.

Schilling, Terence G., and Patricia M. 1987. *Intelligent Drawings: Managing CAD and Information Systems in the Design Office.* New York: McGraw-Hill.

Schmidt, R. J., and R. B. MacKay. 1997. *Timber Frame Tension Joinery.* Research Report of the Dept. of Civil and Architectural Engineering, University of Wyoming, (October).

Schön, Donald A. 1983. *The Reflective Practitioner: How Professionals Think in Action.* New York: Basic Books.

Seike, Kiyosi. 1977. *The Art of Japanese Joinery.* New York: Weatherhill/Tankosha.

Sills, Brad. 1995. "The House on the Trestle: On a Mountainous Site, an Arched Truss Carries the Weight of a Rustic Vacation House." *Fine Homebuilding* 98. (October/November): 96–101.

Sloane, Eric. 1967. *An Age of Barns.* New York: Funk & Wagnalls.

Smulski, S. 1997. "Controlling Indoor Moisture Sources in Wood-Frame Houses." *Wood Design Focus* 8(4): 19–24.

———. 1991. "Glue-Laminated Timbers: Designing and Building with Engineered Beams. *Fine Homebuilding* 71. (December): 55–59.

———. 1996. "Lumber Grade Stamps." *Fine Homebuilding* 103: 70–73.

———. 1990. "Preservative Treated Wood." *Fine Homebuilding* 63: 61–65.

———. 1992. "Wood-Destroying Insects." *Journal of Light Construction* 10(12): 35–39.

———. 1973. "Wood Fungi Causes and Cures." *Journal of Light Construction* 11(8): 22–28.

Stade, Franz. 1989. *Die Holz-Konstruktionen.* Reprint, Verlag Leipzig (originally printed 1904). Goseberg, Germany: Volker Hennig.

Sweet, Justin. 1985. *Legal Aspects of Architecture, Engineering, and the Construction Process, Third Edition.* St. Paul, MN: West Publishing Company.

Task Committee on Fasteners, of the Committee on Wood, of the Structural Division. 1996. *Mechanical Connections in Wood Structures, ASCE Manuals and Reports on Engineering Practice No. 84.* New York: ASCE.

Technical Committee 4.4: Insulation and Moisture Retarders. 1985. "Moisture in Building Construction." Pp. 21.1–21.2 in *ASHRAE Handbook of Fundamentals.* Atlanta, GA: ASHRAE.

Tenner, Edward. 1996. *Why Things Bite Back: Technology and the Revenge of Unintended Consequences.* New York: Alfred A. Knopf.

Tenwolde, A. 1994. "Indoor Humidity and the Building Envelope." Pp. 37–41 in *Bugs, Mold and Rot II: Proceedings of Workshop on Control of Humidity for Health, Artifacts, and Buildings.* Washington, D.C.: NIBS.

The Goldstein Partnership. 1993. *Specifications: Montville Public Library.* West Orange, NJ: The Goldstein Partnership.

———. 1991. *Specifications: Palmer Street Community Center.* West Orange, NJ: The Goldstein Partnership.

Thornton, Charles H., et al. 1993. *Exposed Structure in Building Design.* New York: McGraw-Hill.

"Timber Frame Library Takes Shape in Montville." 1994. *Construction Data News* 18(30). (27 June): 1.

Timoshenko, S., and D. H. Young. 1968. *Elements of Strength of Materials, Fifth Edition.* New York: Van Nostrand Reinhold Company.

"Too-Flexible Floors Need Multidisciplinary Attention." 1997. *Engineering News-Record* 238(20). (19 May): 94.

Truss Plate Institute. 1991. *Commentary and Recommendations for Handling, Installing & Bracing Metal Plate Connected Wood Trusses.* Madison, WI: TPI.

Tschebotarioff, Gregory P. (N.d.) *Soil Mechanics, Foundations, and Earth Structures.* New York: McGraw-Hill.

Tufte, Edward R. 1990. *Envisioning Information.* Cheshire, CT: Graphics Press.

———. *The Visual Display of Quantitative Information.* Cheshire, CT: Graphics Press.

———. 1997. *Visual Explanations: Images and Quantities, Evidence and Narrative.* Cheshire, CT: Graphics Press.

Turner, Marshall R. 1977. "Timber Dome Roof over Stadium Spans Record 502 Ft." *Civil Engineering* (August).

U.S. Department of Agriculture, Forest Products Laboratory. 1990. *Wood Engineering Handbook.* Englewood Cliffs, NJ: Prentice-Hall.

———. 1987. *Wood Handbook: Wood as an Engineering Material. Agricultural Handbook 72,* rev. Washington, D.C.: USDA.

Vogel, Robert M. 1971. *Roebling's Delaware & Hudson Canal Aqueducts.* Washington, D.C.: Smithsonian Institution Press.

Waier, Phillip R., ed. 1997. *Building Construction Cost Data 1998, Fifty-Sixth Annual Edition.* Kingston, MA: R.S. Means Co., Inc.

Wallace, Don. E. "Specifying Rough & Finish Carpentry Materials." 1985. *The Construction Specifier.* Reprint issued by WWPA.

Wardell, Charles. 1991. "Composition Panels." *Fine Homebuilding* (April/May): 77–81.

Weiner, Michael A. 1975. *Plant a Tree: A Working Guide to Regreening America.* New York: Macmillan.

Weismantel, Guy E. 1981. *Paint Handbook.* New York: McGraw-Hill.

Western Wood Products Association. 1994. *Western Lumber Grading Rules 95.* Portland, OR: WWPA.

———. 1985. *Wood Frame Design for Commercial/Multifamily Construction.* Portland, OR: WWPA.

Williams, S., M. Knaebe, and W. Fiest. 1996. *Finishes for Exterior Wood.* Madison, WI: Forest Products Society.

Wood Construction Connectors. 1997. Product catalog. Pleasanton, CA: Simpson Strong-Tie Company, Inc.

Wood-Products Information and Research Center. 1995. "Montville Public Library." Pp. 28–33 in *Large Scale Timber Architecture, Series IV: Representative Examples from North America.* Tokyo: Japan: WPIRC.

Wurman, Richard Saul. 1992. *Follow the Yellow Brick Road: Learning to Give, Take, and Use Instructions.* New York: Bantam Books.

Wyatt, David J. 1997. "Improving Specification Language." *The Construction Specifier* 50(8). (August): 13–15.

———. 1997. "Writing Better Alternates." *The Construction Specifier* 50(9). (September): 24–26.

Youngquist, John A. 1995. "The Marriage of Wood and Non-Wood Materials." *Forest Products Journal* 45(10). (October): 25–30.

Zabel, R., and J. Morrell. 1992. *Wood Microbiology Decay and Its Prevention.* San Diego, CA: Academic Press.

Contributors

Robert L. "Ben" Brungraber, Ph.D., PE, Engineer and Operations Director, Benson Woodworking Co., Inc., Alstead Center, NH (Chap. 7)

Eliot W. Goldstein, AIA, Partner, The Goldstein Partnership, Architects, West Orange, NJ (all other chapters)

Phillip C. Pierce, PE, Structural Engineer, Binghamton, NY (Chap. 15)

Fred Severud, PE, Adjunct Associate Professor of Civil Engineering, University of Missouri/Rolla, Consulting Structural Engineer, Salem, MO (Chap. 9)

Stephen Smulski, Ph.D., President, Wood Science Specialists Inc., Shutesbury, MA (Chap. 3)

Software Note

With the exception of those in the contributed chapters, the figures and icons in this book were produced on Apple Macintoshes, primarily a PowerBook 1400CS and a Power Macintosh 8600/300. The manuscript was prepared using *ClarisWorks 5.0* from Apple, with *Equation Editor* by Design Science, Inc. *MiniCad 6.0* and *7.0* from Diehl Graphsoft were used for the computer models and CAD drawings. Organization-type charts were made with *Inspiration 4.0b* from Inspiration Software, Inc. To produce graphs, data was entered into and manipulated within the spreadsheet mode of *ClarisWorks,* then exported to *DeltaGraph 4.0* from DeltaPoint, Inc., for development of the final displays. Some illustrations were produced using *PageMaker 6.0* from Adobe Systems, Inc. The *American Heritage Dictionary* from SoftKey International, Inc., was extremely valuable during the development of the text. Project correspondence was tracked with Apple's *HyperCard 2.3.* Project scheduling was performed using *FastTrack Schedule 5.0* from AEC Software. And last, but certainly not least, the inevitable loose ends were tracked with *In Control 2.0* from Attain Corporation.

Index

About the Author and Contributors

Eliot W. Goldstein, AIA, an award-winning architect and specialist in timber design, is a partner with The Goldstein Partnership of West Orange, New Jersey. The designer of numerous public buildings, he was selected as "one of the top 25 newsmakers of 1995" by *Engineering News-Record* for his timber-trussed Montville Township Public Library, which that journal said "set a new standard for heavy timber structure as architecture . . . and produced an instant landmark."

Robert Lyman (Ben) Brungraber, Ph.D., PE (Chapter 7, Traditional Joinery), is a manager/part-owner of Benson Woodworking Co., Inc., in Alstead Center, New Hampshire, a firm that has designed and built many prominent heavy timber frame structures in the United States, including more than 30 large public projects. He has also repaired both the oldest and the longest covered bridges in Pennsylvania.

Phillip C. Pierce, PE (Chapter 15, Covered Bridges), a recognized structural and design engineer, has led over 100 bridge design/rehabilitation projects and more than 1000 bridge inspections/evaluations, mostly in the northeastern United States. He is the author of several papers, and is a popular public speaker about wooden bridges. He lives in Binghamton, New York.

Fred Severud, PE (Chapter 9, Lateral Bracing), is a structural engineer and former principal of Severud Associates in New York City. He was in charge of the structural design team on the Blue Cross Headquarters, a 50-story office building in Philadelphia, Pennsylvania, and on the Madison Square Garden renovation in New York City, among many other outstanding projects. He is now an adjunct associate professor of civil engineering at the University of Missouri, Rolla. He lives in nearby Salem, Missouri.

Stephen Smulski, Ph.D. (Chapter 3, Preventing Wood Degradation), is a consulting wood scientist and president/owner of Wood Science Specialists Inc., of Shutesbury, Massachusetts. A consultant to makers, sellers, and users of wood products and the editor of *Wood Design Focus A Journal of Contemporary Wood Engineering,* Dr. Smulski has written extensively on moisture-caused problems in wood construction. He is an adjunct assistant professor of building materials and wood technology at the University of Massachusetts at Amherst.